T0297115

CAMBRIDGE LIBRARY COLLECTION

Books of enduring scholarly value

Mathematical Sciences

From its pre-historic roots in simple counting to the algorithms powering modern desktop computers, from the genius of Archimedes to the genius of Einstein, advances in mathematical understanding and numerical techniques have been directly responsible for creating the modern world as we know it. This series will provide a library of the most influential publications and writers on mathematics in its broadest sense. As such, it will show not only the deep roots from which modern science and technology have grown, but also the astonishing breadth of application of mathematical techniques in the humanities and social sciences, and in everyday life.

Mathematical and Physical Papers

Sir George Stokes (1819-1903) established the science of hydrodynamics with his law of viscosity describing the velocity of a small sphere through a viscous fluid. He published no books, but was a prolific lecturer and writer of papers for the Royal Society, the British Association for the Advancement of Science, the Victoria Institute and other mathematical and scientific institutions. These collected papers (issued between 1880 and 1905) are therefore the only readily available record of the work of an outstanding and influential mathematician, who was Lucasian Professor of Mathematics in Cambridge for over fifty years, Master of Pembroke College, President of the Royal Society (1885-90), Associate Secretary of the Royal Commission on the University of Cambridge and a Member of Parliament for the University.

Cambridge University Press has long been a pioneer in the reissuing of out-of-print titles from its own backlist, producing digital reprints of books that are still sought after by scholars and students but could not be reprinted economically using traditional technology. The Cambridge Library Collection extends this activity to a wider range of books which are still of importance to researchers and professionals, either for the source material they contain, or as landmarks in the history of their academic discipline.

Drawing from the world-renowned collections in the Cambridge University Library, and guided by the advice of experts in each subject area, Cambridge University Press is using state-of-the-art scanning machines in its own Printing House to capture the content of each book selected for inclusion. The files are processed to give a consistently clear, crisp image, and the books finished to the high quality standard for which the Press is recognised around the world. The latest print-on-demand technology ensures that the books will remain available indefinitely, and that orders for single or multiple copies can quickly be supplied.

The Cambridge Library Collection will bring back to life books of enduring scholarly value (including out-of-copyright works originally issued by other publishers) across a wide range of disciplines in the humanities and social sciences and in science and technology.

Mathematical and Physical Papers

VOLUME 2

GEORGE GABRIEL STOKES

CAMBRIDGE UNIVERSITY PRESS

Cambridge New York Melbourne Madrid Cape Town Singapore São Paolo Delhi

Published in the United States of America by Cambridge University Press, New York

www.cambridge.org
Information on this title: www.cambridge.org/9781108002639

© in this compilation Cambridge University Press 2009

This edition first published 1883
This digitally printed version 2009

ISBN 978-1-108-00263-9

MATHEMATICAL

AND

PHYSICAL PAPERS.

London: C. J. CLAY, M.A. & SON,
CAMBRIDGE UNIVERSITY PRESS WAREHOUSE,
17, PATERNOSTER ROW.

CAMBRIDGE: DEIGHTON, BELL, AND CO.
LEIPZIG: F. A. BROCKHAUS.

MATHEMATICAL

AND

PHYSICAL PAPERS

GEORGE GABRIEL STOKES, M.A., D.C.L., LL.D., F.R.S.,

FELLOW OF PEMBROKE COLLEGE AND LUCASIAN PROFESSOR OF MATHEMATICS
IN THE UNIVERSITY OF CAMBRIDGE.

Reprinted from the Original Journals and Transactions,
with Additional Notes by the Author.

VOL. II.

CAMBRIDGE:

AT THE UNIVERSITY PRESS.

1883

𝕮𝖆𝖒𝖇𝖗𝖎𝖉𝖌𝖊:

PRINTED BY C. J. CLAY, M.A. AND SON,

AT THE UNIVERSITY PRESS.

CONTENTS.

ERRATUM.

P. 221, in the Number of the Note. *For* IV. *read* VI.

MATHEMATICAL AND PHYSICAL PAPERS.

[From the *Cambridge and Dublin Mathematical Journal*, Vol. III. p. 121, March, 1848.]

NOTES ON HYDRODYNAMICS*.

III.—*On the Dynamical Equations.*

IN reducing to calculation the motion of a system of rigid bodies, or of material points, there are two sorts of equations with which we are concerned; the one expressing the geometrical connexions of the bodies or particles with one another, or with curves or surfaces external to the system, the other expressing the relations between the changes of motion which take place in the system and the forces producing such changes. The equations belonging to these two classes may be called respectively the geometrical, and the dynamical equations. Precisely the same remarks apply to the motion of fluids. The geometrical equations which occur in

---

* The series of "notes on Hydrodynamics" which are printed in Vols. II., III. and IV. of the *Cambridge and Dublin Mathematical Journal*, were written by agreement between Sir William Thomson and myself mainly for the use of Students. As far as my own share in the series is concerned, there is little contained in the "notes" which may not be found elsewhere. Acting however upon the general advice of my friends, I have included my share of the series in the present reprint. It may be convenient to give here the references to the whole series.

1

Hydrodynamics have been already considered by Professor Thomson, in Notes I. and II. The object of the present Note is to form the dynamical equations.

The fundamental hypothesis of Hydrostatics is, that the mutual pressure of two contiguous portions of a fluid, separated by an imaginary plane, is normal to the surface of separation.   This ·hypothesis forms in fact the mathematical definition of a fluid. The equality of pressure in all directions is in reality not an independent hypothesis, but a necessary consequence of the former. A proof of this may be seen at the commencement of Prof. Miller's *Hydrostatics*.   The truth of our fundamental hypothesis, or at least its extreme nearness to the truth, is fully established by experiment. Some of the nicest processes in Physics depend upon it; for example, the determination of specific gravities, the use of the level, the determination of the zenith by reflection from the surface of mercury.

The same hypothesis is usually made in Hydrodynamics. If it be assumed, the equality of pressure in all directions will follow as a necessary consequence. This may be proved nearly as before, the only difference being that now we have to take into account, along with the impressed forces, forces equal and opposite to the effective forces. The verification of our hypothesis is however much more difficult in the case of motion, partly on account of the mathematical difficulties of the subject, partly because the experiments do not usually admit of great accuracy. Still, theory and experiment have been in certain cases sufficiently compared to shew that our hypothesis may be employed with very little error in many important instances. There are however many phenomena which point out the existence of a tangential force in fluids in motion, analogous in some respects to friction in the case of solids, but differing from it in this respect, that whereas in solids friction is exerted at the surface, and between points which move relatively to each other with a finite velocity, in fluids friction is exerted throughout the mass, where the velocity varies continuously from one point to another. Of course it is the same thing to say that in such cases there is a tangential force along with a normal pressure, as to say that the mutual pressure of two adjacent elements of a fluid is no longer normal to their common surface.

The subsidence of the motion in a cup of tea which has been stirred may be mentioned as a familiar instance of friction, or, which is the same, of a deviation from the law of normal pressure ; and the absolute regularity of the surface when it comes to rest, whatever may have been the nature of the previous disturbance, may be considered as a proof that all tangential force vanishes when the motion ceases.

It does not fall in with the object of this Note to enter into the theory of the friction of fluids in motion*, and accordingly the hypothesis of normal pressure will be adopted. The usual notation will be employed, as in the preceding Notes. Consider the elementary parallelepiped of fluid comprised between planes parallel to the coordinate planes and passing through the points whose co-ordinates are x, y, z, and $x + dx$, $y + dy$, $z + dz$. Let X, Y, Z be the accelerating forces acting on the fluid at the point (x, y, z); then, ρ and X being ultimately constant throughout the element, the moving force parallel to x arising from the accelerating forces which act on the element will be ultimately $\rho X\, dx\, dy\, dz$. The difference between the pressures, referred to a unit of surface, at opposite points of the faces $dy\, dz$ is ultimately $dp/dx \cdot dx$, acting in the direction of x negative, and therefore the difference of the total pressures on these faces is ultimately $dp/dx \cdot dx\, dy\, dz$; and the pressures on the other faces act in a direction perpendicular to the axis of x. The effective moving force parallel to x is ultimately $\rho \cdot D^2x/Dt^2 \cdot dx\, dy\, dz$, where, in order to prevent confusion, D is used to denote differentiation when the independent variables are supposed to be t, and three parameters which distinguish one particle of the fluid from another, as for instance the initial coordinates of the particle, while d is reserved to denote differentiation when the independent variables are x, y, z, t. We have therefore, ultimately,

$$\rho \frac{D^2x}{Dt^2}\, dx\, dy\, dz = \left(\rho X - \frac{dp}{dx}\right) dx\, dy\, dz,$$

---

* The reader who feels an interest in the subject may consult a memoir by Navier, *Mémoires de l'Académie*, tom. VI. p. 389; another by Poisson, *Journal de l'École Polytechnique*, Cahier XX. p. 139; an abstract of a memoir by M. de Saint-Venant, *Comptes Rendus*, tom. XVII. (Nov. 1843) p. 1240; and a paper in the *Cambridge Philosophical Transactions*, Vol. VIII. p. 287. [*Ante*, Vol. I. p. 75.]

with similar equations for y and z. Dividing by $\rho \, dx \, dy \, dz$, transposing, and taking the limit, we get

$$\frac{1}{\rho}\frac{dp}{dx} = X - \frac{D^2 x}{Dt^2}, \quad \frac{1}{\rho}\frac{dp}{dy} = Y - \frac{D^2 y}{Dt^2}, \quad \frac{1}{\rho}\frac{dp}{dz} = Z - \frac{D^2 z}{Dt^2} \dots \dots (1).$$

These are the dynamical equations which must be satisfied at every point in the interior of the fluid mass; but they are not at present in a convenient shape, inasmuch as they contain differential coefficients taken on two different suppositions. It will be convenient to express them in terms of differential coefficients taken on the second supposition, that is, that x, y, z, t are the independent variables. Now $Dx/Dt = u$, and on the second supposition u is a function of t, x, y, z, each of which is a function of t on the first supposition. We have, therefore, by Differential Calculus,

$$\frac{Du}{Dt} \text{ or } \frac{D^2 x}{Dt^2} = \frac{du}{dt} + \frac{du}{dx}\frac{Dx}{Dt} + \frac{du}{dy}\frac{Dy}{Dt} + \frac{du}{dz}\frac{Dz}{Dt} ;$$

or, since by the definitions of u, v, w,

$$\frac{Dx}{Dt} = u, \qquad \frac{Dy}{Dt} = v, \qquad \frac{Dz}{Dt} = w,$$

we have $$\frac{D^2 x}{Dt^2} = \frac{du}{dt} + u\frac{du}{dx} + v\frac{du}{dy} + w\frac{du}{dz},$$

with similar equations for y and z.

Substituting in (1), we have

$$\left.\begin{aligned}
\frac{1}{\rho}\frac{dp}{dx} &= X - \frac{du}{dt} - u\frac{du}{dx} - v\frac{du}{dy} - w\frac{du}{dz} \\[4pt]
\frac{1}{\rho}\frac{dp}{dy} &= Y - \frac{dv}{dt} - u\frac{dv}{dx} - v\frac{dv}{dy} - w\frac{dv}{dz} \\[4pt]
\frac{1}{\rho}\frac{dp}{dz} &= Z - \frac{dw}{dt} - u\frac{dw}{dx} - v\frac{dw}{dy} - w\frac{dw}{dz}
\end{aligned}\right\} \dots\dots\dots\dots(2),$$

which is the usual form of the equations.

The equations (1) or (2), which are physically considered the same, determine completely, so far as Dynamics alone are concerned, the motion of each particle of the fluid. Hence any other purely dynamical equation which we might set down would be identically satisfied by (1) or (2). Thus, if we were to consider the fluid which at the time t is contained within a closed surface S, and set down the last three equations of equilibrium of a rigid body between the pressures exerted on S, the moving forces due to the accelerating forces acting on the contained fluid, and the effective moving forces reversed, we should not thereby obtain any new equation. The surface S may be either finite or infinitesimal, as, for example, the surface of the elementary parallelepiped with which we started. Thus we should fall into error if we were to set down these three equations for the parallelepiped, and think that we had thereby obtained three new independent equations.

If the fluid considered be homogeneous and incompressible, ρ is a constant. If it be heterogeneous and incompressible, ρ is a function of x, y, z, t, and we have the additional equation $D\rho/Dt = 0$, or

$$\frac{d\rho}{dt} + u\frac{d\rho}{dx} + v\frac{d\rho}{dy} + w\frac{d\rho}{dz} = 0 \dots\dots\dots\dots(3),$$

which expresses the fact of the incompressibility. If the fluid be elastic and homogeneous, and at the same temperature θ throughout, and if moreover the change of temperature due to condensation and rarefaction be neglected, we shall have

$$p = k\rho\,(1 + \alpha\theta) \dots\dots\dots\dots\dots\dots(4),$$

where k is a given constant, depending on the nature of the gas, and α a known constant which is the same for all gases [nearly]. The numerical value of α, as determined by experiment, is ·00366, θ being supposed to refer to the centigrade thermometer.

If the condensations and rarefactions of the fluid be rapid, we may without inconsistency take account of the increase of temperature produced by compression, while we neglect the communication of heat from one part of the mass to another. The only important problem coming under this class is that of sound. If we suppose the changes in pressure and density small, and neglect the squares of small quantities, we have, putting p_1, ρ_1 for the values of p, ρ in equilibrium,

$$\frac{p - p_1}{p_1} = K\frac{\rho - \rho_1}{\rho_1} \dots\dots\dots\dots\dots\dots(5),$$

K being a constant which, as is well known, expresses the ratio of the specific heat of the gas considered under a constant pressure to its specific heat when the volume is constant. We are not, however, obliged to consider specific heat at all; but we may if we please regard K merely as the value of $d \log p/d \log \rho$ for $\rho = \rho_1$, p being that function of ρ which it is in the case of a mass of air suddenly compressed or dilated. In whichever point of view we regard K, the observation of the velocity of sound forms the best mode of determining its numerical value.

It will be observed that in the proof given of equations (1) it has been supposed that the pressure exerted by the fluid outside the parallelepiped was exerted wholly on the fluid forming the parallelepiped, and not partly on this portion of fluid and partly on the fluid at the other side of the parallelepiped. Now, the pressure arising directly from molecular forces, this imposes a restriction on the diminution of the parallelepiped, namely that its edges shall not become less than the radius of the sphere of activity of the molecular forces. Consequently we cannot, mathematically speaking, suppose the parallelepiped to be indefinitely diminished. It is known, however, that the molecular forces are insensible at sensible distances, so that we may suppose the parallelepiped to become so small that the values of the forces, &c., for any point of it, do not sensibly differ from their values for one of the corners, and that all summations with respect to such elements may be replaced without sensible error by integrations; so that the values of the several unknown quantities obtained from our equations by differentiation, integration, &c. are sensibly correct, so far as this cause of error is concerned; and that is all that we can ever attain to in the mathematical expression of physical laws. The same remarks apply as to the bearing on our reasoning of the supposition of the existence of ultimate molecules, a question into which we are not in the least called upon to enter.

There remains yet to be considered what may be called the dynamical equation of the bounding surface.

Consider, first, the case of a fluid in contact with the surface of a solid, which may be either at rest or in motion. Let P be a point in the surface, about which the curvature is not infinitely great, ω an element of the surface about P, PN a normal at P, directed into the fluid, and let $PN = h$. Through N draw a plane A perpendicular to PN, and project ω on this plane by a circumscribing cylindrical surface. Suppose h greater than the radius r

of the sphere of activity of the molecular forces, and likewise large enough to allow the plane A not to cut the perimeter of ω. For the reason already mentioned r will be neglected, and therefore no restriction imposed on h on the first account. Let Π be the pressure sustained by the solid, referred to a unit of surface, Π having the value belonging to the point P, and let p' be the pressure of the fluid at N. Consider the element of fluid comprised between ω, its projection on the plane A, and the projecting cylindrical surface. The forces acting on this element are, first, the pressure of the fluid on the base, which acts in the direction NP, and is ultimately equal to $p'\omega$; secondly, the pressure of the solid, which ultimately acts along PN and is equal to $\Pi\omega$; thirdly, the pressure of the fluid on the cylindrical surface, which acts everywhere in a direction perpendicular to PN; and, lastly, the moving forces due to the accelerating forces acting on the fluid; and this whole system of forces is in equilibrium with forces equal and opposite to the effective moving forces. Now the moving forces due to the accelerating forces acting on the fluid, and the effective moving forces, are both of the order ωh, and therefore, whatever may be their directions, vanish in the limit compared with the force $p'\omega$, if we suppose, as we may, that h vanishes in the limit. Hence we get from the equation of the forces parallel to PN, passing to the limit,

$$p = \Pi \dots\dots\dots\dots\dots\dots\dots\dots\dots\dots\dots (6),$$

p being the limiting value of p', or the result obtained by substituting in the general expression for the pressure the coordinates of the point P for x, y, z.

It should be observed that, in proving this equation, the forces on which capillary phenomena depend have not been taken into account. And in fact it is only when such forces are neglected that equation (6) is true.

In the case of a liquid with a free surface, or more generally in the case of two fluids in contact, it may be proved, just as before, that equation (6) holds good at any point in the surface, p, Π being the results obtained on substituting the coordinates of the point considered for the general coordinates in the general expressions for the pressure in the two fluids respectively. In this case, as before, capillary attraction is supposed to be neglected.

[From the *Philosophical Magazine*, Vol. XXXII. p. 343, May, 1848.]

ON THE CONSTITUTION OF THE LUMINIFEROUS ETHER.

THE phenomenon of aberration may be reconciled with the undulatory theory of light, as I have already shown (*Phil. Mag.*, Vol. XXVII. p. 9*), without making the violent supposition that the ether passes freely through the earth in its motion round the sun, but supposing, on the contrary, that the ether close to the surface of the earth is at rest relatively to the earth. This explanation requires us to suppose the motion of the ether to be such, that the expression usually denoted by $udx + vdy + wdz$ is an exact differential. It becomes an interesting question to inquire on what physical properties of the ether this sort of motion can be explained. Is it sufficient to consider the ether as an ordinary fluid, or must we have recourse to some property which does not exist in ordinary fluids, or, to speak more correctly, the existence of which has not been made manifest in such fluids by any phenomenon hitherto observed? I have already attempted to offer an explanation on the latter supposition (*Phil. Mag.*, Vol. XXIX. p. 6†).

In my paper last referred to, I have expressed my belief that the motion for which $udx + $ &c. is an exact differential, which would take place if the ether were like an ordinary fluid, would be unstable; I now propose to prove the same mathematically, though by an indirect method.

Even if we supposed light to arise from vibrations of the ether accompanied by condensations and rarefactions, analogous to the vibrations of the air in the case of sound, since such vibrations would be propagated with about 10,000 times the velocity of the earth,

* *Ante*, Vol. I. p. 134.   † *Ante*, Vol. I. p. 153.

we might without sensible error neglect the condensation of the ether in the motion which we are considering. Suppose, then, a sphere to be moving uniformly in a homogeneous incompressible fluid, the motion being such that the square of the velocity may be neglected. There are many obvious phenomena which clearly point out the existence of a tangential force in fluids in motion, analogous in many respects to friction in the case of solids. When this force is taken into account, the equations of motions become (*Cambridge Philosophical Transactions*, Vol. VIII. p. 297*)

$$\frac{dp}{dx} = - \rho \frac{du}{dt} + \mu \left(\frac{d^2u}{dx^2} + \frac{d^2u}{dy^2} + \frac{d^2u}{dz^2}\right) \dots\dots\dots(1),$$

with similar equations for y and z. In these equations the square of the velocity is omitted, according to the supposition made above, ρ is considered constant, and the fluid is supposed not to be acted on by external forces. We have also the equation of continuity

$$\frac{du}{dx} + \frac{dv}{dy} + \frac{dw}{dz} = 0 \dots\dots\dots(2),$$

and the conditions, (1) that the fluid at the surface of the sphere shall be at rest relatively to the surface, (2) that the velocity shall vanish at an infinite distance.

For my present purpose it is not requisite that the equations such as (1) should be known to be true experimentally; if they were even known to be false they would be sufficient, for they may be conceived to be true without mathematical absurdity. My argument is this. If the motion for which $udx + \dots$ is an exact differential, which would be obtained from the common equations, were stable, the motion which would be obtained from equations (1) would approach indefinitely, as μ vanished, to one for which $udx + \dots$ was an exact differential, and therefore, for anything proved to the contrary, the latter motion might be stable; but if, on the contrary, the motion obtained from (1) should turn out totally different from one for which $udx + \dots$ is an exact differential, the latter kind of motion must necessarily be unstable.

Conceive a velocity equal and opposite to that of the sphere impressed both on the sphere and on the fluid. It is easy to prove that $udx + \ldots$ will or will not be an exact differential after the velocity is impressed, according as it was or was not such before. The sphere is thus reduced to rest, and the problem becomes one of steady motion. The solution which I am about to give is extracted from some researches in which I am engaged, but which are not at present published. It would occupy far too much room in this Magazine to enter into the mode of obtaining the solution : but this is not necessary; for it will probably be allowed that there is but one solution of the equations in the case proposed, as indeed readily follows from physical considerations, so that it will be sufficient to give the result, which may be verified by differentiation.

Let the centre of the sphere be taken for origin; let the direction of the real motion of the sphere make with the axes angles whose cosines are l, m, n, and let v be the real velocity of the sphere; so that when the problem is reduced to one of steady motion, the fluid at a distance from the sphere is moving in the opposite direction with a velocity v. Let a be the sphere's radius : then we have to satisfy the general equations (1) and (2) with the particular conditions

$$u = 0, \quad v = 0, \quad w = 0, \text{ when } r = a \ldots\ldots\ldots\ldots\ldots\ldots(3) ;$$

$$u = -lv, \quad v = -mv, \quad w = -nv, \text{ when } r = \infty \ldots\ldots\ldots\ldots(4),$$

r being the distance of the point considered from the centre of the sphere. It will be found that all the equations are satisfied by the following values,

$$p = \Pi + \frac{3}{2} \mu v \frac{a}{r^3} (lx + my + nz),$$

$$u = \frac{3}{4} v \left(\frac{a}{r^3} - \frac{a^3}{r^5} \right) x (lx + my + nz) + lv \left(\frac{1}{4} \frac{a^3}{r^3} + \frac{3}{4} \frac{a}{r} - 1 \right),$$

with symmetrical expressions for v and w. Π is here an arbitrary constant, which evidently expresses the value of p at an infinite distance. Now the motion defined by the above expressions does not tend, as μ vanishes, to become one for which $udx + \ldots$ is an exact differential, and therefore the motion which would be obtained by supposing $udx + \ldots$ an exact differential, and applying to the ether the common equations of hydrodynamics, would be unstable. The proof supposes the motion in question to be steady; but such it may be proved to be, if the velocity of the earth be regarded as uniform, and an equal and opposite velocity be conceived impressed both on the earth and on the ether. Hence the stars would appear to be displaced in a manner different from that expressed by the well-known law of aberration.

When, however, we take account of a tangential force in the ether, depending, not on relative velocities, or at least not on relative velocities only, but on relative displacements, it then becomes possible, as I have shewn (*Phil. Mag.*, Vol. XXIX. p. 6), to explain not only the perfect regularity of the motion, but also the circumstance that $udx + \ldots$ is an exact differential, at least for the ether which occupies free space; for as regards the motion of the ether which penetrates the air, whether about the limits of the atmosphere or elsewhere, I do not think it prudent, in the present state of our knowledge, to enter into speculation; I prefer resting in the supposition that $udx + \ldots$ *is* an exact differential. According to this explanation, any nascent irregularity of motion, any nascent deviation from the motion for which $udx + \ldots$ is an exact differential, is carried off into space, with the velocity of light, by transversal vibrations, which as such are identical in their physical nature with light, but which do not necessarily produce the sensation of light, either because they are too feeble, as they probably would be, or because their lengths of wave, if the vibrations take place in regular series, fall beyond the limits of the visible spectrum, or because they are discontinuous, and the sensation of light may require the succession of a number of similar vibrations. It is certainly curious that the astronomical phenomenon of the aberration of light should afford an argument in support of the theory of transversal vibrations.

Undoubtedly it does violence to the ideas that we should have been likely to form à *priori* of the nature of the ether, to assert that it must be regarded as an elastic solid in treating of the vibrations of light. When, however, we consider the wonderful simplicity of the explanations of the phenomena of polarization when we adopt the theory of transversal vibrations, and the difficulty, which to me at least appears quite insurmountable, of explaining these phenomena by any vibrations due to the condensation and rarefaction of an elastic fluid such as air, it seems reasonable to suspend our judgement, and be content to learn from phenomena the existence of forces which we should not beforehand have expected. The explanations which I had in view are those which belong to the geometrical part of the theory; but the deduction, from dynamical calculations, of the laws which in the geometrical theory take the place of observed facts must not be overlooked, although here the evidence is of a much more complicated character.

The following illustration is advanced, not so much as explaining the real nature of the ether, as for the sake of offering a plausible mode of conceiving how the apparently opposite properties of solidity and fluidity which we must attribute to the ether may be reconciled.

Suppose a small quantity of glue dissolved in a little water, so as to form a stiff jelly. This jelly forms in fact an elastic solid : it may be constrained, and it will resist constraint, and return to its original form when the constraining force is removed, by virtue of its elasticity ; but if we constrain it too far it will break. Suppose now the quantity of water in which the glue is dissolved to be doubled, trebled, and so on, till at last we have a pint or a quart of glue water. The jelly will thus become thinner and thinner, and the amount of constraining force which it can bear without being dislocated will become less and less. At last it will become so far fluid as to mend itself again as soon as it is dislocated. Yet there seems hardly sufficient reason for supposing that at a certain stage of the dilution the tangential force whereby it resists constraint ceases all of a sudden. In order that the medium should not be dislocated, and therefore should have to be treated as an elastic solid, it is only necessary that the amount of constraint should be very small. The medium would however be what we should call a fluid, as regards the motion of solid bodies through it. The velocity of propagation of normal vibrations in our medium would be nearly the same as that of sound in water ; the velocity of propagation of transversal vibrations, depending as it does on the tangential elasticity, would become very small. Conceive now a medium having similar properties, but incomparably rarer than air, and we have a medium such as we may conceive the ether to be, a fluid as regards the motion of the earth and planets through it, an elastic solid as regards the small vibrations which constitute light. Perhaps we should get nearer to the true nature of the ether by conceiving a medium bearing the same relation to air that thin jelly or glue water bears to pure water. The sluggish transversal vibrations of our thin jelly are, in the case of the ether, replaced by vibrations propagated with a velocity of nearly 200,000 miles in a second : we should expect, à priori, the velocity of propagation of normal vibrations to be incomparably greater. This is just the conclusion to which we are led quite independently, from dynamical principles of the greatest generality, combined with the observed phenomena of optics *.

* See the introduction to an admirable memoir by Green, "On the laws of the Reflexion and Refraction of Light at the common surface of two non-crystallized media." *Cambridge Philosophical Transactions*, Vol. VII. p. 1.

[From the *Philosophical Transactions* for 1848, p. 227.]

ON THE THEORY OF CERTAIN BANDS SEEN IN THE SPECTRUM.

[Read May 25, 1848.]

SOME months ago Professor Powell communicated to me an account of a new case of interference which he had discovered in the course of some experiments on a fluid prism, requesting at the same time my consideration of the theory. As the phenomenon is fully described in Professor Powell's memoir, and is briefly noticed in Art. 1 of this paper, it is unnecessary here to allude to it. It struck me that the theory of the phenomenon was almost identical with that of the bands seen when a spectrum is viewed by an eye, half the pupil of which is covered by a plate of glass or mica. The latter phenomenon has formed the subject of numerous experiments by Sir David Brewster, who has discovered a very remarkable polarity, or apparent polarity, in the bands. The theory of these bands has been considered by the Astronomer Royal in two memoirs " On the Theoretical Explanation of an apparent new Polarity of Light," printed in the *Philosophical Transactions* for 1840 (Part II.) and 1841 (Part I.). In the latter of these Mr Airy has considered the case in which the spectrum is viewed in focus, which is the most interesting case, as being that in which the bands are best seen, and which is likewise far simpler than the case in which the spectrum is viewed out of focus. Indeed, from the mode of approximation adopted, the former memoir can hardly be considered to belong to the bands which formed the subject of Sir David Brewster's experiments, although the memoir no doubt contains the theory of a possible system of bands. On going over the theory of the bands seen when the spectrum is viewed in focus, after the receipt of

Professor Powell's letter, I was led to perceive that the intensity
of the light could be expressed in finite terms. This saves the
trouble of Mr Airy's quadratures, and allows the results to be
discussed with great facility. The law, too, of the variation of
the intensity with the thickness of the plate is very remarkable,
on account of its discontinuity. These reasons have induced me
to lay my investigation before the Royal Society, even though
the remarkable polarity of the bands has been already explained
by the Astronomer Royal. The observation of these bands seems
likely to become of great importance in the determination of the
refractive indices, and more especially the laws of dispersion, of
minerals and other substances which cannot be formed into prisms
which would exhibit the fixed lines of the spectrum.

SECTION I.

Explanation of the formation of the bands on the imperfect theory
of Interferences. Mode of calculating the number of bands
seen in a given part of the spectrum.

1. The phenomenon of which it is the principal object of the
following paper to investigate the theory, is briefly as follows.
Light introduced into a room through a horizontal slit is allowed
to pass through a hollow glass prism containing fluid, with its
refracting edge horizontal, and the spectrum is viewed through
a small telescope with its object-glass close to the prism. On
inserting into the fluid a transparent plate with its lower edge
horizontal, the spectrum is seen traversed from end to end by
very numerous dark bands, which are parallel to the fixed lines.
Under favourable circumstances the dark bands are intensely
black; but in certain cases, to be considered presently, no bands
whatsoever are seen. When the plate is cut from a doubly refracting crystal, there are in general two systems of bands seen
together; and when the light is analysed each system disappears
in turn at every quarter revolution of the analyser.

2. It is not difficult to see that the theory of these bands
must be almost identical with that of the bands described by
Sir David Brewster in the *Report of the Seventh Meeting of the*

British Association, and elsewhere, and explained by Mr Airy in the first part of the *Philosophical Transactions* for 1841. To make this apparent, conceive an eye to view a spectrum through a small glass vessel with parallel faces filled with fluid. The vessel would not alter the appearance of the spectrum. Now conceive a transparent plate bounded by parallel surfaces inserted into the fluid, the plane of the plate being perpendicular to the axis of the eye, and its edge parallel to the fixed lines of the spectrum, and opposite to the centre of the pupil. Then we should have bands of the same nature as those described by Sir David Brewster, the only difference being that in the present case the retardation on which the existence of the bands depends is the difference of the retardations due to the plate itself, and to a plate of equal thickness of the fluid, instead of the absolute retardation of the plate, or more strictly, the difference of retardations of the solid plate and of a plate of equal thickness of air, contained between the produced parts of the bounding planes of the solid plate. In Professor Powell's experiment the fluid fills the double office of the fluid in the glass vessel and of the prism producing the spectrum in the imaginary experiment just described.

It might be expected that the remarkable polarity discovered by Sir David Brewster in the bands which he has described, would also be exhibited with Professor Powell's apparatus. This anticipation is confirmed by experiment. With the arrangement of the apparatus already mentioned, it was found that with certain pairs of media, one being the fluid and the other the retarding plate, no bands were visible. These media were made to exhibit bands by using fluid enough to cover the plate to a certain depth, and stopping by a screen the light which would otherwise have passed through the thin end of the prism underneath the plate.

3. Although the explanation of the polarity of the bands depends on diffraction, it may be well to account for their formation on the imperfect theory of interferences, in which it is supposed that light consists of rays which follow the courses assigned to them by geometrical optics. It will thus readily appear that the number of bands formed with a given plate and fluid, and in a given part of the spectrum, has nothing to do with the

form or magnitude of the aperture, whatever it be, which limits
the pencil that ultimately falls on the retina. Moreover, it seems
desirable to exhibit in its simplest shape the mode of calculating
the number of bands seen in any given case, more especially as
these calculations seem likely to be of importance in the determination of refractive indices.

4. Before the insertion of the plate, the wave of light belonging to a particular colour, and to a particular point of the slit,
or at least a certain portion of it limited by the boundaries of
the fluid, after being refracted at the two surfaces of the prism
enters the object-glass with an unbroken front. The front is here
called unbroken, because the modification which the wave suffers
at its edges is not contemplated. According to geometrical optics,
the light after entering the object-glass is brought to a point near
the principal focus, spherical aberration being neglected; according to the undulatory theory, it forms a small, but slightly diffused image of the point from which it came. The succession of
these images due to the several points of the slit forms the image
of the slit for the colour considered, and the succession of coloured
images forms the spectrum, the waves for the different colours
covering almost exactly the same portion of the object-glass, but
differing from one another in direction.

Apart from all theory, it is certain that the image of a point or
line of homogeneous light seen with a small aperture is diffused.
As the aperture is gradually widened the extent of diffusion decreases continuously, and at last becomes insensible. The perfect
continuity, however, of the phenomenon shows that the true
and complete explanation, whatever it may be, of the narrow
image seen with a broad aperture, ought also to explain the diffused image seen with a narrow aperture. The undulatory theory
explains perfectly both the one and the other, and even pre-
dicts the distribution of the illumination in the image seen
with an aperture of given form, which is what no other theory
has ever attempted.

As an instance of the effect of diffusion in an image, may
be mentioned the observed fact that the definition of a telescope is impaired by contracting the aperture. With a moderate aperture, however, the diffusion is so slight as not to prevent

fine objects, such as the fixed lines of the spectrum, from being well seen.

For the present, however, let us suppose the light entering the telescope to consist of rays which are brought accurately to a focus, but which nevertheless interfere. When the plate is inserted into the fluid the front of a wave entering the object-glass will no longer be unbroken, but will present as it were a *fault*, in consequence of the retardation produced by the plate. Let R be this retardation measured by actual length in air, ρ the retardation measured by phase, M the retardation measured by the number of waves' lengths, so that

$$\rho = \frac{2\pi}{\lambda} R, \quad M = \frac{1}{\lambda} R;$$

then when M is an odd multiple of $\frac{1}{2}$, the vibrations produced by the two streams, when brought to the same focus, will oppose each other, and there will be a minimum of illumination; but when M is an even multiple of $\frac{1}{2}$ the two streams will combine, and the illumination will be a maximum. Now M changes in passing from one colour to another in consequence of the variations both of R and of λ; and since the different colours occupy different angular positions in the field of view, the spectrum will be seen traversed by dark and bright bands. It is nearly thus that Mr Talbot has explained the bands seen when a spectrum is viewed through a hole in a card which is half covered with a plate of glass or mica, with its edge parallel to the fixed lines of the spectrum. Mr Talbot however does not appear to have noticed the polarity of the bands.

Let h, k be the breadths of the interfering streams; then we may take

$$h \sin \frac{2\pi}{\lambda} vt, \quad k \sin \left(\frac{2\pi}{\lambda} vt - \rho \right)$$

to represent the vibrations produced at the focus by the two streams respectively, which gives for the intensity I,

$$I = (h + k \cos \rho)^2 + (k \sin \rho)^2 = h^2 + k^2 + 2hk \cos \rho \ \ldots\ldots\ldots(1),$$

which varies between the limits $(h - k)^2$ and $(h + k)^2$.

5. Although the preceding explanation is imperfect, for the reason already mentioned, and does not account for the polarity,

it is evident that if bands are formed at all in this way, the
number seen in a given part of the spectrum will be determined
correctly by the imperfect theory; for everything will recur, so
far as interference is concerned, when M is decreased or increased
by 1, and not before. This points out an easy mode of determining the number of bands seen in a given part of the spectrum.
For the sake of avoiding a multiplicity of cases, let an acceleration be reckoned as a negative retardation, and suppose R positive
when the stream which passes nearer to the edge of the prism is
retarded relatively to the other. From the known refractive
indices of the plate and fluid, and from the circumstances of the
experiment, calculate the values of R for each of the fixed lines
B, C......H of the spectrum, or for any of them that may be
selected, and thence the values of M, by dividing by the known
values of λ. Set down the results with their proper signs opposite
to the letters B, C... denoting the rays to which they respectively
refer, and then form a table of differences by subtracting the
value of M for B from the value for C, the value for C from the
value for D, and so on. Let N be the number found in the table
of differences corresponding to any interval, as for example from
F to G ; then the numerical value of N, that is to say, N or $-N$,
according as N is positive or negative, gives the number of bands
seen between F and G. For anything that appears from the
imperfect theory of the bands given in the preceding article, it
would seem that the sign of N was of no consequence. It will
presently be seen, however, that the sign is of great importance :
it will be found in fact that the sign $+$ indicates that the second
arrangement mentioned in Art. 2 must be employed; that is to
say, the plate must be made to intercept light from the thin end
of the prism, while the sign $-$ indicates that the first arrangement is required. It is hardly necessary to remark that, if N
should be fractional, we must, instead of the number of bands,
speak of the number of band-intervals and the fraction of an
interval.

Although the number of bands depends on nothing but the
values of N, the values of M are not without physical interest.
For M expresses, as we have seen, the number of waves' lengths
whereby one of the interfering streams is before or behind the
other. Mr Airy speaks of the formation of rings with the light of

a spirit-lamp when the retardation of one of the interfering streams is as much as fifty or sixty waves' lengths. But in some of Professor Powell's experiments, bands were seen which must have been produced by retardations of several hundred waves' lengths. This exalts our ideas of the regularity which must be attributed to the undulations.

6. It appears then that the calculation of the number of bands is reduced to that of the retardation R. As the calculation of R is frequently required in physical optics, it will not be necessary to enter into much detail on this point. The mode of performing the calculation, according to the circumstances of the experiment, will best be explained by a few examples.

Suppose the retarding plate to belong to an ordinary medium, and to be placed so as to intercept light from the thin end of the prism, and to have its plane equally inclined to the faces of the prism. Suppose the prism turned till one of the fixed lines, as F, is seen at a minimum deviation; then the colours about F are incident perpendicularly on the plate; and all the colours may without material error be supposed to be incident perpendicularly, since the directions of the different colours are only separated by the dispersion accompanying the first refraction into the fluid, and near the normal a small change in the angle of incidence produces only a very small change in the retardation. The dispersion accompanying the first refraction into the fluid has been spoken of as if the light were refracted from air directly into the fluid, which is allowable, since the glass sides of the hollow prism, being bounded by parallel surfaces, may be dispensed with in the explanation. Let T be the thickness of the plate, μ the refractive index of the fluid, μ' that of the plate; then

$$R = (\mu' - \mu)\ T \dotfill (2).$$

If the plate had been placed so as to intercept light from the thick end of the prism, we should have had $-R = (\mu' - \mu)\ T$, which would have agreed with (2) if we had supposed T negative. For the future T will be reckoned positive when the plate intercepts light from the thin end of the prism, and negative when it intercepts light from the thick end, so that the same formulæ will apply to both of the arrangements mentioned in Art. 2.

If we put $\mu = 1$, the formula (2) will apply to the experiment in which a plate of glass or mica is held so as to cover half the pupil of the eye when viewing a spectrum formed in any manner, the plate being held perpendicularly to the axis of the eye. The effect of the small obliquity of incidence of some of the colours is supposed to be neglected.

The number of bands which would be determined by means of the formula (2) would not be absolutely exact, unless we suppose the observation taken by receiving each fixed line in succession at a perpendicular incidence. This may be effected in the following manner. Suppose that we want to count the number of bands between F and G, move the plate by turning it round a horizontal axis till the bands about F are seen stationary; then begin to count from F, and before stopping at G incline the plate a little till the bands about G are seen stationary, estimating the fractions of an interval at F and G, if the bands are not too close. The result will be strictly the number given by the formula (2). The difference, however, between this result and that which would be obtained by keeping the plate fixed would be barely sensible. If the latter mode of observation should be thought easier or more accurate, the exact formula which would replace (2) would be easily obtained.

7. Suppose now the nearer face of the retarding plate made to rest on the nearer inner face of the hollow prism, and suppose one of the fixed lines, as F, to be viewed at a minimum deviation. Let ϕ, ϕ' be the angles of incidence and refraction at the first surface of the fluid, i, i' those at the surface of the plate, 2ϵ the angle of the prism. Since the deviation of F is a minimum, the angle of refraction ϕ'_F for F is equal to ϵ, and the angle of incidence ϕ is given by $\sin \phi = \mu_F \sin \phi'_F$, and ϕ is the angle of incidence for all the colours, the incident light being supposed white. The angle of refraction ϕ' for any fixed line is given by the equation $\sin \phi' = 1/\mu . \sin \phi = \mu_F/\mu . \sin \epsilon$; then $i = 2\epsilon - \phi'$, and i' is known from the equation

$$\mu' \sin i'' = \mu \sin i \dots\dots\dots\dots (3).$$

The retardation is given by either of the formulæ

$$R = \mu' T \frac{\sin (i - i'')}{\sin i} \dots\dots\dots\dots (4),$$

$$R = T (\mu' \cos i'' - \mu \cos i) \dots\dots\dots (5).$$

These formulæ might be deduced from that given in Airy's Tract, modified so as to suit the case in which the plate is immersed in a fluid; but either of them may be immediately proved independently by referring everything to the wave's front and not the ray.

By multiplying and dividing the second side of (5) by cos i, and employing (3), we get

$$R = T \sec i \,.\, (\mu' - \mu) - T\mu' \sec i \text{ versin } (i - i') \ldots\ldots\ldots(6).$$

When the refractive indices of the plate and fluid are nearly equal, the last term in this equation may be considered insensible, so that it is not necessary to calculate i' at all.

8. The formulæ (2), (4), (5), (6) are of course applicable to the ordinary ray of a plate cut from a uniaxal crystal. If the plate be cut in a direction parallel to the axis, and if moreover the lower edge be parallel to the axis, so that the axis is parallel to the refracting edge of the prism, the formulæ will apply to both rays. If μ_o, μ_e be the principal indices of refraction referring to the ordinary and extraordinary rays respectively, μ' in the case last supposed must be replaced by μ_o for the bands polarized in a plane perpendicular to the plane of incidence, and by μ_e for the bands polarized in the plane of incidence. In the case of a plate cut from a biaxal crystal in such a direction that one of the principal axes, or axes of elasticity, is parallel to the refracting edge, the same formulæ will apply to that system of bands which is polarized in the plane of incidence.

If the plate be cut from a biaxal crystal in a direction perpendicular to one of the principal axes, and be held in the vertical position, the formula (2) will apply to both systems of bands, if the small effect of the obliquity be neglected. The formula would be exact if the observations were taken by receiving each fixed line in succession at a perpendicular incidence.

If the plate be cut from a uniaxal crystal in a direction perpendicular to the axis, and be held obliquely, we have for the extraordinary bands, which are polarized in a plane perpendicular to the plane of incidence,

$$R = T \left(\frac{\mu_o}{\mu_e} \sqrt{\mu_e^{\,2} - \mu^2 \sin^2 i} - \mu \cos i \right) \ldots\ldots\ldots\ldots(7),$$

which is the same as the formula in Airy's Tract, only modified so as to suit the case in which the plate is immersed in fluid, and expressed in terms of refractive indices instead of velocities. If we take a subsidiary angle j, determined by the equation

$$\sin j = \frac{\mu}{\mu_o} \sin i \quad\dots\dots\dots\dots\dots\dots(8),$$

the formula (7) becomes

$$R = T \left(\mu_o \cos j - \mu \cos i\right)\dots\dots\dots\dots\dots(9),$$

which is of the same form as (5), and may be adapted to logarithmic calculation if required by assuming $\mu_o/\mu = \tan \theta$. The preceding formula will apply to the extraordinary bands formed by a plate cut from a biaxal crystal perpendicular to a principal axis, and inclined in a principal plane, the extraordinary bands being understood to mean those which are polarized in a plane perpendicular to the plane of incidence. In this application we must take for μ_e, μ_o those two of the three principal indices of refraction which are symmetrically related to the axis normal to the plate, and to the axis parallel to the plate, and lying in the plane of incidence, respectively; while in applying the formula (4), (5) or (6) to the other system of bands, the third principal index must be substituted for μ'.

It is hardly necessary to consider the formula which would apply to the general case, which would be rather complicated.

9.   If a plate cut from a uniaxal crystal in a direction perpendicular to the axis be placed in the fluid in an inclined position, and be then gradually made to approach the vertical position, the breadths of the bands belonging to the two systems will become more and more nearly equal, and the two systems will at last coalesce. This statement indeed is not absolutely exact, because the whole spectrum cannot be viewed at once by light which passes along the axis of the crystal, on account of the dispersion accompanying the first refraction, but it is very nearly exact. With quartz it is true there would be two systems of bands seen even in the vertical position, on account of the peculiar optical properties of that substance; but the breadths of the bands belonging to the two systems would be so nearly equal, that it would require a plate of about one-fifth of an inch thickness to give a difference of one in the number of bands seen in the whole

spectrum in the case of the two systems respectively. If the plate should be thick enough to exhibit both systems, the light would of course have to be circularly analyzed to show one system by itself.

SECTION II.—*Investigation of the intensity of the light on the complete theory of undulations, including the explanation of the apparent polarity of the bands.*

10. The explanation of the formation of the bands on the imperfect theory of interferences considered in the preceding section is essentially defective in this respect, that it supposes an annihilation of light when two interfering streams are in opposition; whereas it is a most important principle that *light is never lost by interference.* This statement may require a little explanation, without which it might seem to contradict received ideas. It is usual in fact to speak of light as *destroyed* by interference. Although this is true, in the sense intended, the expression is perhaps not very happily chosen. Suppose a portion of light coming from a luminous point, and passing through a moderately small aperture, to be allowed to fall on a screen. We know that there would be no sensible illumination on the screen except almost immediately in front of the aperture. Conceive now the aperture divided into a great number of small elements, and suppose the same quantity of light as before to pass through each element, the only difference being that now the vibrations in the portions passing through the several elements are supposed to have no relation to each other. The light would now be diffused over a comparatively large portion of the screen, so that a point P which was formerly in darkness might now be strongly illuminated. The disturbance at P is in both cases the aggregate of the disturbances due to the several elements of the aperture; but in the first case the aggregate is insensible on account of interference. It is only in this sense that light is *destroyed* by interference, for the total illumination on the screen is the same in the two cases; the effect of interference has been, not to annihilate any light, but only to alter the "distribution of the illumination," so that the light, instead of being diffused over the screen, is concentrated in front of the aperture.

Now in the case of the bands considered in Section I., if we suppose the plate extremely thin, the bands will be very broad; and the displacement of illumination due to the retardation being small compared with the breadth of a band, it is evident, without calculation, that at most only faint bands can be formed. This particular example is sufficient to show the inadequacy of the imperfect theory, and the necessity of an exact investigation.

11. Suppose first that a point of homogeneous light is viewed through a telescope. Suppose the object-glass limited by a screen in which there is formed a rectangular aperture of length $2l$. Suppose a portion of the incident light retarded, by passing through a plate bounded by parallel surfaces, and having its edge parallel to the length of the aperture. Suppose the unretarded stream to occupy a breadth h of the aperture at one side, the retarded stream to occupy a breadth k at the other, while an interval of breadth $2g$ exists between the streams. In the apparatus mentioned in Section I., the object-glass is not limited by a screen, but the interfering streams of light are limited by the dimensions of the fluid prism, which comes to the same thing. The object of supposing an interval to exist between the interfering streams, is to examine the effect of the gap which exists between the streams when the retarding plate is inclined. In the investigation the effect of diffraction before the light reaches the object-glass of the telescope is neglected.

Let O be the image of the luminous point, as determined by geometrical optics, f the focal length of the object-glass, or rather the distance of O from the object-glass, which will be a little greater than the focal length when the luminous point is not very distant. Let C be a point in the object-glass, situated in the middle of the interval between the two streams, and let the intensity be required at a point M, near O, situated in a plane passing through O and perpendicular to OC. The intensity at any point of this plane will of course be sensibly the same as if the plane were drawn perpendicular to the axis of the telescope instead of being perpendicular to OC. Take OC for the axis of z, the axes of x and y being situated in the plane just mentioned, and that of y being parallel to the length of the aperture. Let p, q be the co-ordinates of M; x, y, z those of a point P in the front of a wave which has just passed through the object-glass, and which forms part of a sphere with O for its centre. Let c be the coefficient of vibration at the distance of the object-glass; then we may take

$$\frac{c}{\lambda} \cdot \frac{1}{PM} \sin \frac{2\pi}{\lambda} (vt - PM)\, dx\, dy \dots\dots\dots (a),$$

to represent the disturbance at M due to the element $dx\,dy$ of the aperture at P, P being supposed to be situated in the unretarded stream, which will be supposed to lie at the negative side of the axis of x. In the expression (a), it is assumed that the proper multiplier of c/PM is $1/\lambda$. This may be shown to be a necessary consequence of the principle mentioned in the preceding article, that light is never lost by interference; and this principle follows directly from the principle of *vis viva*. In proving that λ^{-1} is the proper multiplier, it is not in the least necessary to enter into the consideration of the law of the variation of intensity in a secondary wave, as the angular distance from the normal to the primary wave varies; the result depends merely on the assumption that in the immediate neighbourhood of the normal the intensity may be regarded as sensibly constant.

In the expression (a) we have

$$PM = \sqrt{\{z^2 + (x - p)^2 + (y - q)^2\}} = \sqrt{\{f^2 + p^2 + q^2 - 2px - 2qy\}}$$

$$= f - \frac{1}{f}(px + qy),\ \text{nearly,}$$

if we write f for $\sqrt{(f^2 + p^2 + q^2)}$. It will be sufficient to replace $1/PM$ outside the circular function by $1/f$. We may omit the constant f under the circular function, which comes to the same thing as changing the origin of t. We thus get for the disturbance at M due to the unretarded stream,

$$\frac{c}{\lambda f} \int_{-(g+h)}^{-g} \int_{-l}^{l} \sin \frac{2\pi}{\lambda} \left\{ vt + \frac{1}{f}(px + qy) \right\} dx\, dy,$$

or on performing the integrations and reducing,

$$\frac{2chl}{\lambda f} \cdot \frac{\lambda f}{2\pi ql} \sin \frac{2\pi ql}{\lambda f} \cdot \frac{\lambda f}{\pi ph} \sin \frac{\pi ph}{\lambda f} \cdot \sin \frac{2\pi}{\lambda} \left(vt - \frac{pg}{f} - \frac{ph}{2f} \right) \dots(b).$$

For the retarded stream, the only difference is that we must subtract R from vt, and that the limits of x are g and $g + k$. We thus get for the disturbance at M due to this stream,

$$\frac{2ckl}{\lambda f} \cdot \frac{\lambda f}{2\pi ql} \sin \frac{2\pi ql}{\lambda f} \cdot \frac{\lambda f}{\pi pk} \sin \frac{\pi pk}{\lambda f} \cdot \sin \frac{2\pi}{\lambda} \left(vt - R + \frac{pg}{f} + \frac{pk}{2f} \right) \dots(c).$$

If we put for shortness τ for the quantity under the last circular function in (b), the expressions (b), (c) may be put under the forms $u \sin \tau$, $v \sin (\tau - \alpha)$, respectively ; and if I be the intensity, I will be measured by the sum of the squares of the coefficients of $\sin \tau$ and $\cos \tau$ in the expression

$$u \sin \tau + v \sin (\tau - \alpha),$$

so that

$$I = u^2 + v^2 + 2uv \cos \alpha,$$

which becomes, on putting for u, v and α, their values, and putting

$$\left\{ \frac{\lambda f}{2\pi q l} \sin \frac{2\pi q l}{\lambda f} \right\}^2 = Q \dots\dots\dots\dots(10),$$

$$I = Q \cdot \frac{4c^2 l^2}{\pi^2 p^2} \left\{ \left(\sin \frac{\pi p h}{\lambda f} \right)^2 + \left(\sin \frac{\pi p k}{\lambda f} \right)^2 \right.$$

$$\left. + 2 \sin \frac{\pi p h}{\lambda f} \cdot \sin \frac{\pi p k}{\lambda f} \cos \left[\rho - \frac{\pi p}{\lambda f} (4g + h + k) \right] \right\} \dots(11).$$

12. Suppose now that instead of a point we have a line of homogeneous light, the line being parallel to the axis of y. The luminous line is supposed to be a narrow slit, through which light enters in all directions, and which is viewed in focus. Consequently each element of the line must be regarded as an independent source of light. Hence the illumination on the object-glass due to a portion of the line which subtends the small angle β at the distance of the object-glass varies as β, and may be represented by $A\beta$. Let the former origin O be referred to a new origin O' situated in the plane xy, and in the image of the line ; and let η, q' be the ordinates of O, M referred to O', so that $q = q' - \eta$. In order that the luminous point considered in the last article may represent an element of the luminous line considered in the present, we must replace c^2 by $A d\beta$ or $A f^{-1} d\eta$; and in order to get the aggregate illumination due to the whole line, we must integrate from a large negative to a large positive value of η, the largeness being estimated by comparison with $\lambda f / l$. Now the angle $2\pi q l / \lambda f$ changes by π when q changes by $\lambda f / 2l$, which is therefore the breadth, in the direction of y, of one of the diffraction bands which would be seen with a luminous point. Since l is supposed not to be extremely small, but on the contrary moderately large, the whole system of diffraction bands would occupy but a very small portion of the field of view in the direction of y, so that we may without

sensible error suppose the limits of η to be $-\infty$ and $+\infty$. We have then

$$\int_{-\infty}^{\infty} Q d\eta = \int_{-\infty}^{\infty} \left\{ \frac{\lambda f}{2\pi l \left(q' - \eta\right)} \sin \frac{2\pi l \left(q' - \eta\right)}{\lambda f} \right\}^2 d\eta = \frac{\lambda f}{2\pi l} \int_{-\infty}^{\infty} \left(\frac{\sin \theta}{\theta} \right)^2 d\theta,$$

by taking the quantity under the circular function in place of η for the independent variable. Now it is known that the value of the last integral is π, as will also presently appear, and therefore we have for the intensity I at any point,

$$I = \frac{2 A \lambda l}{\pi^2 p^2} \left\{ \left(\sin \frac{\pi p h}{\lambda f} \right)^2 + \left(\sin \frac{\pi p k}{\lambda f} \right)^2 \right.$$

$$\left. + 2 \sin \frac{\pi p h}{\lambda f} . \sin \frac{\pi p k}{\lambda f} . \cos \left[\rho - \frac{\pi p}{\lambda f} \left(4g + h + k \right) \right] \right\} \dots (12),$$

which is independent of q', as of course it ought to be.

13. Suppose now that instead of a line of homogeneous light we have a line of white light, the component parts of which have been separated, whether by refraction or by diffraction is immaterial, so that the different colours occupy different angular positions in the field of view. Let $B\beta\psi$ be the illumination on the object-glass due to a length of the line which subtends the small angle β, and to a portion of the spectrum which subtends the small angle ψ at the centre of the object-glass. In the axis of x take a new origin O'', and let ξ, p' be the abscissæ of O', M reckoned from O'', so that $p = p' - \xi$. In order that (12) may express the intensity at M due to an elementary portion of the spectrum, we must replace A by $Bd\psi$, or $Bf^{-1}d\xi$; and in order to find the aggregate illumination at M, we must integrate so as to include all values of ξ which are sufficiently near to p' to contribute sensibly to the illumination at M. It would not have been correct to integrate using the displacement instead of the intensity, because the different colours cannot interfere. Suppose the angular extent, in the direction of x, of the system of diffraction bands which would be seen with homogeneous light, or at least the angular extent of the brighter part of the system, to be small compared with that of the spectrum. Then we may neglect the variations of B and of λ in the integration, considering only those of ξ and ρ, and we may suppose the changes of ρ proportional to those of ξ; and we may moreover suppose the limits of ξ to be $-\infty$ and $+\infty$. Let ρ' be the value of ρ, and $-\varpi$ that of $d\rho/d\xi$, when $\xi = p'$, so that we may

put $\rho = \rho' + \varpi\,(p' - \xi)$; and take p instead of ξ for the independent variable. Then putting for shortness

$$\frac{\pi h}{\lambda f} = h_{,}, \qquad \frac{\pi k}{\lambda f} = k_{,}, \qquad \varpi - \frac{\pi}{\lambda f}(4g + h + k) = g_{,}, \quad \dots\dots(13),$$

we have for the intensity,

$$I = \frac{2B\lambda l}{\pi^2 f}\int_{-\infty}^{\infty}\{\sin^2 h_{,}p + \sin^2 k_{,}p + 2\sin h_{,}p\,.\sin k_{,}p\,.\cos(\rho' - g_{,}p)\}\,\frac{dp}{p^2}.$$

Now $$\int_{-\infty}^{\infty}\sin^2 h_{,}p\,\frac{dp}{p^2} = h_{,}\int_{-\infty}^{\infty}\sin^2\theta\,\frac{d\theta}{\theta^2} = \pi h_{,}.$$

Similarly, $$\int_{-\infty}^{\infty}\sin^2 k_{,}p\,.\,\frac{dp}{p^2} = \pi k_{,}.$$

Moreover, if we replace

$$\cos(\rho' - g_{,}p) \text{ by } \cos\rho'\,.\cos g_{,}p + \sin\rho'\,.\sin g_{,}p,$$

the integral containing $\sin\rho'$ will disappear, because the positive and negative elements will destroy each other, and we have only to find w, where

$$w = \int_{-\infty}^{\infty}\sin h_{,}p\,.\sin k_{,}p\,.\cos g_{,}p\,.\frac{dp}{p^2}.$$

Now we get by differentiating under the integral sign,

$$\frac{dw}{dg_{,}} = -\int_{-\infty}^{\infty}\sin h_{,}p\,.\sin k_{,}p\,.\sin g_{,}p\,.\frac{dp}{p}$$

$$= \frac{1}{4}\int_{-\infty}^{\infty}\{\sin(g_{,} + h_{,} + k_{,})\,p + \sin(g_{,} - h_{,} - k_{,})\,p$$

$$- \sin(g_{,} + h_{,} - k_{,})\,p - \sin(g_{,} + k_{,} - h_{,})\,p\}\,\frac{dp}{p}.$$

But it is well known that

$$\int_{-\infty}^{\infty}\frac{\sin sp}{p}\,dp = \pi, \text{ or } = -\pi,$$

according as s is positive or negative. If then we use $F(s)$ to denote a discontinuous function of s which is equal to $+1$ or -1 according as s is positive or negative, we get

$$\frac{dw}{dg_{,}} = \frac{\pi}{4}\{F(g_{,}+h_{,}+k_{,}) + F(g_{,}-h_{,}-k_{,}) - F(g_{,}+h_{,}-k_{,}) - F(g_{,}+k_{,}-h_{,})\}.$$

This equation gives

$$\frac{dw}{dg_{,}} = 0, \text{ from } g_{,} = -\infty \text{ to } g_{,} = -(h_{,}+k_{,})$$

$$= \frac{\pi}{2}, \text{ from } g_{,} = -(h_{,} + k_{,}) \text{ to } g_{,} = -(h_{,} \sim k_{,})$$

$$= 0, \text{ from } g_{,} = -(h_{,} \sim k_{,}) \text{ to } g_{,} = +(h_{,} \sim k_{,})$$

$$= -\frac{\pi}{2}, \text{ from } g_{,} = h_{,} \sim k_{,} \text{ to } g_{,} = h_{,} + k_{,}$$

$$= 0, \text{ from } g_{,} = h_{,} + k_{,} \text{ to } g_{,} = \infty.$$

Now w vanishes when $g_{,}$ is infinite, on account of the fluctuation of the factor $\cos g_{,}p$ under the integral sign, whence we get by integrating the value of $dw/dg_{,}$ given above, and correcting the integral so as to vanish for $g_{,} = -\infty$,

$w = 0$, from $g_{,} = -\infty$ to $g_{,} = -(h_{,} + k_{,})$;

$w = \frac{\pi}{2}(h_{,} + k_{,} + g_{,})$, from $g_{,} = -(h_{,} + k_{,})$ to $g_{,} = -(h_{,} \sim k_{,})$;

$w = \pi k_{,}$ or $= \pi h_{,}$, (according as $h_{,} > k_{,}$ or $h_{,} < k_{,}$)
$$\text{from } g = -(h_{,} \sim k_{,}) \text{ to } g_{,} = +(h_{,} \sim k_{,}) ;$$

$w = \frac{\pi}{2}(h_{,} + k_{,} - g_{,})$, from $g_{,} = h' \sim k_{,}$ to $g_{,} = h_{,} + k_{,}$;

$w = 0$, from $g_{,} = h_{,} + k_{,}$ to $g_{,} = \infty$.

Substituting in the expression for the intensity, and putting in (13) $g_{,} = \pi g'/\lambda f$, so that

$$g' = \frac{\varpi \lambda f}{\pi} - 4g - h - k \dots\dots\dots\dots(14),$$

we get

$$I = \frac{2Bl}{f^2}(h + k) \dots\dots\dots\dots\dots(15),$$

when the numerical value of g' exceeds $h + k$;

$$I = \frac{2Bl}{f^2}\{h + k + (h + k - \sqrt{g'^2})\cos \rho'\}\dots\dots\dots(16),$$

when the numerical value of g' lies between $h + k$ and $h \sim k$;

$$I = \frac{2Bl}{f^2}(h + k + 2h\cos \rho'), \text{ or } = \frac{2Bl}{f^2}(h + k + 2k\cos \rho') \dots(17),$$

according as h or k is the smaller of the two, when the numerical value of g' is less than $h \sim k$.

The discontinuity of the law of intensity is very remarkable.

By supposing $g_i = 0$, $k_i = h_i$ in the expression for w, and observing that these suppositions reduce w to

$$\int_{-\infty}^{\infty} \sin^2 h_i p \cdot \frac{dp}{p^2},$$

we get

$$\int_{-\infty}^{\infty} \left(\frac{\sin h_i p}{p}\right)^2 dp = \pi h_i,$$

a result already employed. This result would of course have been obtained more readily by differentiating with respect to h_i.

14. The preceding investigation will apply, with a very trifling modification, to Sir David Brewster's experiment, in which the retarding plate, instead of being placed in front of the object-glass of a telescope, is held close to the eye. In this case the eye itself takes the place of the telescope; and if we suppose the whole refraction to take place at the surface of the cornea, which will not be far from the truth, we must replace f by the diameter of the eye, and ψ by the angular extent of the portion of the spectrum considered, diminished in the ratio of m to 1, m being the refractive index of the cornea. When a telescope is used in this experiment, the retarding plate being still held close to the eye, it is still the naked eye, and not the telescope, which must be assimilated to the telescope considered in the investigation; the only difference is that ψ must be taken to refer to the magnified, and not the unmagnified spectrum.

Let the axis of x be always reckoned positive in the direction in which the blue end of the spectrum is seen, so that in the image formed at the focus of the object-glass or on the retina, according as the retarding plate is placed in front of the object-glass or in front of the eye, the blue is to the negative side of the red. Although the plate has been supposed at the positive side, there will thus be no loss of generality, for should the plate be at the negative side it will only be requisite to change the sign of ρ.

First, suppose ρ to decrease algebraically in passing from the red to the blue. This will be the case in Sir David Brewster's experiment when the retarding plate is held at the side on which the red is seen. It will be the case in Professor Powell's experiment when the first of the arrangements mentioned in Art. 2 is employed, and the value of N in the table of differences mentioned in Art. 5 is positive, or when the second arrangement is employed and N is negative. In this case ϖ is negative, and therefore $g' < - (h + k)$, and therefore (15) is the expression for the intensity. This expression indicates a uniform intensity, so that there are no bands at all.

Secondly, suppose ρ to increase algebraically in passing from the red to the blue. This will be the case in Sir David Brewster's experiment when the retarding plate is held at the side on which the blue is seen. It will be the case in Professor Powell's experiment when the first arrangement is employed and N is negative, or when the second arrangement is employed and N is positive. In this case ϖ is positive; and since ϖ varies as the thickness of the plate, g' may be made to assume any value from $- (4g + h + k)$ to $+ \infty$ by altering the thickness of the plate. Hence, provided the thickness lie within certain limits, the expression for the intensity will be (16) or (17). Since these expressions have the same form as (1), the magnitude only of the coefficient of $\cos \rho'$, as compared with the constant term, being different, it is evident that the number of bands and the places of the minima are given correctly by the imperfect theory considered in Section I.

15. The plate being placed as in the preceding paragraph, suppose first that the breadths h, k of the interfering streams are equal, and that the streams are contiguous, so that $g = 0$. Then the expression (17) may be dispensed with, since it only holds good when $g' = 0$, in which case it agrees with (16). Let T_0 be the value of the thickness T for which $g' = 0$. Then $T = 0$ corresponds to $g' = - (h + k)$, $T = T_0$ to $g' = 0$, and $T = 2T_0$ to $g' = h + k$; and for values of T equidistant from T_0, the values of g' are equal in magnitude but of opposite signs. Hence, provided T be less than $2T_0$, there are dark and bright bands formed, the vividness of the bands being so much the greater as T is more nearly equal to T_0, for which particular value the minima are absolutely black.

Secondly, suppose the breadths h, k of the two streams to be equal as before, but suppose the streams separated by an interval $2g$; then the only difference is that $g' = - (h + k)$ corresponds to a positive value, T_2 suppose, of T. If T be less than T_2, or greater than $2T_0 - T_2$, there are no bands; but if T lie between T_2 and $2T_0 - T_2$ bands are formed, which are most vivid when $T = T_0$, in which case the minima are perfectly black.

Thirdly, suppose the breadths h, k of the interfering streams unequal, and suppose, as before, that the streams are separated by an interval $2g$; then $g' = -(h+k)$ corresponds to a positive value, T_2 suppose, of T: $g' = -(h \sim k)$ corresponds to another positive value, T_1 suppose, of T, T_1 lying between T_2 and T_0, T_0 being, as before, the value of T which gives $y' = 0$. As T increases from T_0, g' becomes positive and increases from 0, and becomes equal to $h \sim k$ when $T = 2T_0 - T_1$, and to $h + k$ when $T = 2T_0 - T_2$. When $T < T_2$ there are no bands. As T increases to T_1 bands become visible, and increase in vividness till $T = T_1$, when the ratio of the minimum intensity to the maximum becomes that of $h - k$ to $h + 3k$, or of $k - h$ to $k + 3h$, according as h or k is the greater of the two, h, k. As T increases to $2T_0 - T_1$, the vividness of the bands remains unchanged; and as T increases from $2T_0 - T_1$ to $2T_0 - T_2$, the vividness decreases by the same steps as it before increased. When $T = 2T_0 - T_2$, the bands cease to exist, and no bands are formed for a greater value of T.

Although in discussing the intensity of the bands the aperture has been supposed to remain fixed, and the thickness of the plate to alter, it is evident that we might have supposed the thickness of the plate to remain the same and the aperture to alter. Since $\varpi \propto T$, the vividness of the bands, as measured by the ratio of the maximum to the minimum intensity, will remain the same when T varies as the aperture. This consideration, combined with the previous discussion, renders unnecessary the discussion of the effect of altering the aperture. It will be observed that, as a general rule, fine bands require a comparatively broad aperture in order that they may be well formed, while broad bands require a narrow aperture.

16. The particular thickness T_0 may be conveniently called the *best thickness*. This term is to a certain extent conventional, since when h and k are unequal the thickness may range from T_1 to $2T_0 - T_1$ without any change being produced in the vividness of the bands. The best thickness is determined by the equation

$$\varpi = -\frac{d\rho}{d\xi} = \frac{\pi}{\lambda f}.(4g + h + k).$$

Now in passing from one band to its consecutive, ρ changes by 2π, and ξ by e, if e be the linear breadth of a band; and for this small change of ξ we may suppose the changes of ρ and ξ proportional, or put $-d\rho/d\xi = 2\pi/e$. Hence the best aperture for a given thickness is that for which

$$4g + h + k = \frac{2\lambda f}{e}.$$

If $g = 0$ and $k = h$, this equation becomes $h = \lambda f/e$.

The difference of distances of a point in the plane xy whose coordinates are ξ, 0 from the centres of the portions of the object-glass which are covered by the interfering streams, is nearly

$$\frac{\xi}{f}\left(g + \frac{1}{2}h\right) - \frac{\xi}{f}\left(-g - \frac{1}{2}k\right), \quad \text{or} \quad \frac{\xi}{2f}(4g + h + k);$$

and if δ be the change of ξ when this difference changes by λ,

$$4g + h + k = \frac{2\lambda f}{\delta}.$$

Hence, when the thickness of the plate is equal to the best thickness, $e = \delta$, or the interval between the bands seen in the spectrum is equal to the interval between the bands formed by the interference of two streams of light, of the colour considered, coming from a luminous line seen in focus, and entering the object-glass through two very narrow slits parallel to the axis of y, and situated in the middle of the two interfering streams respectively. This affords a ready mode of remembering and calculating the best thickness of plate for a given aperture, or the best aperture for a given thickness of plate.

17. According to the preceding explanation, no bands would be formed in Sir David Brewster's experiment when the plate was held on the side of the spectrum on which the red was seen. Mr Airy has endeavoured to explain the existence of bands under such circumstances[*]. Mr Airy appears to speak doubtfully of his explanation, and in fact to offer it as little more than a conjecture to account for an observed phenomenon. In the experiments of Mr Talbot and Mr Airy, bands appear to have been seen when the retarding plate was held at the red side of the spectrum; whereas Sir David Brewster has stated that he has repeatedly looked for the bands under these circumstances and has never been able to

---

[1] *Philosophical Transactions* for 1841, Part I. p. 6.

find the least trace of them; and he considers the bands seen by Mr Talbot and Mr Airy in this case to be of the nature of Newton's rings. While so much uncertainty exists as to the experimental circumstances under which the bands are seen when the retarding plate is held at the red side of the spectrum, if indeed they are seen at all, it does not seem to be desirable to enter into speculations as to the cause of their existence.

[From the *Cambridge and Dublin Mathematical Journal*, Vol. III. p. 209 (*November*, 1848)].

NOTES ON HYDRODYNAMICS.

IV.—*Demonstration of a Fundamental Theorem.*

THEOREM. Let the accelerating forces X, Y, Z, acting on the fluid, be such that $Xdx + Ydy + Zdz$ is the exact differential dV of a function of the coordinates. The function V may also contain the time t explicitly, but the differential is taken on the supposition that t is constant. Suppose the fluid to be either homogeneous and incompressible, or homogeneous and elastic, and of the same temperature throughout, except in so far as the temperature is altered by sudden condensation or rarefaction, so that the pressure is a function of the density. Then if, either for the whole fluid mass, or for a certain portion of it, the motion is at one instant such that $udx + vdy + wdz$ is an exact differential, that expression will always remain an exact differential, in the first case throughout the whole mass, in the second case throughout the portion considered, a portion which will in general continually change its position in space as the motion goes on. In particular, the proposition is true when the motion begins from rest.

Two demonstrations of this important theorem will here be given. The first is taken from a memoir by M. Cauchy, " Mémoire sur la Théorie des Ondes, &c." (*Mém. des Savans Étrangers*, Tom. I. (1827), p. 40). M. Cauchy has obtained three first integrals of the equations of motion for the case in which $Xdx + Ydy + Zdz$ is an exact differential, and in which the pressure is a function of the density; a case which embraces almost all the problems of any interest in this subject. M. Cauchy, it is true, has only considered an incompressible fluid, in accordance with the problem he had in hand, but his method applies to the more general case in which the pressure is a function of the density. The theorem considered follows as a particular consequence from M. Cauchy's integrals. As however the equations employed in obtaining these integrals are rather long, and the integrals themselves do not seem to lead to any result of much interest except the theorem enunciated at the beginning of this article*, I have given another demonstration of the theorem, which is taken from the *Cambridge Philosophical Transactions* (Vol. VIII. p. 307†). A new proof of the theorem for the case of an incompressible fluid will be given by Professor Thomson in this *Journal*.

FIRST DEMONSTRATION. Let the time t and the initial coordinates a, b, c be taken for the independent variables; and let $\int \dfrac{dp}{\rho} = P$, p being by hypothesis a function of ρ. Since we have, by the Differential Calculus,

$$\frac{dP}{da} = \frac{dP}{dx}\frac{dx}{da} + \frac{dP}{dy}\frac{dy}{da} + \frac{dP}{dz}\frac{dz}{da},$$

with similar equations for b and c, we get from equations (1), p. 124 (Notes on Hydrodynamics, No. III.) [*Ante*, p. 4],

$$\left.\begin{aligned}
\frac{dV}{da} - \frac{dP}{da} &= \frac{d^2x}{dt^2}\frac{dx}{da} + \frac{d^2y}{dt^2}\frac{dy}{da} + \frac{d^2z}{dt^2}\frac{dz}{da} \\
\frac{dV}{db} - \frac{dP}{db} &= \frac{d^2x}{dt^2}\frac{dx}{db} + \frac{d^2y}{dt^2}\frac{dy}{db} + \frac{d^2z}{dt^2}\frac{dz}{db} \\
\frac{dV}{dc} - \frac{dP}{dc} &= \frac{d^2x}{dt^2}\frac{dx}{dc} + \frac{d^2y}{dt^2}\frac{dy}{dc} + \frac{d^2z}{dt^2}\frac{dz}{dc}
\end{aligned}\right\} \quad \ldots\ldots\ldots\ldots (1).$$

In these equations d^2x/dt^2, dx/da, &c. have been written for D^2x/Dt^2, Dx/Da, &c., since the context will sufficiently explain the sense in which the differential coefficients are taken. By differentiating the first of equations (1) with respect to b, the second with respect

* [See however the note at p. 47.]

† [*Ante*, Vol. I. p. 108. Although given already in nearly the same form, the demonstration is here retained, to avoid breaking the continuity of the present article.]

to a, and subtracting, we get, after putting for dx/dt, dy/dt, dz/dt their values u, v, w,

$$\frac{d^2u}{dtdb}\frac{dx}{da} - \frac{d^2u}{dtda}\frac{dx}{db} + \frac{d^2v}{dtdb}\frac{dy}{da} - \frac{d^2v}{dtda}\frac{dy}{db} + \frac{d^2w}{dtdb}\frac{dz}{da}$$

$$- \frac{d^2w}{dtda}\frac{dz}{db} = 0 \ldots\ldots (2).$$

By treating the second and third, and then the third and first of equations (1) as the first and second have been treated, we should get two more equations, which with (2) would form a symmetrical system. Now it is easily seen, on taking account of the equations $dx/dt = u$, &c., that the first side of (2) is the differential coefficient with respect to t of

$$\frac{du}{db}\frac{dx}{da} - \frac{du}{da}\frac{dx}{db} + \frac{dv}{db}\frac{dy}{da} - \frac{dv}{da}\frac{dy}{db} + \frac{dw}{db}\frac{dz}{da} - \frac{dw}{da}\frac{dz}{db} \ldots\ldots (3),$$

the differential coefficient in question being of course of the kind denoted by D in No. III. of these Notes. Hence the expression (3) is constant for the same particle. Let u_0, v_0, w_0 be the initial velocities of the particle which at the time t is situated at the point (x, y, z); then if we observe that $x = a$, $y = b$, $z = c$, when $t = 0$, we shall get from (2) and the two other equations of that system,

$$\left.\begin{array}{l}\dfrac{du}{db}\dfrac{dx}{da} - \dfrac{du}{da}\dfrac{dx}{db} + \dfrac{dv}{db}\dfrac{dy}{da} - \dfrac{dv}{da}\dfrac{dy}{db} + \dfrac{dw}{db}\dfrac{dz}{da} - \dfrac{dw}{da}\dfrac{dz}{db} = \dfrac{du_0}{db} - \dfrac{dv_0}{da} \\[2ex] \dfrac{du}{dc}\dfrac{dx}{db} - \dfrac{du}{db}\dfrac{dx}{dc} + \dfrac{dv}{dc}\dfrac{dy}{db} - \dfrac{dv}{db}\dfrac{dy}{dc} + \dfrac{dw}{dc}\dfrac{dz}{db} - \dfrac{dw}{db}\dfrac{dz}{dc} = \dfrac{dv_0}{dc} - \dfrac{dw_0}{db} \\[2ex] \dfrac{du}{da}\dfrac{dx}{dc} - \dfrac{du}{dc}\dfrac{dx}{da} + \dfrac{dv}{da}\dfrac{dy}{dc} - \dfrac{dv}{dc}\dfrac{dy}{da} + \dfrac{dw}{da}\dfrac{dz}{dc} - \dfrac{dw}{dc}\dfrac{dz}{da} = \dfrac{dw_0}{da} - \dfrac{du_0}{dc}\end{array}\right\} \ldots (4).$$

These are the three first integrals of the equations of motion already mentioned. If we replace the differential coefficients of u, v and w, taken with respect to a, b and c, by differential coefficients of the same quantities taken with respect to x, y and z, and differential coefficients of x, y and z taken with respect to a, b and c, the first sides of equations (4) become

$$\left.\begin{aligned}
&\left(\frac{du}{dy}-\frac{dv}{dx}\right)\left(\frac{dy}{db}\frac{dx}{da}-\frac{dy}{da}\frac{dx}{db}\right)+\left(\frac{dv}{dz}-\frac{dw}{dy}\right)\left(\frac{dz}{db}\frac{dy}{da}-\frac{dz}{da}\frac{dy}{db}\right)\\
&\qquad+\left(\frac{dw}{dx}-\frac{du}{dz}\right)\left(\frac{dx}{db}\frac{dz}{da}-\frac{dx}{da}\frac{dz}{db}\right)\\
&\left(\frac{du}{dy}-\frac{dv}{dx}\right)\left(\frac{dy}{dc}\frac{dx}{db}-\frac{dy}{db}\frac{dx}{dc}\right)+\left(\frac{dv}{dz}-\frac{dw}{dy}\right)\left(\frac{dz}{dc}\frac{dy}{db}-\frac{dz}{db}\frac{dy}{dc}\right)\\
&\qquad+\left(\frac{dw}{dx}-\frac{du}{dz}\right)\left(\frac{dx}{dc}\frac{dz}{db}-\frac{dx}{db}\frac{dz}{dc}\right)\\
&\left(\frac{du}{dy}-\frac{dv}{dx}\right)\left(\frac{dy}{da}\frac{dx}{dc}-\frac{dy}{dc}\frac{dx}{da}\right)+\left(\frac{dv}{dz}-\frac{dw}{dy}\right)\left(\frac{dz}{da}\frac{dy}{dc}-\frac{dz}{dc}\frac{dy}{da}\right)\\
&\qquad+\left(\frac{dw}{dx}-\frac{du}{dz}\right)\left(\frac{dx}{da}\frac{dz}{dc}-\frac{dx}{dc}\frac{dz}{da}\right)
\end{aligned}\right\}\dots(5).$$

Having put the first sides of equations (4) under the form (5), we may solve the equations, regarding

$$\frac{du}{dy}-\frac{dv}{dx},\ \frac{dv}{dz}-\frac{dw}{dy},\ \frac{dw}{dx}-\frac{du}{dz}$$

as the unknown quantities. For this purpose multiply equations (4) by dz/dc, dz/da, dz/db, and add; then the second and third unknown quantities will disappear. Again, multiply by dx/dc, dx/da, dx/db, and add; then the third and first will disappear. Lastly, multiply by dy/dc, dy/da, dy/db, and add; then the first and second will disappear. Putting for shortness

$$\frac{dx}{da}\frac{dy}{db}\frac{dz}{dc}-\frac{dx}{da}\frac{dy}{dc}\frac{dz}{db}+\frac{dx}{db}\frac{dy}{dc}\frac{dz}{da}-\frac{dx}{db}\frac{dy}{da}\frac{dz}{dc}$$

$$+\frac{dx}{dc}\frac{dy}{da}\frac{dz}{db}-\frac{dx}{dc}\frac{dy}{db}\frac{dz}{da}=R\ \dots\dots(6),$$

we thus get

$$\left.\begin{aligned}
\frac{du}{dy}-\frac{dv}{dx}&=\frac{1}{R}\left\{\frac{dz}{dc}\left(\frac{du_0}{db}-\frac{dv_0}{da}\right)+\frac{dz}{da}\left(\frac{dv_0}{dc}-\frac{dw_0}{db}\right)+\frac{dz}{db}\left(\frac{dw_0}{da}-\frac{du_0}{dc}\right)\right\}\\
\frac{dv}{dz}-\frac{dw}{dy}&=\frac{1}{R}\left\{\frac{dx}{dc}\left(\frac{du_0}{db}-\frac{dv_0}{da}\right)+\frac{dx}{da}\left(\frac{dv_0}{dc}-\frac{dw_0}{db}\right)+\frac{dx}{db}\left(\frac{dw_0}{da}-\frac{du_0}{dc}\right)\right\}\\
\frac{dw}{dx}-\frac{du}{dz}&=\frac{1}{R}\left\{\frac{dy}{c}\left(\frac{du_0}{ab}-\frac{dv_0}{da}\right)+\frac{dy}{da}\left(\frac{dv_0}{dc}-\frac{dw_0}{db}\right)+\frac{dy}{ab}\left(\frac{dw_0}{da}-\frac{du_0}{dc}\right)\right\}
\end{aligned}\right\}\dots(7).$$

Consider the element of fluid which at first occupied the rectangular parallelepiped formed by planes drawn parallel to the coordinate planes through the points (a, b, c) and $(a + da, b + db, c + dc)$. At the time t the element occupies a space bounded by six curved surfaces, which in the limit becomes an oblique-angled parallelepiped. The coordinates of the particle which at first was situated at the point (a, b, c) are x, y, z at the time t; and the coordinates of the extremities of the three edges of the oblique-angled parallelepiped which meet in the point (x, y, z) are

$$x + \frac{dx}{da} da, \qquad y + \frac{dy}{da} da, \qquad z + \frac{dz}{da} da;$$

$$x + \frac{dx}{db} db, \qquad y + \frac{dy}{db} db, \qquad z + \frac{dz}{db} db;$$

$$x + \frac{dx}{dc} dc, \qquad y + \frac{dy}{dc} dc, \qquad z + \frac{dz}{dc} dc.$$

Consequently, by a formula in analytical geometry, the volume of the element which at first was $da\,db\,dc$ is $R\,da\,db\,dc$ at the time t. Hence if ρ_0 be the initial density,

$$R = \frac{\rho_0}{\rho} \dots\dots\dots\dots\dots\dots(8).$$

From the mode in which this equation has been obtained, it is evident that it can be no other than the equation of continuity expressed in terms of a, b, c and t as independent variables, and integrated with respect to t.

The preceding equations are true independently of any particular supposition respecting the motion. If the initial motion be such that $u_0 da + v_0 db + w_0 dc$ is an exact differential, and in particular if the motion begin from rest, we shall have

$$\frac{du_0}{db} - \frac{dv_0}{da} = 0, \qquad \frac{dv_0}{dc} - \frac{dw_0}{db} = 0, \qquad \frac{dw_0}{da} - \frac{du_0}{dc} = 0;$$

and since by (8) R cannot vanish, it follows from (7) that at any time t

$$\frac{du}{dy} - \frac{dv}{dx} = 0, \qquad \frac{dv}{dz} - \frac{dw}{dy} = 0, \qquad \frac{dw}{dx} - \frac{du}{dz} = 0,$$

or $u\,dx + v\,dy + w\,dz$ is an exact differential.

Since any instant may be taken for the origin of the time, and t may be either negative or positive, it is evident that for a given portion of the fluid $u\,dx + v\,dy + w\,dz$ cannot cease to be an exact differential if it is once such, and cannot become an exact differential, not having been such previously.

SECOND DEMONSTRATION. The equations of motion in their usual form are

$$\left. \begin{aligned} \frac{1}{\rho}\frac{dp}{dx} &= X - \frac{du}{dt} - u\frac{du}{dx} - v\frac{du}{dy} - w\frac{du}{dz} \\ \frac{1}{\rho}\frac{dp}{dy} &= Y - \frac{dv}{dt} - u\frac{dv}{dx} - v\frac{dv}{dy} - w\frac{dv}{dz} \\ \frac{1}{\rho}\frac{dp}{dz} &= Z - \frac{dw}{dt} - u\frac{dw}{dx} - v\frac{dw}{dy} - w\frac{dw}{dz} \end{aligned} \right\} \dots\dots\dots(9).$$

Differentiating the first of these equations with respect to y and the second with respect to x, subtracting, and observing that by hypothesis p is a function of ρ, and $Xdx + Ydy + Zdz$ is an exact differential, we have

$$\left(\frac{d}{dt} + u\frac{d}{dx} + v\frac{d}{dy} + w\frac{d}{dz}\right)\left(\frac{du}{dy} - \frac{dv}{dx}\right) + \frac{du}{dy}\frac{du}{dx} + \frac{dv}{dy}\frac{du}{dy}$$

$$+ \frac{dw}{dy}\frac{du}{dz} - \frac{du}{dx}\frac{dv}{dx} - \frac{dv}{dx}\frac{dv}{dy} - \frac{dw}{dx}\frac{dv}{dz} = 0 \dots\dots(10).$$

According to the notation before employed,

$$\frac{d}{dt} + u\frac{d}{dx} + v\frac{d}{dy} + w\frac{d}{dz}$$

means the same as D/Dt. Let

$$\frac{dw}{dy} - \frac{dv}{dz} = 2\omega', \quad \frac{du}{dz} - \frac{dw}{dx} = 2\omega'', \quad \frac{dv}{dx} - \frac{du}{dy} = 2\omega''' \dots(11);$$

then the last six terms of (10) become, on adding and subtracting $\frac{du}{dz}\frac{dv}{dz}^{*}$,

$$2\frac{du}{dz}\omega' + 2\frac{dv}{dz}\omega'' - 2\left(\frac{du}{dx} + \frac{dv}{dy}\right)\omega'''$$

---

* $\frac{dw}{dx}\ \frac{dw}{dy}$ would have done as well.

We thus get from (10), and the other two equations which would be formed in a similar manner from (9),

$$
\left.
\begin{aligned}
\frac{D\omega'''}{Dt} &= \frac{du}{dz}\,\omega' + \frac{dv}{dz}\,\omega'' - \left(\frac{du}{dx} + \frac{dv}{dy}\right)\omega''' \\[2mm]
\frac{D\omega'}{dt} &= \frac{dv}{dx}\,\omega'' + \frac{dw}{dx}\,\omega''' - \left(\frac{dv}{dy} + \frac{dw}{dz}\right)\omega' \\[2mm]
\frac{D\omega''}{Dt} &= \frac{dw}{dy}\,\omega''' + \frac{du}{dy}\,\omega' - \left(\frac{dw}{dz} + \frac{du}{dx}\right)\omega''
\end{aligned}
\right\} \dots\dots(12).
$$

Now the motion at any instant varying continuously from one point of the fluid to another, the coefficients of ω', ω'', ω''' on the second sides of equations (12) cannot become infinite. Suppose that when $t = 0$ either there is no motion, or the motion is such that $u\,dx + v\,dy + w\,dz$ is an exact differential. This may be the case either throughout the whole fluid mass or throughout a limited portion of it. Then ω', ω'', ω''' vanish when $t = 0$. Let L be a superior limit to the numerical values of the coefficients of ω', ω'', ω''' on the second sides of equations (12) from the time 0 to the time t: then evidently ω', ω'', ω''' cannot increase faster than if they satisfied the equations

$$
\left.
\begin{aligned}
-\frac{D\omega'''}{Dt} &= L\,(\omega' + \omega'' + \omega''') \\[2mm]
\frac{D\omega'}{Dt} &= L\,(\omega' + \omega'' + \omega''') \\[2mm]
\frac{D\omega''}{Dt} &= L\,(\omega' + \omega'' + \omega''')
\end{aligned}
\right\} \dots\dots\dots\dots(13),
$$

instead of (12), vanishing in this case also when $t = 0$. By integrating equations (13), and determining the arbitrary constants by the conditions that ω', ω'', ω''' shall vanish when $t = 0$, we should find the general values of ω', ω'', and ω''' to be zero. We need not, however, take the trouble of integrating the equations; for, putting for shortness

$$\omega' + \omega'' + \omega''' = \Omega,$$

we get, by adding together the right and left-hand sides respectively of equations (13),

$$\frac{D\Omega}{Dt} = L\Omega.$$

The integral of this equation is $\Omega = Ce^{\iota t}$; and since $\Omega = 0$ when $t = 0$, $C = 0$; therefore the general value of Ω is zero. But Ω is the sum of the three quantities ω', ω'', ω''', which evidently cannot be negative, and therefore the general values of ω', ω'', ω''' are each zero. Since, then, ω', ω'', ω''' would have to be equal to zero, even if they satisfied equations (13), they must *a fortiori* be equal to zero in the actual case, since they satisfy equations (12), which proves the theorem enunciated.

It is evident that it is for a given mass of fluid, not for the fluid occupying a given portion of space, that the proposition is true, since equations (12) contain the differential coefficients $D\omega'/Dt$, &c. and not $d\omega'/dt$, &c. It is plain also that the same demonstration will apply to negative values of t.

If the motion should either be produced at first, or modified during its course, by impulsive pressures applied to the surface of the fluid, which of course can only be the case when the fluid is incompressible, the proposition will still be true. In fact, the change of motion produced by impulsive pressures is merely the limit of the change of motion produced by finite pressures, when the intensity of the pressures is supposed to increase and the duration of their action to decrease indefinitely. The proposition may however be proved directly in the case of impulsive forces by using the equations of impulsive motion. If q be the impulsive pressure, u_0, v_0, w_0 the velocities just before, u, v, w the velocities just after impact, it is very easy to prove that the equations of impulsive motion are

$$\frac{1}{\rho}\frac{dq}{dx} = -(u - u_0), \quad \frac{1}{\rho}\frac{dq}{dy} = -(v - v_0), \quad \frac{1}{\rho}\frac{dq}{dz} = -(w - w_0) \dots(14).$$

No forces appear in these equations, because finite forces disappear from equations of impulsive motion, and there are no forces which bear to finite forces, like gravity, acting all over the mass, the same relation that impulsive bear to finite pressures applied at the surface; and the impulsive pressures applied at the surface will appear, not in the general equations which hold good throughout the mass, but in the particular equations which have to be satisfied at the surface. The equations (14) are applicable to a heterogeneous, as well as to a homogeneous liquid. They must be combined with the equation of continuity of a liquid, (equation (6), p. 286 of the preceding volume.) In the

44   NOTES ON HYDRODYNAMICS.

case under consideration, however, ρ is constant; and therefore
from (14)
$$(u - u_0)\,dx + (v - v_0)\,dy + (w - w_0)\,dz$$
is an exact differential $d\,(-q/\rho)$; and therefore if u_0, v_0, w_0 be
zero, or if they be such that $u_0 dx + v_0 dy + w_0 dz$ is an exact differential $d\phi_0$, $u dx + v dy + w dz$ will also be an exact differential
$d\,(\phi_0 - q/\rho)$.

When $u dx + v dy + w dz$ is an exact differential $d\phi$, the expression for dP obtained from equations (9) is immediately integrable,
and we get
$$P = V - \frac{d\phi}{dt} - \tfrac{1}{2}\left\{ \left(\frac{d\phi}{dx}\right) + \left(\frac{d\phi}{dy}\right)^2 + \left(\frac{d\phi}{dz}\right)^2 \right\} \,\ldots\ldots\,(15),$$
supposing the arbitrary function of t introduced by integration
to be included in ϕ.

M. Cauchy's proof of the theorem just considered does not
seem to have attracted the attention which it deserves. It does
not even appear to have been present to Poisson's mind when
he wrote his *Traité de Mécanique*. The demonstration which
Poisson has given[*] is in fact liable to serious objections[†]. Poisson
indeed was not satisfied as to the generality of the theorem. It
is not easy to understand the objections which he has raised[‡],
which after all do not apply to M. Cauchy's demonstration, in
which no expansions are employed. As Poisson gives no hint
where to find the "examples" in which he says the theorem
fails, if indeed he ever published them, we are left to conjecture.
In speaking of the developments of u, v, w in infinite series of
exponentials or circular functions, suited to particular problems,
by which all the equations of the problem are satisfied, he remarks that one special character of such expansions is, not always
to satisfy the equations which are deduced from those of motion
by new differentiations. It is true that the equations which
would *apparently* be obtained by differentiation would not always
be satisfied; for the differential coefficients of the expanded
functions cannot in general be obtained by direct differentiation,
that is by differentiating under the sign of summation, but must

* *Traité de Mécanique,* tom. II. p. 688 (2nd edition).
† See *Cambridge Philosophical Transactions*, Vol. VIII. p. 305. [*Ante*, Vol. I. p. 110.]
‡ *Traité de Mécanique*, tom. II. p. 690.

be got from formulæ applicable to the particular expansions*.
Poisson appears to have met with some contradiction, from
whence he concluded that the theorem was not universally true,
the contradiction probably having arisen from his having dif-
ferentiated under the sign of summation in a case in which it
is not allowable to do so.

It has been objected to the application of the theorem proved
in this note to the case in which the motion begins from rest,
that we are not at liberty to call $udx + vdy + wdz$ an exact differential when u, v, and w vanish with t, unless it be proved that
if u_1, v_1, w_1 be the results obtained by dividing u, v, w by the
lowest power of t occurring as a factor in u, v, w, and then putting
$t = 0$, $u_1dx + v_1dy + w_1dz$ is an exact differential. Whether we call
$udx + vdy + wdz$ in all cases an exact differential when u, v and w
vanish, is a matter of definition, although reasons might be assigned which would induce us to allow of the application of the
term in all such cases : the demonstration of the theorem is not
at all affected. Indeed, in enunciating and demonstrating the
theorem there is no occasion to employ the term *exact differential*
at all. The theorem might have been enunciated as follows.
If the three quantities $du/dy - dv/dx$, &c. are numerically equal
to zero when $t = 0$, they will remain numerically equal to zero
throughout the motion. This theorem having been established,
it follows as a result that when u, v, and w vanish with t,
$u_1dx + v_1dy + w_1dz$ is an exact differential.

The theorem has been shewn to be a rigorous consequence
of the hypothesis of the absence of all tangential force in fluids
in motion. It now becomes a question, How far is the theorem
practically true, or nearly true ; or in what cases would it lead
to results altogether at variance with observation ?

As a general rule it may be answered that the theorem will
lead to results nearly agreeing with observation when the motion
of the particles which are moving is continually beginning from
rest, or nearly from rest, or is as good as if it were continually
beginning from rest ; while the theorem will practically fail when
the velocity of a given particle, or rather its velocity relatively

---

* See a paper "On the Critical Values of the sums of Periodic Series," *Cambridge
Philosophical Transactions*, Vol. VIII. Part 5. [*Ante*, Vol. I. p. 236.]

to other particles, takes place for a long continuance in one
direction.

Thus, when a wave of sound is propagated through air, a new
set of particles is continually coming into motion ; or the motion,
considered with reference to the individual particles, is continually
beginning from rest. When a wave is propagated along the
surface of water, although the motion of the water at a distance
from the wave is not mathematically zero, it is insensible, so that
the set of particles which have got any sensible motion is continually changing. When a series of waves of sound is propagated in air, as for example the series of waves coming from
a musical instrument, or when a series of waves is propagated
along the surface of water, it is true that the motion is not
continually beginning from rest, but it is as good as if it were
continually beginning from rest. For if at any instant the disturbing cause were to cease for a little, and then go on again,
the particles would be reduced to rest, or nearly to rest, when
the first series of waves had passed over them, and they would
begin to move afresh when the second series reached them. Again,
in the case of the simultaneous small oscillations of solids and
fluids, when the forward and backward oscillations are alike, equal
velocities in opposite directions are continually impressed on the
particles at intervals of time separated by half the time of a complete oscillation. In such cases the theorem would generally lead
to results agreeing nearly with observation.

If however water coming from a reservoir where it was sensibly at rest were to flow down a long canal, or through a long
pipe, the tendency of friction being always the same way, the
motion would soon altogether differ from one for which
$udx + vdy + wdz$ was an exact differential. The same would
be the case when a solid moves continually onwards in a fluid.
Even in the case of an oscillating solid, when the forward and
backward oscillations are not similar, as for example when a
cone oscillates in the direction of its axis, it may be conceived that the tendency of friction to alter the motion of
the fluid in the forward oscillation may not be compensated in
the backward oscillation; so that, even if the internal friction
be very small, the motion of the fluid after several oscillations
may differ widely from what it would have been had there been

absolutely no friction. I do not expect that there would be this wide difference; but still the actual motion would probably not agree so well with the theoretical, as in those cases in which the forward and backward oscillations are alike. By the theoretical motion is of course meant that which would be obtained from the common theory, in which friction is not taken into account.

It appears from experiments on pendulums that the effect of the internal friction in air and other gases is greater than might have been anticipated. In Dubuat's experiments on spheres oscillating in air the spheres were large, and the alteration in the time of oscillation due to the resistance of the air, as determined by his experiments, agrees very nearly with the result obtained from the common theory. Other philosophers, however, having operated on smaller spheres, have found a considerable discrepancy, which is so much the greater as the sphere employed is smaller. It appears, moreover, from the experiments of Colonel Sabine, that the resistance depends materially upon the nature of the gas. Thus it is much greater, in proportion to the density, in hydrogen than in air.

NOTE REFERRED TO AT P. 37.

[It may be noticed that two of Helmholtz's fundamental propositions respecting vortex motion[*] follow immediately from Cauchy's integrals; or rather, two propositions the same as those of Helmholtz merely generalized so as to include elastic fluids follow from Cauchy's equations similarly generalized.

On substituting in (7) for R the expression given by (8), and introducing the notation of angular velocities, as in (11), equations (7) become

$$\omega' = \frac{\rho}{\rho_0}\left(\frac{dx}{da}\omega_0' + \frac{dx}{db}\omega_0'' + \frac{dx}{dc}\omega_0'''\right)$$

$$\omega'' = \frac{\rho}{\rho_0}\left(\frac{dy}{da}\omega_0' + \frac{dy}{db}\omega_0'' + \frac{dy}{dc}\omega_0'''\right) \quad\quad \ldots\ldots\ldots\ldots (a).$$

$$\omega''' = \frac{\rho}{\rho_0}\left(\frac{dz}{da}\omega_0' + \frac{dz}{db}\omega_0'' + \frac{dz}{dc}\omega_0'''\right)$$

[*] Crelle's *Journal*, Vol. LV. p. 25.

We see at once from these equations that if

$$\frac{da}{\omega_0{}'} = \frac{db}{\omega_0{}''} = \frac{dc}{\omega_0{}'''} \quad\text{...................... (b),}$$

then
$$\frac{dx}{\omega'} = \frac{dy}{\omega''} = \frac{dz}{\omega'''} \quad\text{.................. .... (c),}$$

but (b) are the differential equations of the system of vortex lines at the time 0, and (c), as being of the form

$$\frac{dx}{P} = \frac{dy}{Q} = \frac{dz}{R},$$

are the differential equations of the loci of the particles at the time t which at the time 0 formed the vortex lines respectively. But when we further take account of the values of P, Q, R, as exhibited in (c), we see that (c) are also the differential equations of the system of vortex lines at the time t. Therefore the same loci of particles which at one moment are vortex lines remain vortex lines throughout the motion.

Let Ω_0 be the resultant angular velocity at the time 0 of a particle P_0 which at the time t is at P, and has Ω for its angular velocity; let ds_0 drawn from P_0 be an element of the vortex line at time 0 passing through P_0, and ds the element of the vortex line passing through P at the time t which consists of the same set of particles. Then each member of equations (b) is equal to ds_0/Ω_0, and each member of equations (c) equal to ds/Ω. Hence we get from any one of equations (a)

$$\frac{\rho\, ds}{\Omega} = \frac{\rho_0 ds_0}{\Omega_0}$$

Let A_0 be the area of a perpendicular section, at P_0, of a vortex thread containing the vortex line passing through P_0 at the time 0, a vortex thread meaning the portion of fluid contained within an elementary tube made up of vortex lines; then by what precedes the same set of particles will at the time t constitute a vortex thread passing through P; let A be a perpendicular section of it passing through P at the time t, and draw two other perpendicular sections passing respectively through the other extremities of the elements ds_0 and ds. Then if we suppose, as we are at liberty to do, that the linear dimensions of A_0 are indefinitely small compared with the length ds_0, we see at once that the elements of volume comprised between the tube and the pair of sections at the time 0 and at the time t respectively contain ultimately the same particles, and therefore

$$\rho A\, ds = \rho_0 A_0 ds_0,$$

whence $\qquad\qquad \Omega A = \Omega_0 A_0,$

or the angular velocity of any given particle varies inversely as the area of a perpendicular section through it of the vortex thread to which it belongs, and that, whether the fluid be incompressible or elastic.

When these results are deduced from Cauchy's integrals, the state of the fluid at any time is compared directly with its state at any other time; in Helmholtz's method the state at the time t is compared with the state at the time $t + dt$, and so on step by step.

A remaining proposition of Helmholtz's, that along a vortex line the angular velocity varies at any given time inversely as the perpendicular section of the vortex thread, has no immediate relation to Cauchy's integrals, inasmuch as it relates to a comparison of the state of the fluid at different points at the same moment. It may however be convenient to the reader that the demonstration, which is very brief, should be reproduced here.

We have at once from (11)

$$\frac{d\omega'}{dx} + \frac{d\omega''}{dy} + \frac{d\omega'''}{dz} = 0,$$

and consequently

$$\iiint \left(\frac{d\omega'}{dx} + \frac{d\omega''}{dy} + \frac{d\omega'''}{dz}\right) dx\, dy\, dz = 0,$$

where the integration extends over any arbitrary portion of the fluid. This equation gives

$$\iint \omega' dy dz + \iint \omega'' dz dx + \iint \omega''' dx dy = 0,$$

where the double integrals extend over the surface of the space in question. The latter equation again becomes by a well-known transformation

$$\iint \Omega \cos \theta\, dS = 0,$$

where dS is an element of the surface of the space, and θ the

4

angle between the instantaneous axis and the normal to the surface drawn outwards.

Let now the space considered be the portion of a vortex thread comprised between any two perpendicular sections, of which let A and A' denote the areas. All along the side of the tube $\theta = 90^\circ$, and at the two ends $\theta = 180^\circ$ and $= 0^\circ$, respectively, and therefore if Ω' denotes the angular velocity at the second extremity of the portion of the vortex thread considered

$$\Omega A = \Omega' A',$$

which proves the theorem.]

[From the *Philosophical Magazine*, Vol. XXXIII., p. 349 (*November*, 1848.)]

On a Difficulty in the Theory of Sound.

THE theoretical determination of the velocity of sound has recently been the occasion of a discussion between Professor Challis and the Astronomer Royal. It is not my intention to enter into the controversy, but merely to consider a very remarkable difficulty which Professor Challis has noticed in connexion with a known first integral of the accurate equations of motion for the case of plane waves.

The difficulty alluded to is to be found at page 496 of the preceding volume of this Magazine*. In what follows I shall use Professor Challis's notation.

* [The following quotation will suffice to put the reader in possession of the apparent contradiction discovered by Professor Challis. It should be stated that the investigation relates to plane waves, propagated in the direction of z, and that the pressure is supposed to vary as the density.

"The function f being quite arbitrary, we may give it a particular form. Let, therefore,

$$w = m \sin \frac{2\pi}{\lambda} \{z - (a+w) t\}.$$

This equation shows that at any time t_1 we shall have $w = 0$ at points on the axis of z, for which

$$z - (a+w) t_1 = \frac{n\lambda}{2},$$

or

$$z = at_1 + \frac{n\lambda}{2}.$$

At the same time w will have the value $\pm m$ at points of the axis for which

$$z - (a+m) t_1 = \left(\frac{n}{2} - \tfrac{1}{4}\right) \lambda,$$

or

$$z = at_1 + \frac{n\lambda}{2} + mt_1 - \frac{\lambda}{4}.$$

Without entering into the consideration of the mode in which Poisson obtained the particular integral

$$w = f\{z - (a + w)\, t\} \quad\dots\dots\dots\dots\dots(1),$$

it may easily be shown, by actual differentiation and substitution, that the integral does satisfy our equations. The function f being arbitrary, we may assign to it any form we please, as representing a particular possible motion, and may employ the result, *so long as no step tacitly assumed in the course of our reasoning fails.* The interpretation of the integral (1) will be rendered more easy by the consideration of a curve. In Fig. 1 let oz be the axis of z, and let the ordinate of the curve represent the values of w for $t = 0$. The equation (1) merely asserts that whatever value the

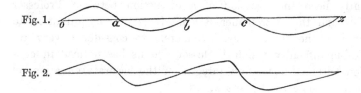

Fig. 1.

Fig. 2.

velocity w may have at any particular point when $t = 0$, the same value will it have at the time t at a point in advance of the former by the space $(a + w)\, t$. Take any point P in the curve of Fig. 1, and from it draw, in the positive direction, the right line PP' parallel to the axis of z, and equal to $(a + w)\, t$. The locus of all the points P' will be the velocity-curve for the time t. This curve is represented in Fig. 2, except that the displacement at common to all points of the original curve is omitted, in order that the modification in the form of the curve may be more easily perceived. This comes to the same thing as drawing PP' equal to wt instead of $(a + w)\, t$. Of course in this way P' will lie on the positive or negative side of P, according as P lies above or below the axis of z. It is evident that in the neighbourhood of the points a, c the curve becomes more and more steep as t increases, while in the neigh-

Here it is observable that no relation exists between the points of no velocity and the points of maximum velocity. As m, t_1, and λ are arbitrary constants, we may even have

$$m t_1 - \frac{\lambda}{4} = 0,$$

in which case the points of no velocity are also points of maximum velocity."]

bourhood of the points o, b, z its inclination becomes more and more gentle.

The same result may easily be obtained analytically. In Fig. 1, take two points, infinitely close to each other, whose abscissæ are z and $z + dz$; the ordinates will be w and

$$w + \frac{dw}{dz}\, dz.$$

After the time t these same ordinates will belong to points whose abscissæ will have become (in Fig. 2) $z + wt$ and

$$z + dz + \left(w + \frac{dw}{dz}\, dz\right) t.$$

Hence the horizontal distance between the points, which was dz, will have become

$$\left(1 + \frac{dw}{dz}\, t\right) dz\,;$$

and therefore the tangent of the inclination, which was dw/dz, will have become

$$\frac{\dfrac{dw}{dz}}{1 + \dfrac{dw}{dz}\, t} \quad\dots\dots\dots\dots\dots\dots\dots\dots \text{(A)}.$$

At those points of the original curve' at which the tangent is horizontal, $dw/dz = 0$, and therefore the tangent will constantly remain horizontal at the corresponding points of the altered curve. For the points for which dw/dz is positive, the denominator of the expression (A) increases with t, and therefore the inclination of the curve continually decreases. But when dw/dz is negative, the denominator of (A) decreases as t increases, so that the curve becomes steeper and steeper. At last, for a sufficiently large value of t, the denominator of (A) becomes infinite for some value of z. Now the very formation of the differential equations of motion with which we start, tacitly supposes that we have to deal with finite and continuous functions; and therefore in the case under consideration we must not, without limitation, push our results beyond the least value of t which renders (A) infinite. This value is evidently the reciprocal, taken positively, of the greatest negative value of dw/dz; w here, as in the whole of this paragraph, denoting the velocity when $t = 0$.

By the term *continuous function*, I here understand a function whose value does not alter *per saltum*, and not (as the term is sometimes used) a function which preserves the same algebraical expression. Indeed, it seems to me to be of the utmost importance, in considering the application of partial differential equations to physical, and even to geometrical problems, to contemplate functions apart from all idea of algebraical expression.

In the example considered by Professor Challis,

$$w = m \sin \frac{2\pi}{\lambda} \{z - (a + w)\, t\},$$

where m may be supposed positive; and we get by differentiating and putting $t = 0$,

$$\frac{dw}{dz} = \frac{2\pi m}{\lambda} \cos \frac{2\pi z}{\lambda},$$

the greatest negative value of which is $- 2\pi m/\lambda$; so that the greatest value of t for which we are at liberty to use our results without limitation is $\lambda/2\pi m$, whereas the contradiction arrived at by Professor Challis is obtained by extending the result to a larger value of t, namely $\lambda/4m$.

Of course, after the instant at which the expression (A) becomes infinite, some motion or other will go on, and we might wish to know what the nature of that motion was. Perhaps the most natural supposition to make for trial is, that a surface of discontinuity is formed, in passing across which there is an abrupt change of density and velocity. The existence of such a surface will presently be shown to be possible*, on the two suppositions that the pressure is equal in all directions about the same point, and that it varies as the density. I have however convinced myself, by a train of reasoning which I do not think it worth while to give, inasmuch as the result is merely negative, that even on the supposition of the existence of a surface of discontinuity, it is not possible to satisfy all the conditions of the problem by means of a single function of the form $f\{z - (a + w)\, t\}$. Apparently, something like reflexion must take place. Be that as it may, it is evident that the change which now takes place in the nature of the motion, beginning with the particle (or rather plane of particles) for which (A) first becomes infinite, cannot influence a

---

* [Not so: see the substituted paragraph at the end.]

particle at a finite distance from the former until after the expiration of a finite time. Consequently even after the change in the nature of the motion, our original expressions are applicable, at least for a certain time, to a certain portion of the fluid. It was for this reason that I inserted the words "without limitation," in saying that we are not at liberty to use our original results without limitation beyond a certain value of t. The full discussion of the motion which would take place after the change above alluded to, if possible at all, would probably require more pains than the result would be worth.

[So long as the motion is continuous, and none of the differential coefficients involved become infinite, the two principles of the conservation of mass and what may be called the conservation of momentum, applied to each infinitesimal slice of the fluid, are not only necessary but also sufficient for the complete determination of the motion, the functional relation existing between the pressure and density being of course supposed known.   Hence any other principle known to be true, such for example as that of the conservation of energy, must be virtually contained in the former. It was accordingly a not unnatural mistake to make to suppose that in the limit, when we imagine the motion to become discontinuous, the same two principles of conservation of mass and of momentum applied to each infinitesimal slice of the fluid should still be sufficient, even though one such slice might contain a surface of discontinuity.   It was however pointed out to me by Sir William Thomson, and afterwards independently by Lord Rayleigh, that the discontinuous motion supposed above involves a violation of the principle of the conservation of energy.   In fact, the equation of energy, applied to the fluid in the immediate neighbourhood of the surface of discontinuity, and combined with the two equations deduced from the two principles first mentioned, leads in the case of $p \propto \rho$ to

$$2\rho\rho' \log \rho/\rho' = \rho^2 - \rho'^2,$$

where ρ, ρ' are the densities at the two sides of the supposed surface of discontinuity; but this equation has no real root except $\rho = \rho'$.]

[From the *Transactions of the Cambridge Philosophical Society*, Vol. VIII. p. 642.]

ON THE FORMATION OF THE CENTRAL SPOT OF NEWTON'S RINGS BEYOND THE CRITICAL ANGLE.

[Read *December* 11, 1848.]

WHEN Newton's Rings are formed between the under surface of a prism and the upper surface of a lens, or of another prism with a slightly convex face, there is no difficulty in increasing the angle of incidence on the under surface of the first prism till it exceeds the critical angle. On viewing the rings formed in this manner, it is found that they disappear on passing the critical angle, but that the central black spot remains. The most obvious way of accounting for the formation of the spot under these circumstances is, perhaps, to suppose that the forces which the material particles exert on the ether extend to a small, but sensible distance from the surface of a refracting medium; so that in the case under consideration the two pieces of glass are, in the immediate neighbourhood of the point of contact, as good as a single uninterrupted medium, and therefore no reflection takes place at the surfaces. This mode of explanation is however liable to one serious objection. So long as the angle of incidence falls short of the critical angle, the central spot is perfectly explained, along with the rest of the system of which it forms a part, by ordinary reflection and refraction. As the angle of incidence gradually increases, passing through the critical angle, the appearance of the central spot changes gradually, and but slightly. To account then for the existence of this spot by ordinary reflection and refraction so long as the angle of incidence falls short of the critical angle, but by the finite extent of the sphere of action of the molecular forces when the angle of incidence exceeds the critical angle, would be to give a discontinuous explanation to a continuous phenomenon. If we adopt the latter mode of explanation in the one case we must adopt it in the other, and thus separate the theory of the central spot from that of the rings, which to all appearance belong to the same system; although the admitted theory of the rings fully accounts likewise for the existence of the spot, and not only for its existence, but also for some remarkable modifications which it undergoes in certain circumstances*.

Accordingly the existence of the central spot beyond the critical angle has been attributed by Dr Lloyd, without hesitation, to the disturbance in the second medium which takes the place of that which, when the angle of incidence is less than the critical angle, constitutes the refracted light†. The expression for the intensity of the light, whether reflected or transmitted, has not however been hitherto given, so far as I am aware. The object of the present paper is to supply this deficiency.

In explaining on dynamical principles the total internal reflection of light, mathematicians have been led to an expression for the disturbance in the second medium involving an exponential, which contains in its index the perpendicular distance of the point considered from the surface. It follows from this expression that the disturbance is insensible at the distance of a small multiple of the length of a wave from the surface. This circumstance is all that need be attended to, so far as the refracted light is concerned, in explaining total internal reflection; but in considering the theory of the central spot in Newton's Rings, it is precisely the superficial disturbance just mentioned that must be taken into account. In the present paper I have not adopted any special dynamical theory: I have preferred deducing my results from Fresnel's formulæ for the intensities of reflected and refracted polarized light, which in the case considered became imaginary, interpreting these imaginary expressions, as has been done by Professor O'Brien‡,

---

* I allude especially to the phenomena described by Mr Airy in a paper printed in the fourth volume of the *Cambridge Philosophical Transactions*, p. 409.

† Report on the present state of Physical Optics. *Reports of the British Association*, Vol. III. p. 310.

‡ *Cambridge Philosophical Transactions*, Vol. VIII. p. 20.

in the way in which general dynamical considerations show that they ought to be interpreted.

By means of these expressions, it is easy to calculate the intensity of the central spot. I have only considered the case in which the first and third media are of the same nature : the more general case does not seem to be of any particular interest. Some conclusions follow from the expression for the intensity, relative to a slight tinge of colour about the edge of the spot, and to a difference in the size of the spot according as it is seen by light polarized in, or by light polarized perpendicularly to the plane of incidence, which agree with experiment.

1. Let a plane wave of light be incident, either externally or internally, on the surface of an ordinary refracting medium, suppose glass. Regard the surface as plane, and take it for the plane xy; and refer the media to the rectangular axes of x, y, z, the positive part of the last being situated in the second medium, or that into which the refraction takes place. Let l, m, n be the cosines of the angles at which the normal to the incident wave, measured in the direction of propagation, is inclined to the axes; so that $m = 0$ if we take, as we are at liberty to do, the axis of y parallel to the trace of the incident wave on the reflecting surface. Let V, $V_{,}$, V' denote the incident, reflected, and refracted vibrations, estimated either by displacements or by velocities, it does not signify which; and let a, $a_{,}$, a' denote the coefficients of vibration. Then we have the following possible system of vibrations :

$$V = a \, \cos\frac{2\pi}{\lambda} \, (vt - lx - nz),$$
$$V_{,} = a_{,} \cos\frac{2\pi}{\lambda} \, (vt - lx + nz),$$
$$V' = a' \cos\frac{2\pi}{\lambda'} (v't - l'x - n'z),$$
\hspace{2cm} \right\} \; \dots\dots\dots (A).

In these expressions v, v' are the velocities of propagation, and λ, λ' the lengths of wave, in the first and second media; so that v, v', and the velocity of propagation in vacuum, are proportional to λ, λ', and the length of wave in vacuum: l is the sine, and n the cosine of the angle of incidence, l' the sine, and n' the cosine of the angle of refraction, these quantities being connected by the equations

$$\frac{l}{v} = \frac{l'}{v'}, \quad n = \sqrt{1 - l^2}, \quad n' = \sqrt{1 - l'^2} \ldots\ldots\ldots (1).$$

2. The system of vibrations (A) is supposed to satisfy certain linear differential equations of motion belonging to the two media, and likewise certain linear equations of condition at the surface of separation, for which $z = 0$. These equations lead to certain relations between a, $a_{,}$, and a', by virtue of which the ratios of $a_{,}$ and a' to a are certain functions of l, v, and v', and it might·be also of λ. The equations, being satisfied identically, will continue to be satisfied when l' becomes greater than 1, and consequently n' imaginary, which may happen, provided $v' > v$; but the interpretation before given to the equations (A) and (1) fails.

When n' becomes imaginary, and equal to $\nu' \sqrt{(-1)}$, ν' being equal to $\sqrt{(l'^2 - 1)}$, z instead of appearing under a circular function in the third of equations (A), appears in one of the exponentials $\epsilon^{\pm k'\nu'z}$, k' being equal to $2\pi/\lambda'$. By changing the sign of $\sqrt{(-1)}$ we should get a second system of equations (A), satisfying, like the first system, all the equations of the problem; and we should get two new systems by writing $vt + \lambda/4$ for vt. By combining these four systems by addition and subtraction, which is allowable on account of the linearity of our equations, we should be able to get rid of the imaginary quantities, and likewise of the exponential $\epsilon^{+k'\nu'z}$, which does not correspond to the problem, inasmuch as it relates to a disturbance which increases indefinitely in going from the surface of separation into the second medium, and which could only be produced by a disturbing cause existing in the second medium, whereas none such is supposed to exist.

3. The analytical process will be a good deal simplified by replacing the expressions (A) by the following symbolical expressions for the disturbance, where k is put for $2\pi/\lambda$, so that $kv = k'v'$;

$$\left.\begin{array}{l} V = a\epsilon^{k(vt - lx - nz)\sqrt{-1}}, \\ V_{,} = a_{,}\epsilon^{k(vt - lx + nz)\sqrt{-1}}, \\ V' = a'\epsilon^{k(v't - l'x - n'z)\sqrt{-1}} \end{array}\right\} \ldots\ldots\ldots\ldots\ldots (B).$$

In these expressions, if each exponential of the form $\epsilon^{P\sqrt{(-1)}}$ be replaced by $\cos P + \sqrt{(-1)}\sin P$, the real part of the expressions will agree with (A), and therefore will satisfy the equations of the problem. The coefficients of $\sqrt{(-1)}$ in the imaginary part will be derived from the real part by writing $t + \lambda/4v$ for t, and therefore will form a system satisfying the same equations, since the form of these equations is supposed in no way to depend on the origin of the time; and since the equations are linear they will be satisfied by the complete expressions (B).

Suppose now l' to become greater than 1, so that n' becomes $\pm \nu' \sqrt{(-1)}$. Whichever sign we take, the real and imaginary parts of the expressions (B), which must separately satisfy the equations of motion and the equations of condition, will represent two possible systems of waves; but the upper sign does not correspond to the problem, for the reason already mentioned, so that we must use the lower sign. At the same time that n' becomes $\nu' \sqrt{(-1)}$, let a, $a_{,}$, a' become

$$\rho \epsilon^{\theta\sqrt{-1}}, \qquad \rho_{,}\epsilon^{\theta,\sqrt{-1}}, \qquad \rho'\epsilon^{\theta'\sqrt{-1}}, \text{ respectively :}$$

then we have the symbolical system

$$\left. \begin{aligned} V &= \rho\epsilon^{-\theta\sqrt{-1}} . \epsilon^{k(vt-lx-nz)\sqrt{-1}}, \\ V_{,} &= \rho_{,}\epsilon^{-\theta,\sqrt{-1}} . \epsilon^{k(vt-lx+nz)\sqrt{-1}}, \\ V' &= \rho'\epsilon^{-\theta'\sqrt{-1}} . \epsilon^{-k'\nu'z} . \epsilon^{k'(v't-l'x)\sqrt{-1}}, \end{aligned} \right\} \dots\dots\dots\dots(C),$$

of which the real part

$$\left. \begin{aligned} V &= \rho \cos\{k(vt-lx-nz)-\theta\}, \\ V_{,} &= \rho_{,} \cos\{k(vt-lx+nz)-\theta_{,}\}, \\ V' &= \rho'\epsilon^{-k'\nu'z} \cos\{k'(v't-l'x)-\theta'\}, \end{aligned} \right\} \dots\dots\dots (D)$$

forms the system required.

As I shall frequently have occasion to allude to a disturbance of the kind expressed by the last of equations (D), it will be convenient to have a name for it, and I shall accordingly call it a *superficial undulation*.

4. The interpretation of our results is not yet complete, inasmuch as it remains to consider what is meant by V'. When the vibrations are perpendicular to the plane of incidence there is no difficulty. In this case, whether the angle of incidence be greater or less than the critical angle, V' denotes a displacement, or

else a velocity, perpendicular to the plane of incidence. When the vibrations are in the plane of incidence, and the angle of incidence is less than the critical angle, V' denotes a displacement or velocity in the direction of a line lying in the plane xz, and inclined at angles $\pi - i'$, $-(\frac{1}{2}\pi - i')$ to the axes of x, z, i' being the angle of refraction. But when the angle of incidence exceeds the critical angle there is no such thing as an angle of refraction, and the preceding interpretation fails. Instead therefore of considering the whole vibration V', consider its resolved parts V_x', V_z' in the direction of the axes of x, z. Then when the angle of incidence is less than the critical angle, we have

$$V_x' = -n'V' = -\cos i'' \cdot V'; \quad V_z' = l'V' = \sin i' \cdot V',$$

V' being given by (A), and being reckoned positive in that direction which makes an acute angle with the positive part of the axis of z. When the angle of incidence exceeds the critical angle, we must first replace the coefficient of V' in V_x', namely $-n'$, by $\nu'\epsilon^{\frac{1}{2}\pi \sqrt{-1}}$, and then, retaining ν' for the coefficient, add $\frac{1}{2}\pi$ to the phase, according to what was explained in the preceding article.

Hence, when the vibrations take place in the plane of incidence, and the angle of incidence exceeds the critical angle, V' in (D) must be interpreted to mean an expression from which the vibrations in the directions of x, z may be obtained by multiplying by ν', l' respectively, and increasing the phase in the former case by $\frac{1}{2}\pi$. Consequently, so far as depends on the third of equations (D), the particles of ether in the second medium describe small ellipses lying in the plane of incidence, the semi-axes of the ellipses being in the directions of x, z, and being proportional to ν', l', and the direction of revolution being the same as that in which the incident ray would have to revolve in order to diminish the angle of incidence.

Although the elliptic paths of the particles lie in the plane of incidence, that does not prevent the superficial vibration just considered from being of the nature of transversal vibrations. For it is easy to see that the equation

$$\frac{dV_x'}{dx} + \frac{dV_z'}{dz} = 0$$

is satisfied; and this equation expresses the condition that there is no change of density, which is the distinguishing characteristic of transversal vibrations.

5.   When the vibrations of the incident light take place in the plane of incidence, it appears from investigation that the equations of condition relative to the surface of separation of the two media cannot be satisfied by means of a system of incident, reflected and refracted waves, in which the vibrations are transversal.   If the media be capable of transmitting normal vibrations with velocities comparable with those of transversal vibrations, there will be produced, in addition to the waves already mentioned, a series of reflected and a series of refracted waves in which the vibrations are normal, provided the angle of incidence be less than either of the two critical angles corresponding to the reflected and refracted normal vibrations respectively.   It has been shown however by Green, in a most satisfactory manner, that it is necessary to suppose the velocities of propagation of normal vibrations to be incomparably greater than those of transversal vibrations, which comes to the same thing as regarding the ether as sensibly incompressible ; so that the two critical angles mentioned above must be considered evanescent*.   Consequently the reflected and refracted normal waves are replaced by undulations of the kind which I have called superficial.   Now the existence of these superficial undulations does not affect the interpretation which has been given to the expressions (A) when the angle of incidence becomes greater than the critical angle corresponding to the refracted transversal wave ;   in fact, so far as regards that interpretation, it is immaterial whether the expressions (A) satisfy the linear equations of motion and condition alone, or in conjunction with other terms referring to the normal waves, or rather to the superficial undulations which are their representatives.   The expressions (D) however will not represent the whole of the disturbance in the two media, but only that part of it which relates to the transversal waves, and to the superficial undulation which is the representative of the refracted tranversal wave.

6.   Suppose now that in the expressions (A) n becomes imaginary, n' remaining real, or that n and n' both become imaginary.

---

* *Cambridge Philosophical Transactions*, Vol. VII. p. 2.

The former case occurs in the theory of Newton's Rings when the angle of incidence on the surface of the second medium becomes greater than the critical angle, and we are considering the superficial undulation incident on the third medium : the latter case would occur if the third medium as well as the second were of lower refractive power than the first, and the angle of incidence on the surface of the second were greater than either of the critical angles corresponding to refraction out of the first into the second, or out of the first into the third. Consider the case in which n becomes imaginary, n' remaining real ; and let $\sqrt{(l^2 - 1)} = \nu$. Then it may be shown as before that we must put $-\nu\sqrt{(-1)}$, and not $\nu\sqrt{(-1)}$, for n ; and using ρ, θ in the same sense as before, we get the symbolical system,

$$
\left.
\begin{aligned}
V &= \rho\epsilon^{-\theta\sqrt{-1}} . \epsilon^{-kvz} . \epsilon^{k(vt-lx)\sqrt{-1}}, \\
V_{\prime} &= \rho_{\prime}\epsilon^{-\theta_{\prime}\sqrt{-1}} . \epsilon^{kvz} . \epsilon^{k(vt-lx)\sqrt{-1}}, \\
V' &= \rho'\epsilon^{-\theta'\sqrt{-1}} . \epsilon^{k'(v't-l'x-n'z)\sqrt{-1}},
\end{aligned}
\right\} \quad\dots\dots\dots\dots \text{(E)},
$$

to which corresponds the real system

$$
\left.
\begin{aligned}
V &= \rho\epsilon^{-kvz} \cos\{k(vt-lx)-\theta\}, \\
V_{\prime} &= \rho_{\prime}\epsilon^{kvz} \cos\{k(vt-lx)-\theta_{\prime}\}, \\
V' &= \rho'\cos\{k'(v't-l'x-n'z)-\theta'\},
\end{aligned}
\right\} \quad\dots\dots\dots\dots \text{(F)}.
$$

When the vibrations take place in the plane of incidence, V and $V_{\prime}$ in these expressions must be interpreted in the same way as before. As far as regards the incident and reflected superficial undulations, the particles of ether in the first medium will describe small ellipses lying in the plane of incidence. The ellipses will be similar and similarly situated in the two cases; but the direction of revolution will be in the case of the incident undulation the same as that in which the refracted ray would have to turn in order to diminish the angle of refraction, whereas in the reflected undulation it will be the opposite.

It is unnecessary to write down the formulæ which apply to the case in which n and n' both become imaginary.

7. If we choose to employ real expressions, such as (D) and (F), we have this general rule. When any one of the undulations, incident, reflected, or refracted, becomes superficial, remove z from under the circular function, and insert the exponential

$\epsilon^{-k\nu z}$, $\epsilon^{k\nu z}$, or $\epsilon^{-k'\nu'z}$, according as the incident, reflected, or refracted undulation is considered. At the same time put the coefficients, which become imaginary, under the form

$$\rho \left\{\cos \theta \pm \sqrt{(-1)} \sin \theta\right\},$$

the double sign corresponding to the substitution of

$$\pm \nu \sqrt{(-1)}, \text{ or } \pm \nu' \sqrt{(-1)} \text{ for } n \text{ or } n',$$

retain the modulus ρ for coefficient, and subtract θ from the phase.

It will however be far more convenient to employ symbolical expressions such as (B). These expressions will remain applicable without any change when n or n' becomes imaginary: it will only be necessary to observe to take

$$\pm \nu \sqrt{(-1)}, \text{ or } \pm \nu' \sqrt{(-1)}$$

with the negative sign. If we had chosen to employ the expressions (B) with the opposite sign in the index, which would have done equally well, it would then have been necessary to take the positive sign.

8.   We are now prepared to enter on the regular calculation of the intensity of the central spot; but before doing so it will be proper to consider how far we are justified in omitting the consideration of the superficial undulations which, when the vibrations are in the plane of incidence, are the representatives of normal vibrations. These undulations may conveniently be called *normal superficial undulations,* to distinguish them from the superficial undulations expressed by the third of equations (D), or the first and second of equations (F), which may be called *transversal.* The former name however might, without warning, be calculated to carry a false impression; for the undulations spoken of are not propagated by way of condensation and rarefaction; the disturbance is in fact precisely the same as that which exists near the surface of deep water when a series of oscillatory waves is propagated along it, although the cause of the propagation is extremely different in the two cases.

Now in the ordinary theory of Newton's Rings, no account is taken of the normal superficial undulations which may be supposed to exist; and the result so obtained from theory agrees very

well with observation. When the angle of incidence passes through
the critical angle, although a material change takes place in the
nature of the refracted transversal undulation, no such change
takes place in the case of the normal superficial undulations: the
critical angle is in fact nothing particular as regards these undulations. Consequently, we should expect the result obtained from
theory when the normal superficial undulations are left out of consideration to agree as well with experiment beyond the critical
angle as within it.

9. It is however one thing to show why we are justified in
expecting a near accordance between the simplified theory and
experiment, beyond the critical angle, in consequence of the
observed accordance within that angle; it is another thing to show
why a near accordance ought to be expected both in the one case
and in the other. The following considerations will show that the
effect of the normal superficial undulations on the observed
phenomena is most probably very slight.

At the point of contact of the first and third media, the reflection and refraction will take place as if the second medium were
removed, so that the first and third were in contact throughout.
Now Fresnel's expressions satisfy the condition of giving the same
intensity for the reflected and refracted light whether we suppose
the refraction to take place directly out of the first medium into
the third, or take into account the infinite number of reflections
which take place when the second medium is interposed, and then
suppose the thickness of the interposed medium to vanish. Consequently the expression we shall obtain for the intensity by neglecting the normal superficial undulations will be strictly correct
for the point of contact, Fresnel's expressions being supposed correct, and of course will be sensibly correct for some distance round
that point. Again, the expression for the refracted normal superficial undulation will contain in the index of the exponential
$-klz$, in place of $-k\sqrt{(l^2 - v^2/v'^2)}\, z$, which occurs in the expression for the refracted transversal superficial undulation; and therefore the former kind of undulation will decrease much more rapidly,
in receding from the surface, than the latter, so that the effect
of the former will be insensible at a distance from the point of
contact at which the effect of the latter is still important. If we
cembine these two considerations, we can hardly suppose the

S. II.   5

effect of the normal superficial undulations at intermediate points to be of any material importance.

10. The phenomenon of Newton's Rings is the only one in which I see at present any chance of rendering these undulations sensible in experiment : for the only way in which I can conceive them to be rendered sensible is, by their again producing transversal vibrations; and in consequence of the rapid diminution of the disturbance on receding from the surface, this can only happen when there exists a second reflecting surface in close proximity with the first. It is not my intention to pursue the subject further at present, but merely to do for angles of incidence greater than the critical angle what has long ago been done for smaller angles, in which case light is refracted in the ordinary way. Before quitting the subject however I would observe that, for the reasons already mentioned, the near accordance of observation with the expression for the intensity obtained when the normal superficial undulations are not taken into consideration cannot be regarded as any valid argument against the existence of such undulations.

11. Let Newton's Rings be formed between a prism and a lens, or a second prism, of the same kind of glass. Suppose the incident light polarized, either in the plane of incidence, or in a plane perpendicular to the plane of incidence. Let the coefficient of vibration in the incident light be taken for unity; and, according to the notation employed in Airy's *Tract*, let the coefficient be multiplied by b for reflection and by c for refraction when light passes from glass into air, and by e for reflection and f for refraction when light passes from air into glass. In the case contemplated b, c, e, f become imaginary, but that will be taken into account further on. Then the incident vibration will be represented symbolically by

$$\epsilon^{k\,(vt-lx-nz)\,\sqrt{-1}},$$

according to the notation already employed; and the reflected and refracted vibrations will be represented by

$$b\epsilon^{k(vt-lx+nz)\sqrt{-1}},$$

$$c\epsilon^{-k'v'z}.\,\epsilon^{k'\,(v't-l'x)\,\sqrt{-1}}.$$

Consider a point at which the distance of the pieces of glass is D; and, as in the usual investigation, regard the plate of air about that point as bounded by parallel planes. When the superficial undulation represented by the last of the preceding expressions is incident on the second surface, the coefficient of vibration will become cq, q being put for shortness in place of $\epsilon^{-k'\nu'D}$; and the reflected and refracted vibrations will be represented by

$$cqe\epsilon^{k'\nu'z} . \epsilon^{k'(\nu't - l'x)\sqrt{-1}},$$

$$cqfe^{k(\nu t - lx - nz)\sqrt{-1}},$$

z being now measured from the lower surface. It is evident that each time that the undulation passes from one surface to the other the coefficient of vibration will be multiplied by q, while the phase will remain the same. Taking account of the infinite series of reflections, we get for the symbolical expression for the reflected vibration

$$\{b + cefq^2 (1 + e^2q^2 + e^4q^4 + \ldots)\} \epsilon^{k(\nu t - lx + nz)\sqrt{-1}}.$$

Summing the geometric series, we get for the coefficient of the exponential

$$b + \frac{cefq^2}{1 - e^2q^2}.$$

Now it follows from Fresnel's expressions that

$$b = -e, \quad cf = 1 - e^2 *,$$

These substitutions being made in the coefficient, we get for the symbolical expression for the reflected vibration

$$\frac{(1 - q^2) b}{1 - q^2b^2} \epsilon^{k(\nu t - lx + nz)\sqrt{-1}} \ldots\ldots\ldots\ldots\ldots(G).$$

Let the coefficient, which is imaginary, be put under the form $\rho \{\cos \psi + \sqrt{(-1)} \sin \psi\}$; then the real part of the whole expression, namely

$$\rho \cos \{k (\nu t - lx + nz) + \psi\},$$

will represent the vibration in the reflected light, so that ρ^2 is the intensity, and ψ the acceleration of phase.

12. Let i be the angle of incidence on the first surface of the plate of air, μ the refractive index of glass; and let λ now denote the length of wave in air. Then in the expression for q

$$k'\nu' = \frac{2\pi}{\lambda} \sqrt{\mu^2 \sin^2 i - 1}.$$

---

* I have proved these equations in a very simple manner, without any reference to Fresnel's formulæ, in a paper which will appear in the next number of the *Cambridge and Dublin Mathematical Journal* [p. 89 of the present volume].

In the expression for b we must, according to Art. 2, take the imaginary expression for $\cos i'$ with the negative sign. We thus get for light polarized in the plane of incidence (Airy's *Tract*, p. 362, 2nd edition*), changing the sign of $\sqrt{-1}$,

$$b = \cos 2\theta + \sqrt{-1} \sin 2\theta,$$

where

$$\tan \theta = \frac{\sqrt{\mu^2 \sin^2 i - 1}}{\mu \cos i} \quad\dots\dots\dots\dots\dots(2).$$

Putting C for the coefficient in the expression (G), we have

$$C = \frac{1 - q^2}{b^{-1} - q^2 b} = \frac{1 - q^2}{(1 - q^2)\cos 2\theta - \sqrt{-1}\,(1 + q^2)\sin 2\theta}$$

$$= \frac{(1 - q^2)\{(1 - q^2)\cos 2\theta + \sqrt{-1}\,(1 + q^2)\sin 2\theta\}}{(1 - q^2)^2 + 4q^2 \sin^2 2\theta} ;$$

whence

$$\tan \psi = \frac{1 + q^2}{1 - q^2}\tan 2\theta \dots\dots\dots\dots\dots(3),$$

$$\rho^2 = \frac{(1 - q^2)^2}{(1 - q^2)^2 + 4q^2 \sin^2 2\theta} \dots\dots\dots\dots(4),$$

where

$$q = \epsilon^{-\frac{2\pi D}{\lambda}\sqrt{\mu^2 \sin^2 i - 1}} \dots\dots\dots\dots\dots(5).$$

If we take ρ positive, as it will be convenient to do, we must take ψ so that $\cos \psi$ and $\cos 2\theta$ may have the same sign. Hence from (3) $\sin \psi$ must be positive, since $\sin 2\theta$ is positive, inasmuch as θ lies between 0 and $\frac{1}{2}\pi$. Hence, of the two angles lying between $-\pi$ and π which satisfy (2), we must take that which lies between 0 and π.

For light polarized perpendicularly to the plane of incidence, we have merely to substitute ϕ for θ in the equations (3) and (4), where

$$\tan \phi = \frac{\mu \sqrt{\mu^2 \sin^2 i - 1}}{\cos i} \dots\dots\dots\dots\dots(6).$$

The value of q does not depend on the nature of the polarization.

---

* Mr Airy speaks of "vibrations perpendicular to the plane of incidence," and "vibrations parallel to the plane of incidence," adopting the theory of Fresnel; but there is nothing in this paper which requires us to enter into the question whether the vibrations in plane polarized light are in or perpendicular to the plane of polarization.

13. For the transmitted light we have an expression similar to (G), with $-nz$ in place of nz, and a different coefficient $C_{\prime}$, where

$$C_{\prime} = cqf(1 + e^2q^2 + e^4q^4 + \ldots) = \frac{qcf}{1 - e^2q^2} = \frac{q(1 - b^2)}{1 - q^2b^2} = \frac{q(b^{-1} - b)}{b^{-1} - q^2b}.$$

When the light is polarized in the plane of incidence we have

$$C_{\prime} = \frac{-\sqrt{-1} \cdot 2q \sin 2\theta}{(1 - q^2)\cos 2\theta - \sqrt{-1}(1 + q^2)\sin 2\theta}$$

$$= \frac{2q \sin 2\theta \{(1 + q^2)\sin 2\theta - \sqrt{-1}(1 - q^2)\cos 2\theta\}}{(1 - q^2)^2 + 4q^2 \sin^2 2\theta} \quad \ldots\ldots(7);$$

so that if $\psi_{\prime}$ and $\rho_{\prime}$ refer to the transmitted light we have

$$\tan \psi_{\prime} = -\frac{1 - q^2}{1 + q^2} \cot 2\theta \ldots\ldots\ldots\ldots\ldots\ldots(8),$$

$$\rho_{\prime}^2 = \frac{4q^2 \sin^2 2\theta}{(1 - q^2)^2 + 4q^2 \sin^2 2\theta} \quad \ldots\ldots\ldots\ldots\ldots(9).$$

If we take $\rho_{\prime}$ positive, as it will be supposed to be, we must take $\psi_{\prime}$ such that $\cos \psi_{\prime}$ may be positive; and therefore, of the two angles lying between $-\pi$ and π which satisfy (8), we must choose that which lies between $-\frac{1}{2}\pi$ and $+\frac{1}{2}\pi$. Hence, since from (3) and (8) $\psi_{\prime}$ is of the form $\psi + \frac{1}{2}\pi + n\pi$, n being an integer, we must take $\psi_{\prime} = \psi - \frac{1}{2}\pi$.

For light polarized perpendicularly to the plane of incidence we have only to put ϕ for θ. It follows from (4) and (9) that the sum of the intensities of the reflected and transmitted light is equal to unity, as of course ought to be the case. This renders it unnecessary to discuss the expression for the intensity of the transmitted light.

14. Taking the expression (4) for the intensity of the reflected light, consider first how it varies on receding from the point of contact.

As the point of contact $D = 0$, and therefore from (5) $q = 1$, and therefore $\rho^2 = 0$, or there is absolute darkness. On receding from the point of contact q decreases, but slowly at first, inasmuch as D varies as r^2, r being the distance from the point of contact. It follows from (4) that the intensity ρ^2 varies ultimately as r^4, so that it increases at first with extreme slowness. Consequently the darkness is, as far as sense can decide, perfect for some distance round the point of contact. Further on q decreases more rapidly, and soon becomes insensible. Consequently the intensity decreases, at first rapidly, and then slowly again as it approaches its limiting value 1, to which it soon becomes sensibly equal. All this agrees with observation.

15. Consider next the variation of intensity as depending on the colour. The change in θ and ϕ in passing from one colour to another is but small, and need not here be taken into account: the quantity whose variation it is important to consider is q. Now it follows from (5) that q changes the more rapidly in receding from the point of contact the smaller be λ. Consequently the spot must be smaller for blue light than for red; and therefore towards the edge of the spot seen by reflection, that is beyond the edge of the central portion of it, which is black, there is a predominance of the colours at the blue end of the spectrum; and towards the edge of the bright spot seen by transmission the colours at the red end predominate. The tint is more conspicuous in the transmitted, than in the reflected light, in consequence of the quantity of white light reflected about the edge of the spot. The separation of colours is however but slight, compared with what takes place in dispersion or diffraction, for two reasons. First, the point of minimum intensity is the same for all the colours, and the only reason why there is any tint produced is, that the intensity approaches more rapidly to its limiting value 1 in the case of the blue than in the case of the red. Secondly, the same fraction of the incident light is reflected at points for which $D \propto \lambda$, and therefore $r \propto \sqrt{\lambda}$; and therefore, on this account also, the separation of colours is less than in diffraction, where the colours are arranged according to the values of λ, or in dispersion, where they are arranged according to values of λ^{-2} nearly. These conclusions agree with observation. A faint blueish tint may be perceived about the dark spot seen by reflection; and the fainter portions of the bright spot seen by transmission are of a decided reddish brown.

16. Let us now consider the dependance of the size of the spot on the nature of the polarization. Let s be the ratio of the

intensity of the transmitted light to that of the reflected; s_1, s_2, the particular values of s belonging to light polarized in the plane of incidence and to light polarized perpendicularly to the plane of incidence respectively; then

$$s_1 = \frac{4q^2 \sin^2 2\theta}{(1-q^2)^2}, \quad s_2 = \frac{4q^2 \sin^2 2\phi}{(1-q^2)^2},$$

$$\frac{s_1}{s_2} = \left(\frac{\sin 2\theta}{\sin 2\phi}\right)^2 = \{(\mu^2+1)\sin^2 i - 1\}^2 \ldots\ldots\ldots (10).$$

Now according as s is greater or less, the spot is more or less conspicuous; that is, conspicuous in regard to extent, and intensity at some distance from the point of contact; for in the immediate neighbourhood of that point the light is in all cases wholly transmitted. Very near the critical angle we have from (10) $s_2 = \mu^4 s_1$, and therefore the spot is much more conspicuous for light polarized perpendicularly to the plane of incidence than for light polarized in that plane. As i increases the spots seen in the two cases become more and more nearly equal in magnitude: they become exactly alike when $i = \iota$, where

$$\sin^2 \iota = \frac{2}{1+\mu^2}.$$

When i becomes greater than ι the order of magnitude is reversed; and the spots become more and more unequal as i increases. When $i = 90^\circ$ we have $s_1 = \mu^4 s_2$, so that the inequality becomes very great. This however must be understood with reference to relative, not absolute magnitude; for when the angle of incidence becomes very great both spots become very small.

I have verified these conclusions by viewing the spot through a rhomb of Iceland spar, with its principal plane either parallel or perpendicular to the plane of incidence, as well as by using a doubly refracting prism; but I have not attempted to determine experimentally the angle of incidence at which the spots are exactly equal. Indeed, it could not be determined in this way with any precision, because the difference between the spots is insensible through a considerable range of incidence.

17.   It is worthy of remark that the angle of incidence ι at which the spots are equal, is exactly that at which the difference of acceleration of phase of the oppositely polarized pencils, which arises from total internal reflection, is a maximum.

When $i = \iota$ we have

$$\sin 2\theta = \sin 2\phi = \frac{2\mu}{\mu^2 + 1} \; ; \; \text{ whence } \cot\theta = \tan\phi = \mu \,\dots\dots\, (11) \; ;$$

and

$$\rho^2 = \frac{(1+\mu^2)^2 (1-q^2)^2}{(1+\mu^2)^2 (1-q^2)^2 + 16\mu^2 q^2} \, ,$$

where

$$q = \epsilon^{-\frac{2\pi D}{\lambda}\sqrt{\frac{\mu^2-1}{\mu^2+1}}} \,\dots\dots\dots\dots\dots\,(12).$$

If we determine in succession the angles θ, ζ, η from the equations $\cot\theta = \mu$, $\tan\zeta = q$, $\tan\eta = \sin 2\theta \tan 2\zeta$,

we have $\quad \rho_{\prime}^2 = 1 - \rho^2 = \frac{1}{2} \text{ versin } 2\eta.$

The expression for the intensity may be adapted to numerical computation in the same way for any angle of incidence, except that θ or ϕ must be determined by (2) or (6) instead of (11), and q by (5) instead of (12).

18.   When light is incident at the critical angle, which I shall denote by γ, the expression for the intensity takes the form $0/0$. Putting for shortness $\sqrt{(\mu^2 \sin^2 i - 1)} = w$, we have ultimately

$$q = 1 - \frac{2\pi D}{\lambda} w, \quad \tan\theta = \theta = \frac{w}{\mu\cos i} = \frac{w}{\sqrt{\mu^2 - 1}}, \quad \phi = \mu^2\theta \; ;$$

and we get in the limit

$$\rho^2 = \frac{\left(\frac{\pi D}{\lambda}\right)^2}{\left(\frac{\pi D}{\lambda}\right)^2 + \frac{1}{\mu^2 - 1}} , \quad \text{or} = \frac{\left(\frac{\pi D}{\lambda}\right)^2}{\left(\frac{\pi D}{\lambda}\right)^2 + \frac{\mu^4}{\mu^2 - 1}} \,\dots\dots\,(13),$$

according as the light is polarized in or perpendicularly to the plane of incidence. The same formulæ may be obtained from the expression given at page 304 of Airy's *Tract*, which gives the intensity when $i < \gamma$, and which like (4) takes the form $0/0$ when i becomes equal to γ, in which case e becomes equal to -1.

19.   When i becomes equal to γ, the infinite series of Art. 11 ceases to be convergent: in fact, its several terms become ultimately equal to each other, while at the same time the coefficient by which the series is multiplied vanishes, so that the whole takes the form $0 \times \infty$. The same remark applies to the series at page

303 of Airy's *Tract*. If we had included the coefficient in each term of the series, we should have got series which ceased to be convergent at the same time that their several terms vanished. Now the sum of such a series may depend altogether on the point of view in which it is regarded as a limit. Take for example the convergent infinite series

$$f(x, y) = x \sin y + \tfrac{1}{3} x^3 \sin 3y + \tfrac{1}{5} x^5 \sin 5y + \ldots = \tfrac{1}{2} \tan^{-1} \frac{2x \sin y}{1 - x^2},$$

where x is less than 1, and may be supposed positive. When x becomes 1 and y vanishes $f(x, y)$ becomes indeterminate, and its limiting value depends altogether upon the order in which we suppose x and y to receive their limiting values, or more generally upon the arbitrary relation which we conceive imposed upon the otherwise independent variables x and y as they approach their limiting values together. Thus, if we suppose y first to vanish, and then x to become 1, we have $f(x, y) = 0$; but if we suppose x first to become 1, and then y to vanish, $f(x, y)$ becomes $\pm \pi/4$, + or − according as y vanishes positively or negatively. Hence in the case of such a series a mode of approximating to the value of x or y, which in general was perfectly legitimate, might become inadmissible in the extreme case in which $x = 1$, or nearly $= 1$. Consequently, in the case of Newton's Rings when $i \sim \gamma$ is extremely small, it is no longer safe to neglect the defect of parallelism of the surfaces. Nevertheless, inasmuch as the expression (4), which applies to the case in which $i > \gamma$, and the ordinary expression which applies when $i < \gamma$, alter continuously as i alters, and agree with (13) when $i = \gamma$, we may employ the latter expression in so far as the phenomenon to be explained alters continuously as i alters. Consequently we may apply the expression (13) to the central spot when $i = \gamma$, or nearly $= \gamma$, at least if we do not push the expression beyond values of D corresponding to the limits of the central spot as seen at other angles of incidence. To explain however the precise mode of disappearance of the rings, and to determine their greatest dilatation, we should have to enter on a special investigation in which the inclination of the surfaces should be taken into account.

20.   I have calculated the following Table of the intensity of the transmitted light, taking the intensity of the incident light at 100. The Table is calculated for values of D increasing by $\lambda/4$,

and for three angles of incidence, namely, the critical angle, the angle ι before mentioned, and a considerable angle, for which I have taken 60°. I have supposed $\mu = 1\cdot63$, which is about the refractive index for the brightest part of the spectrum in the case of flint glass. This value of μ gives $\gamma = 37° 51'$, $\iota = 42° 18'$. The numerals I., II. refer to light polarized in and perpendicularly to the plane of incidence respectively.

$\dfrac{4D}{\lambda}$	$i=\gamma$		$i=\iota$	$i=60°$	
	I.	II.	I. and II.	I.	II.
0	100	100	100	100	100
1	49	87	33	16	6
2	20	63	5	1	0
3	10	43	1	0	
4	6	30	0		
5	4	22			
6	3	16			
7	2	12			
8	2	10			
9	1	8			
10	1	6			
11	1	5			
12	1	5			
13	1	4			
⋮		⋮			
26		1			
27		0			

21. A Table such as this would enable us to draw the curve of intensity, or the curve in which the abscissa is proportional to the distance of the point considered from the point of contact, and the ordinate proportional to the intensity. For this purpose it would only be requisite to lay down on the axis of the abscissæ, on the positive and negative sides of the origin, distances proportional to the square roots of the numbers in the first column, and to take for ordinates lengths proportional to the numbers in one of the succeeding columns. To draw the curve of intensity for $i = \iota$ or for $i = 60°$, the table ought to have been calculated with smaller intervals between the values of D; but the law of the decrease of the intensity cannot be accurately observed.

22. From the expression (13) compared with (4), it will be seen that the intensity decreases much more rapidly, at some

distance from the point of contact, when i is considerably greater than γ than when $i = \gamma$ nearly. This agrees with observation. What may be called the *ragged edge* of the bright spot seen by transmission is in fact much broader in the latter case than in the former.

When i becomes equal to $90°$ there is no particular change in the value of q, but the angles θ and ϕ become equal to $90°$, and therefore $\sin 2\theta$ and $\sin 2\phi$ vanish, so that the spot vanishes. Observation shows that the spot becomes very small when i becomes nearly equal to $90°$.

23. Suppose the incident light to be polarized in a plane making an angle α with the plane of incidence. Then at the point of contact the light, being transmitted as if the first and third media formed one uninterrupted medium, will be plane polarized, the plane of polarization being the same as at first. At a sufficient distance from the point of contact there is no sensible quantity of light transmitted. At intermediate distances the transmitted light is in general elliptically polarized, since it follows from (8) and the expression thence derived by writing ϕ for θ that the two streams of light, polarized in and perpendicularly to the plane of incidence respectively, into which the incident light may be conceived to be decomposed, are unequally accelerated or retarded. At the point of contact, where $q = 1$, these two expressions agree in giving $\psi_{\prime} = 0$. Suppose now that the transmitted light is analyzed, so as to extinguish the light which passes through close to the point of contact. Then the centre of the spot will be dark, and beyond a certain distance all round there will be darkness, because no sensible quantity of light was incident on the analyzer; but at intermediate distances a portion of the light incident on the analyzer will be visible. Consequently the appearance will be that of a luminous ring with a perfectly dark centre.

24. Let the coefficient of vibration in the incident light be taken for unity; then the incident vibration may be resolved into two, whose coefficients are $\cos \alpha$, $\sin \alpha$, belonging to light polarized in and perpendicularly to the plane of incidence respectively. The phases of vibration will be accelerated by the angles $\psi_{\prime}$, $\psi_{\prime\prime}$, and the coefficients of vibration will be multiplied by $\rho_{\prime}$, $\rho_{\prime\prime}$, if $\psi_{\prime\prime}$, $\rho_{\prime\prime}$

are what $\psi_{,}$, $\rho_{,}$ in Art. (13) become when ϕ is put for θ. Hence we may take

$$\rho_{,} \cos \alpha \, . \, \cos \left\{ \frac{2\pi}{\lambda} \, (vt - \mu x') + \psi_{,} \right\},$$

$$\rho_{,,} \sin \alpha \, . \, \cos \left\{ \frac{2\pi}{\lambda} \, (vt - \mu x') + \psi_{,,} \right\},$$

to represent the vibrations which compounded together make up the transmitted light, x' being measured in the direction of propagation. The light being analyzed in the way above mentioned, it is only the resolved parts of these vibrations in a direction perpendicular to that of the vibrations in the incident light which are preserved. We thus get, to express the vibration with which we are concerned,

$$\sin \alpha \, \cos \alpha \, \left\{ \rho_{,} \cos \left(\frac{2\pi}{\lambda} \, (vt - \mu x') + \psi_{,} \right) - \rho_{,,} \cos \left(\frac{2\pi}{\lambda} \, (vt - \mu x') + \psi_{,,} \right) \right\},$$

which gives for the intensity (I) at any point of the ring

$$I = \tfrac{1}{4} \sin^2 2\alpha \, \{ (\rho_{,} \cos \psi_{,} - \rho_{,,} \cos \psi_{,,})^2 + (\rho_{,} \sin \psi_{,} - \rho_{,,} \sin \psi_{,,})^2 \} \ldots (14),$$

$$= \tfrac{1}{4} \sin^2 2\alpha \, \{ \rho_{,}{}^2 + \rho_{,,}{}^2 - 2\rho_{,}\rho_{,,} \cos (\psi_{,,} - \psi_{,}) \}.$$

Let P_θ, Q_θ be respectively the real part of the expression at the second side of (7) and the coefficient of $\sqrt{(-1)}$, and let P_ϕ, Q_ϕ be what P_θ, Q_θ become when ϕ is put for θ. Then we may if we please replace (14) by

$$I = \tfrac{1}{4} \sin^2 2\alpha \, \{ (P_\theta - P_\phi)^2 + (Q_\theta - Q_\phi)^2 \} \ldots \ldots \ldots (15).$$

The ring is brightest, for a given angle of incidence, when $\alpha = 45°$. When $i = \iota$, the two kinds of polarized light are transmitted in the same proportion; but it does not therefore follow that the ring vanishes, inasmuch as the change of phase is different in the two cases. In fact, in this case the angles ϕ, θ are complementary; so that $\cot 2\phi$, $\cot 2\theta$ are equal in magnitude but opposite in sign, and therefore from (8) the phase in the one case is accelerated and in the other case retarded by the angle

$$\tan^{-1} \left(\frac{1 - q^2}{1 + q^2} \cot 2\theta \right), \text{ or } \tan^{-1} \left(\frac{1 - q^2}{1 + q^2} \frac{\mu^2 - 1}{2\mu} \right).$$

It follows from (14) that the ring cannot vanish unless $\rho_{,} \cos \psi_{,} = \rho_{,,} \cos \psi_{,,}$, and $\rho_{,} \sin \psi_{,} = \rho_{,,} \sin \psi_{,,}$. This requires that $\rho_i{}^2 = \rho_{ii}{}^2$, which is satisfied only when $i = \iota$, in which case as we have seen the ring does not vanish. Consequently a ring is formed at all angles of incidence; but it should be remembered that the spot, and consequently the ring, vanishes when i becomes $90°$.

25. When $i = \gamma$, the expressions for P_θ, Q_θ, take the form $0/0$, and we find, putting for shortness $\pi D/\lambda = p$,

$$P_\theta = \frac{(\mu^2 - 1)^{-1}}{p^2 + (\mu^2 - 1)^{-1}}, \qquad P_\phi = \frac{\mu^4 (\mu^2 - 1)^{-1}}{p^2 + \mu^4 (\mu^2 - 1)^{-1}},$$

$$Q_\theta = -\frac{p (\mu^2 - 1)^{-\frac{1}{2}}}{p^2 + (\mu^2 - 1)^{-1}}, \quad Q_\phi = -\frac{p\mu^2 (\mu^2 - 1)^{-\frac{1}{2}}}{p^2 + \mu^4 (\mu^2 - 1)^{-1}}.$$

If we take two subsidiary angles χ, ω, determined by the equations

$$\frac{\pi D}{\lambda} \sqrt{\mu^2 - 1} = \tan \chi = \mu^2 \tan \omega,$$

we get

$$P_\theta = \cos^2 \chi, \qquad P_\phi = \cos^2 \omega,$$

$$Q_\theta = -\sin \chi \cos \chi, \quad Q_\phi = -\sin \omega \cos \omega.$$

Substituting in (15) and reducing we get, supposing $\alpha = 45°$,

$$I = \tfrac{1}{8} \text{ versin } (2\chi - 2\omega) \dots\dots\dots\dots\dots\dots (16).$$

When $i = \iota$, $\cos 2\phi = -\cos 2\theta$, $\sin 2\phi = \sin 2\theta$; and therefore $P_\phi = P_\theta$, $Q_\phi = -Q_\theta$, which when $\alpha = 45°$ reduces (15) to $I = Q_\theta{}^2$.

If we determine the angle ϖ from the equation

$$1 - q^2 = 2q \sin 2\theta \tan \varpi, \text{ or } \tan \varpi = \cot 2\zeta . \operatorname{cosec} 2\theta,$$

we get

$$I = \tfrac{1}{4} \sin^2 2\varpi . \cos^2 2\theta \dots\dots\dots\dots\dots\dots (17).$$

In these equations

$$\log_e \tan \zeta = -\frac{2\pi D}{\lambda} \sqrt{\frac{\mu^2 - 1}{\mu^2 + 1}}, \quad \cot \theta = \mu.$$

26. The following Table gives the intensity of the ring for the two angles of incidence $i = \gamma$ and $i = \iota$, and for values of D increasing by $\lambda/10$. The intensity is calculated by the formulæ (16) and (17). The intensity of the incident polarized light is taken at 100, and μ is supposed equal to $1\cdot63$, as before.

$\dfrac{D}{\lambda}$	I $i=\gamma$	I $i=\iota$	$\dfrac{D}{\lambda}$	I $i=\gamma$
·0	0·0	0·0	1·6	1·4
·1	1·3	3·2	1·7	1·2
·2	3·5	5·1	1·8	1·1
·3	4·8	3·6	1·9	1·0
·4	5·1	1·9	2·0	·9
·5	4·9	·9	2·1	·9
·6	4·5	·4	2·2	·8
·7	4·0	·2	2·3	·7
·8	3·6	·1	2·4	·7
·9	3·1	·0	2·5	·6
1·0	2·8		2·6	·6
1·1	2·4		2·7	·5
1·2	2·1		2·8	·5
1·3	1·9		2·9	·5
1·4	1·7		3·0	·4
1·5	1·5			

The column for $i=\gamma$ may be continued with sufficient accuracy, by taking I to vary inversely as the square of the number in the first column.

27. I have seen the ring very distinctly by viewing the light transmitted at an angle of incidence a little greater than the critical angle. In what follows, in speaking of angles of position, I shall consider those positive which are measured in the direction of motion of the hands of a watch, to a person looking at the light. The plane of incidence being about 45° to the positive side of the plane of primitive polarization, the appearance presented as the analyzer (a Nicol's prism) was turned, in the positive direction, through the position in which the light from the centre was extinguished, was as follows. On approaching that position, in addition to the general darkening of the spot, a dark ring was observed to separate itself from the dark field about the spot, and to move towards the centre, where it formed a broad dark patch, surrounded by a rather faint ring of light. On continuing to turn, the ring got brighter, and the central patch ceased to be quite black. The light transmitted near the centre increased in intensity till the dark patch disappeared: the patch did not break up into a dark ring travelling outwards.

On making the analyzer revolve in the contrary direction, the same appearances were of course repeated in a reverse order: a dull central patch was seen, which became darker and darker till it appeared quite black, after which it broke up into a dark ring which travelled outwards till it was lost in the dark field surrounding the spot. The appearance was a good deal disturbed by the imperfect annealing of the prisms. When the plane of incidence was inclined at an angle of about $-45°$ to the plane of primitive polarization, the same appearance as before was presented on reversing the direction of rotation of the analyzer.

28. Although the complete theoretical investigation of the moving dark ring would require a great deal of numerical calculation, a general explanation may very easily be given. At the point of contact the transmitted light is plane polarized, the plane of polarization being the same as at first*. At some distance from the point of contact, although strictly speaking the light is elliptically polarized, it may be represented in a general way by plane polarized light with its plane of polarization further removed than at first from the plane of incidence, in consequence of the larger proportion in which light polarized perpendicularly to the plane of incidence is transmitted, than light polarized in that plane. Consequently the transmitted light may be represented in a general way by plane polarized, with its plane of polarization receding from the plane of incidence on going from the centre outwards. If therefore we suppose the position of the plane of incidence, and the direction of rotation of the analyzer, to be those first mentioned, the plane of polarization of light transmitted by the analyzer will become perpendicular to the plane of polarization of the transmitted light of the spot sooner towards the edge of the spot than in the middle. The locus of the point where the two planes are perpendicular to each other will in fact be a circle, whose radius will contract as the analyzer turns round. When the analyzer has passed the position in which its plane of polarization is perpendicular to that of the light at the centre of the spot, the inclination of the planes of polarization of the analyzer and of the transmitted light of the spot decreases, for a given position of the analyzer, in passing from the centre outwards; and therefore there is formed, not a dark ring travelling outwards as the analyzer turns round, but a dark patch, darkest in the centre, and becoming

---

* The rotation of the plane of polarization due to the refraction at the surfaces at which the light enters the first prism and quits the second is not here mentioned, as it has nothing to do with the phenomenon discussed.

brighter, and therefore less and less conspicuous, as the analyzer turns round. The appearance will of course be the same when the plane of incidence is turned through 90°, so as to be equally inclined to the plane of polarization on the opposite side, provided the direction of rotation of the analyzer be reversed.

29. The investigation of the intensity of the spot formed beyond the critical angle when the third medium is of a different nature from the first, does not seem likely to lead to results of any particular interest. Perhaps the most remarkable case is that in which the second and third media are both of lower refractive power than the first, and the angle of incidence is greater than either of the critical angles for refraction out of the first medium into the second, or out of the first into the third. In this case the light must be wholly reflected; but the acceleration of phase due to the total internal reflection will alter in the neighbourhood of the point of contact. At that point it will be the same as if the third medium occupied the place of the second as well as its own; at a distance sufficient to render the influence of the third medium insensible, it will be the same as if the second medium occupied the place of the third as well as its own. The law of the variation of the acceleration from the one to the other of its extreme values, as the distance from the point of contact varies, would result from the investigation. This law could be put to the test of experiment by examining the nature of the elliptic polarization of the light reflected in the neighbourhood of the point of contact when the incident light is polarized at an azimuth of 45°, or thereabouts. The theoretical investigation does not present the slightest difficulty in principle, but would lead to rather long expressions; and as the experiment would be difficult, and is not likely to be performed, there is no occasion to go into the investigation.

30. In viewing the spot formed between a prism and a lens, I was struck with the sudden, or nearly sudden disappearance of the spot at a considerable angle of incidence. The cause of the disappearance no doubt was that the lens was of lower refractive power than the prism, and that the critical angle was reached which belongs to refraction out of the prism into the lens. Before disappearing, the spot became of a bright sky blue, which

shows that the ratio of the refractive index of the prism to that of the lens was greater for the blue rays than for the red. As the disappearance of the spot can be observed with a good deal of precision, it may be possible to determine in this way the refractive index of a substance of which only a very minute quantity can be obtained. The examination of the refractive index of the globule obtained from a small fragment of a fusible mineral might afford the mineralogist a means of discriminating between one mineral and another. For this purpose a plate, which is what a prism becomes when each base angle becomes 90°, would probably be more convenient than a prism. Of course the observation is possible only when the refractive index of the substance to be examined is less than that of the prism or plate.

[From the *Philosophical Magazine*, Vol. XXXIV. p. 52, (*January*, 1849.)]

ON SOME POINTS IN THE RECEIVED THEORY OF SOUND*.

I PROCEED now to notice the apparent contradiction at which Professor Challis has arrived by considering spherical waves, a contradiction which it is the chief object of this communication to consider. The only reason why I took no notice of it in a former communication was, that it was expressed with such brevity by Professor Challis (Vol. XXXII. p. 497), that I did not perceive how the conclusion that the condensation varies inversely as the square of the distance was arrived at. On mentioning this circumstance to Professor Challis, he kindly explained to me his reasoning, which he has since stated in detail (Vol. XXXIII. p. 463)†.

---

* The beginning and end of this Paper are omitted, as being merely controversial, and of ephemeral interest.

† The objection is put in two slightly different forms in the two Papers. The substance of it may be placed before the reader in a few words.

Conceive a wave of sound of small disturbance to be travelling outwards from a centre, the disturbance being alike in all directions round the centre. Then according to the received theory the condensation is expressed by equation (1), where r is the distance from the centre, and s the condensation. It follows from this equation that any phase of the wave is carried outwards with the velocity of propagation a, and that the condensation varies inversely as the distance from the centre. But if we consider the shell of infinitesimal thickness a comprised between spherical surfaces of radii r and $r+a$ corresponding to given phases, so that these surfaces travel outwards with the velocity a, the excess of matter in the shell over the quantity corresponding to the undisturbed density will vary as the condensation multiplied by the volume, and therefore as $r^2 s$; and as the constancy of mass requires that this excess should be constant, s must vary inversely as r^2 not r.

Or instead of considering only an infinitesimal shell, consider the whole of an outward travelling wave, and for simplicity's sake suppose it to have travelled so far that its thickness is small compared with its mean radius r or at, t being

The whole force of the reasoning rests on the tacit supposition that when a wave is propagated from the centre outwards, any arbitrary portion of the wave, bounded by spherical surfaces concentric with the bounding surfaces of the wave, may be isolated, the rest of the wave being replaced by quiescent fluid; and that being so isolated, it will continue to be propagated outwards as before, all the fluid except the successive portions which form the wave in its successive positions being at rest. At first sight it might seem as if this assumption were merely an application of the principle of the coexistence of small motions, but it is in reality extremely different. The equations are competent to decide whether the isolation be possible or not. The subject may be considered in different ways; they will all be found to lead to the same result.

1. We may evidently without absurdity conceive an outward travelling wave to exist already, without entering into the question of its original generation; and we may suppose the condensation to be given arbitrarily throughout this wave. By an outward travelling wave, I mean one for which the quantity usually denoted by ϕ contains a function of $r - at$, unaccompanied by a function of $r + at$, in which case the expressions for v and s will likewise contain functions of $r - at$ only. Let

$$as = \frac{f'(r - at)}{r} \quad \ldots\ldots\ldots\ldots\ldots\ldots\ldots(1).$$

We are at liberty to suppose $f'(z) = 0$, except from $z = b$ to $z = c$, where b and c are supposed positive; and we may take $f'(z)$ to denote any arbitrary function for which the portion from $z = b$

the time of travelling from the origin to the distance r. Then assuming the expression (1), and putting the factor r outside the sign of integration, as we are at liberty to do in consequence of the supposition made above as to the distance the wave has travelled, we have for the quantity of matter existing at any time in the wave beyond what would occupy the same space in the quiescent state of the fluid,

$$4\pi . a^2 t^2 \times \rho \int f'(r - at)\, dr \div a^2 t$$

very nearly, or $4\pi \rho A t$, putting A for the value of the integral $\int f'(r - at)\, dr$ taken from the inner to the outer boundary of the wave. Hence the matter increases in quantity with the time.

to $z = c$ has been isolated, the rest having been suppressed. Equation (1) gives

$$\phi = -a^2 \int s\, dt = \frac{f(r - at)}{r} + \psi(r), \dots\dots\dots\dots (2),$$

$\psi(r)$ being an arbitrary function of r, to determine which we must substitute the value of ϕ given by (2) in the equation which ϕ has to satisfy, namely

$$\frac{d^2 . r\phi}{dt^2} = a^2 \frac{d^2 . r\phi}{dr^2} \dots\dots\dots\dots\dots (3).$$

This equation gives $\psi(r) = C + D/r$, C and D being arbitrary constants, whence

$$v = \frac{d\phi}{dr} = \frac{f'(r - at)}{r} - \frac{f(r - at)}{r^2} - \frac{D}{r^2} \dots\dots\dots\dots (4).$$

Now the function $f(z)$ is merely defined as an integral of $f'(z)\, dz$, and we may suppose the integral so chosen as to vanish when $z = b$, and therefore when z has any smaller value. Consequently we get from (4), for every point within the sphere which forms the inner boundary of the wave of condensation,

$$v = -\frac{D}{r^2} \dots\dots\dots\dots\dots\dots (5).$$

Again, if we put $f(c) = A$, so that $f(z) = A$ when $z > c$, we have for any point outside the wave of condensation,

$$v = -\frac{A + D}{r^2} \dots\dots\dots\dots\dots (6).$$

The velocities expressed by (5) and (6) are evidently such as could take place in an incompressible fluid. Now Professor Challis's reasoning requires that the fluid be at rest beyond the limits of the wave of condensation, since otherwise the conclusion cannot be drawn that the matter increases with the time. Consequently we must have $D = 0$, $A = 0$; but if $A = 0$ the reasoning at p. 463 evidently falls to the ground.

2. We may if we please consider an outward travelling wave which arose from a disturbance originally confined to a sphere of radius ϵ. At p. 463 Professor Challis has referred to Poisson's expressions relating to this case. It should be observed that Poisson's expressions at page 706 of the *Traité de Mécanique* (second edition) do not apply to the whole wave from $r = at - \epsilon$

to $r = at + \epsilon$, but only to the portion from $r = at - \epsilon$ to $r = at$; the expressions which apply to the remainder are those given near the bottom of page 705. We may of course represent the condensation s by a single function $1/ar \cdot \chi (r - at)$, where

$$\chi (-z) = f' (z), \quad \chi (z) = F' (z),$$

z being positive; and we shall have

$$A = \int_{-\epsilon}^{\epsilon} \chi (z) \, dz = f (\epsilon) - f (0) + F (\epsilon) - F (0).$$

Now Poisson has proved, and moreover expressly stated at page 706, that the functions F, f vanish at the limits of the wave; so that $f (\epsilon) = 0$, $F (\epsilon) = 0$. Also Poisson's equations (6) give in the limiting case for which $z = 0$, $f (0) + F (0) = 0$, so that $A = 0$ as before.

3. We may evidently without absurdity conceive the velocity and condensation to be both given arbitrarily for the instant at which we begin to consider the motion; but then we must take the *complete* integral of (3), and determine the two arbitrary functions which it contains. We are at liberty, for example, to suppose the condensation and velocity when $t = 0$ given by the equations

$$as = \frac{f' (r)}{r}, \quad v = \frac{f' (r)}{r} - \frac{f (r)}{r^2},$$

from $r = b$ to $r = c$, and to suppose them equal to zero for all other values of r; but we are not therefore at liberty to suppress the second arbitrary function in the integral of (3). The problem is only a particular case of that considered by Poisson, and the arbitrary functions are determined by his equations (6) and (8), where, however, it must be observed, that the arbitrary functions which Poisson denotes by f, F must not be confounded with the given function here denoted by f, which latter will appear at the *right-hand* side of equations (8). The solution presents no difficulty in principle, but it is tedious from the great number of cases to be considered, since the form of one of the functions which enter into the result changes whenever the value of $r + at$ or of $r - at$ passes through either b or c, or when that of $r - at$ passes through zero. It would be found that unless $f (b) = 0$, a backward wave sets out from the inner surface of the spherical shell containing the disturbed portion of the fluid; and unless $f(c) = 0$, a similar wave starts from the outer surface. Hence, whenever the disturbance can be propagated in the positive direction only, we must have A, or $f(c) - f(b)$, equal to zero. When a backward wave is formed, it first approaches the centre, which in due time it reaches, and then begins to diverge outwards, so that after the time c/a there is nothing left but an outward travelling wave, of breadth $2c$, in which the fluid is partly rarefied and partly condensed, in such a manner that $\int rs\, dr$ taken throughout the wave, or A, is equal to zero.

It appears, then, that for any outward travelling wave, or for any portion of such a wave which can be isolated, the quantity A is necessarily equal to zero. Consequently the conclusion arrived at, that the mean condensation in such a wave or portion of a wave varies ultimately inversely as the distance from the centre, proves not to be true. It is true, as commonly stated, that the condensation at corresponding points in such a wave in its successive positions varies ultimately inversely as the distance from the centre; it is likewise true, as Professor Challis has argued, that the mean condensation in any portion of the wave which may be isolated varies ultimately inversely as the square of the distance; but these conclusions do not in the slightest degree militate against each other.

If we suppose b to increase indefinitely, the condensation or rarefaction in the wave which travels towards the centre will be a small quantity, of the order b^{-1}, compared with that in the shell. In the limiting case, in which $b = \infty$, the condensation or rarefaction in the backward travelling wave vanishes. If in the equations of paragraph 3 we write $b + x$ for r, $b\sigma(x)$ for $f'(r)$, and then suppose b to become infinite, we shall get $as = \sigma(x)$, $v = \sigma(x)$. Consequently a plane wave in which the relation $v = as$ is satisfied will be propagated in the positive direction only, no matter whether $\int \sigma(x)\, dx$ taken from the beginning to the end of the wave be or be not equal to zero; and therefore any arbitrary portion of such a wave may be conceived to be isolated, and being isolated, will continue to travel in the positive direction only, without sending back any wave which will be propagated in the negative direction. This result follows at once from the equations which apply directly to plane waves; I mean, of course, the approximate equations obtained by neglecting the squares of small quantities. It may be observed, however, that it appears from what has been proved, that it is a property of every plane wave which is the limit of a spherical wave, to have its mean condensation equal to zero; although there is no absurdity in conceiving a plane wave in which that is not the case as already existing, and inquiring in what manner such a wave will be propagated.

There is another way of putting the apparent contradiction arrived at in the case of spherical waves, which Professor Challis has mentioned to me, and has given me permission to publish. Conceive an elastic spherical envelope to exist in an infinite mass of air which is at rest, and conceive it to expand for a certain time, and then to come to rest again, preserving its spherical form and the position of its centre during expansion. We should apparently have a wave consisting of condensation only, without rarefaction, travelling outwards, in which case the conclusion would follow, that the quantity of matter altered with the time.

Now in this or any similar case we have a perfectly definite problem, and our equations are competent to lead to the complete solution, and so make known whether or not a wave will be propagated outwards leaving the fluid about the envelope at rest, and if such a wave be formed, whether it will consist of condensation only, or of condensation accompanied by rarefaction: that condensation will on the whole prevail is evident beforehand, because a certain portion of space which was occupied by the fluid is now occupied by the envelope.

In order to simplify as much as possible the analysis, instead of an expanding envelope, suppose that we have a sphere, of a constant radius b, at the surface of which fluid is supplied in such a manner as to produce a constant velocity V from the centre outwards, the supply lasting from the time 0 to the time τ, and then ceasing. This problem is evidently just as good as the former for the purpose intended, and it has the advantage of leading to a result which may be more easily worked out. On account of the length to which the present article has already run, I am unwilling to go into the detail of the solution; I will merely indicate the process, and state the nature of the result.

Since we have no reason to suspect the existence of a function of the form $F(r + at)$ in the value of ϕ which belongs to the present case, we need not burden our equations with this function, but we may assume as the expression for ϕ

$$\phi = \frac{f(r - at)}{r} \quad \ldots\ldots\ldots\ldots\ldots\ldots\ldots\ldots(7).$$

For we can always, if need be, fall back on the complete integral of (3); and if we find that the particular integral (7) enables us to satisfy all the conditions of the problem, we are certain that we should have arrived at the same result had we used the complete integral all along.　These conditions are

$$\phi = 0 \text{ when } t = 0, \text{ from } r = b \text{ to } r = \infty \ldots\ldots\ldots\ldots(8);$$

for ϕ must be equal to a constant, since there is neither condensation nor velocity, and that constant we are at liberty to suppose equal to zero;

$$\frac{d\phi}{dr} = V \text{ when } r = b, \text{ from } t = 0 \text{ to } t = \tau \ldots\ldots\ldots\ldots\ldots(9);$$

$$\frac{d\phi}{dr} = 0 \text{ when } r = b, \text{ from } t = \tau \text{ to } t = \infty \ldots\ldots\ldots\ldots(10).$$

(8) determines $f(z)$ from $z = b$ to $z = \infty$; (9) determines $f(z)$ from $z = b$ to $z = b - a\tau$; and (10) determines $f(z)$ from $z = b - a\tau$ to $z = -\infty$, and thus the motion is completely determined.

It appears from the result that if we consider any particular value of r there is no condensation till $at = r - b$, when it suddenly commences.　The condensation lasts during the time τ, when it is suddenly exchanged for rarefaction, which decreases indefinitely, tending to 0 as its limit as t tends to ∞.　The sudden commencement of the condensation, and its sudden change into rarefaction, depend of course on the sudden commencement and cessation of the supply of fluid at the surface of the sphere, and have nothing to do with the object for which the problem was investigated. Since there is no isolated wave of condensation travelling outwards, the complete solution of the problem leads to no contradiction, as might have been confidently anticipated.

[From the *Cambridge and Dublin Mathematical Journal*, Vol. IV. p. 1, (*February*, 1849.)]

On the Perfect Blackness of the Central Spot in Newton's Rings, and on the Verification of Fresnel's Formulæ for the Intensities of Reflected and Refracted Rays.

WHEN Newton's rings are formed between two glasses of the same kind, the central spot in the reflected rings is observed to be perfectly black. This result is completely at variance with the theory of emissions, according to which the central spot ought to be half as bright as the brightest part of the bright rings, supposing the incident light homogeneous. On the theory of undulations, the intensity of the light reflected at the middle point depends entirely on the proportions in which light is reflected and refracted at the two surfaces of the plate of air, or other interposed medium, whatever it may be. The perfect blackness of the central spot was first explained by Poisson, in the case of a perpendicular incidence, who shewed that when the infinite series of reflections and refractions is taken into account, the expression for the intensity at the centre vanishes, the formula for the intensity of light reflected at a perpendicular incidence first given by Dr Young being assumed. Fresnel extended this conclusion to all incidences by means of a law discovered experimentally by M. Arago, that light is reflected in the same proportions at the first and second surfaces of a transparent plate*. I have thought of a very simple mode of obtaining M. Arago's law from theory, and at the same

---

* See Dr Lloyd's Report on Physical Optics.—*Reports of the British Association*, Vol. III. p. 344.

time establishing theoretically the loss of half an undulation in internal, or else in external reflection.

This method rests on what may be called the *principle of reversion*, a principle which may be enunciated as follows.

If any material system, in which the forces acting depend only on the positions of the particles, be in motion, if at any instant the velocities of the particles be reversed, the previous motion will be repeated in a reverse order. In other words, whatever were the positions of the particles at the time t before the instant of reversion, the same will they be at an equal interval of time t after reversion; from whence it follows that the velocities of the particles in the two cases will be equal in magnitude and opposite in direction.

Let S be the surface of separation of two media which are both transparent, homogeneous, and uncrystallized. For the present purpose S may be supposed a plane. Let A be a point in the surface S where a ray is incident along IA in the first medium. Let AR, AF be the directions of the reflected and refracted rays, AR' the direction of the reflected ray for a ray incident along FA, and therefore also the direction of the refracted ray for a ray incident along RA. Suppose the vibrations in the incident ray to be either parallel or perpendicular to the plane of incidence. Then the vibrations in the reflected and refracted rays will be in the first case parallel and in the second case perpendicular to the plane of incidence, since everything is symmetrical with respect to that plane. The direction of vibration being determined, it remains to determine the alteration of the coefficient of vibration. Let the maximum vibration in the incident light be taken for unity, and, according to the notation employed in Airy's *Tract*, let the coefficient of vibration be multiplied by b for reflection and by c for refraction at the surface S, and by e for reflection and f for refraction at a parallel surface separating the second medium from a third, of the same nature as the first.

Let x be measured from A negatively backwards along AI, and positively forwards along AR or AF, and let it denote the distance from A of the particle considered multiplied by the refractive index of the medium in which the particle is situated, so that it expresses an equivalent length of path in vacuum. Let λ be the length of a wave, and v the velocity of propagation in vacuum; and for shortness sake let

$$\frac{2\pi}{\lambda}(vt - x) = X.$$

Then $\sin X$, $b \sin X$, $c \sin X$ may be taken to represent respectively the incident, reflected, and refracted rays; and it follows from the principle of reversion, if we suppose it applicable to light, that the reflected and refracted rays reversed will produce the incident ray reversed. Now if in the reversed rays we measure x positively along AI or AR', and negatively along AR or AF, the reflected ray reversed will give rise to the rays represented by

$$b^2 \sin X, \text{ reflected along } AI;$$

$$bc \sin X, \text{ refracted along } AR'; *$$

and the refracted ray reversed will give rise to

$$cf \sin X, \text{ refracted along } AI;$$

$$ce \sin X, \text{ reflected along } AR'.$$

The two rays along AR' superposed must destroy each other, and the two along AI must give a ray represented by $\sin X$. We have therefore

$$bc + ce = 0, \quad b^2 + cf = 1;$$

---

* It does not at once appear whether on reversing a ray we ought or ought not to change the sign of the coefficient; but the following considerations will shew that we must leave the sign unaltered. Let the portion of a wave, in which the displacement of the ether is in the direction which is considered positive, be called the *positive portion*, and the remaining part the *negative portion;* and let the points of separation be called *nodes*. There are evidently two sorts of nodes: the nodes of one sort, which may be called *positive nodes*, being situated in front of the positive portions of the waves, and the nodes of the other sort, which may be called *negative nodes*, being situated behind the positive portions or in front of the negative, the terms *in front* and *behind* referring to the direction of propagation. Now when the angle X vanishes, the particle considered is in a node; and since, at the same time, the expression for the velocity of the particle is positive, the coefficient of $\sin X$ being supposed positive, the node in question is a positive node. When a ray is reversed, we must in the first instance change the sign of the coefficient, since the velocity is reversed; but since the nodes which in the direct ray were positive are negative in the reversed ray, and *vice versâ*, we must moreover add $\pm \pi$ to the phase, which comes to the same thing as changing the sign back again. Thus we must take $b^2 \sin X$, as in the text, and not $- b^2 \sin X$, to represent the ray reflected along AI, and so in other cases.

and therefore, since c is not zero,

$$b = -e \dots\dots\dots\dots\dots\dots\dots\dots(1),$$
$$cf = 1 - b^2 = 1 - e^2 \dots\dots\dots\dots\dots\dots(2).$$

Equation (1) contains at the same time M. Arago's law and the loss of half an undulation; and equations (1) and (2) together explain the perfect blackness of the centre of Newton's rings. (See Airy's *Tract*.)

If the incident light be common light, or polarized light, of any kind except plane polarized for which the plane of polarization either coincides with the plane of incidence or is perpendicular to it, we can resolve the vibrations in and perpendicular to the plane of incidence, and consider the two parts separately.

It may be observed that the principle of reversion is just as applicable to the theory of emissions as to the theory of undulations; and thus the emissionists are called on to explain how two rays incident along RA, FA respectively can fail to produce a ray along AR'. In truth this is not so much a new difficulty as an old difficulty in a new shape; for if any mode could be conceived of explaining interference on the theory of emissions, it would probably explain the non-existence of the ray along AR'.

Although the principle of reversion applies to the theory of emissions, it does not lead, on that theory, to the law of intensity resulting from equations (1) and (2). For the formation of these equations involves the additional principle of superposition, which on the theory of undulations is merely a general dynamical principle applied to the fundamental hypotheses, but which does not apply to the theory of emissions, or at best must be assumed, on that theory, as the expression of a property which we are compelled to attribute to light, although it appears inexplicable.

In forming equations (1) and (2) it has been tacitly assumed that the reflections and refractions were unaccompanied by any change of phase, except the loss of half an undulation, which may be regarded indifferently as a change of phase of 180°, or a change of sign of the coefficient of vibration. In very highly refracting substances, however, such as diamond, it appears that when the incident light is polarized in a plane perpendicular to the plane of incidence, the reflected light does not wholly vanish at the polarizing angle; but as the angle of incidence passes through the polarizing angle, the intensity of the reflected light passes through a small minimum value, and the phase changes rapidly through an angle of nearly 180°. Suppose, for the sake of perfect generality, that all the reflections and refractions are accompanied by changes of phase. While the coefficient of vibration is multiplied by b, c, e, or f, according to the previous notation, let the phase of vibration be accelerated by the angle β, γ, ϵ, or ϕ, a retardation being reckoned as a negative acceleration. Then, if we still take $\sin X$ to represent the incident ray, we must take $b \sin (X + \beta)$, $c \sin (X + \gamma)$ to represent respectively the reflected and the refracted rays. After reversion we must change the signs of β and γ, because, whatever distance a given phase of vibration has receded from A in consequence of the acceleration accompanying reflection or refraction, the same additional distance will it have to get over in returning to A after reversion. We have therefore $b \sin (X - \beta)$, $c \sin (X - \gamma)$ to represent the rays incident along RA, FA, which together produce the ray $\sin X$ along AI. Now the ray along RA alone would produce the rays

$$b^2 \sin X \text{ along } AI, \quad bc \sin (X - \beta + \gamma) \text{ along } AR' ;$$

and the ray along FA alone would produce the rays

$$cf \sin (X - \gamma + \phi) \text{ along } AI, \quad ce \sin (X - \gamma + \epsilon) \text{ along } AR'.$$

We have therefore in the same way as before

$$cf \sin (X - \gamma + \phi) = (1 - b^2) \sin X,$$
$$b \sin (X - \beta + \gamma) + e \sin (X - \gamma + \epsilon) = 0.$$

Now each of these equations has to hold good for general values of X, and therefore, as may very easily be proved, the angles added to X in the two terms must either be equal or must differ by a multiple of 180°. But the addition of any multiple of 360° to the angle in question leaves everything the same as before, and the addition of 180° comes to the same thing as changing the sign of c or f in the first equation, or of b or e in the second. We are therefore at liberty to take

$$\phi = \gamma \quad \dots\dots\dots\dots\dots\dots\dots\dots \text{(3)},$$
$$\beta + \epsilon = 2\gamma \dots\dots\dots\dots\dots\dots\dots \text{(4)} ;$$

and the relations between b, c, e, and f will be the same as before.

Hence M. Arago's law holds good even when reflection and re-fraction are accompanied by a change of phase.

Equations (3) and (4) express the following laws with refer-ence to the changes of phase. *The sum of the accelerations of phase at the two reflections is equal to the sum of the accelerations at the two refractions; and the accelerations at the two refractions are equal to each other.* It will be observed that the accelerations are here supposed to be so measured as to give like signs to c and f, and unlike to b and e.

If we suppose the reflections and refractions accompanied by changes of phase, it is easy to prove, from equations (3) and (4), that when Newton's rings are formed between two transparent media of the same kind, the intensities of the light in the re-flected and transmitted systems are given by the same formulæ as when there are no changes of phase, provided only we replace the retardation $2\pi V/\lambda$ (according to the notation in Airy's *Tract*) by $2\pi V/\lambda - 2\epsilon$, or replace D, the distance of the media, by $D - \lambda\epsilon/2\pi \cos\beta$.

Let us now consider some circumstances which might at first sight be conceived to affect the conclusions arrived at.

When the vibrations of the incident light take place in the plane of incidence, it appears from investigation that the condi-tions at the surface of separation cannot all be satisfied by means of an incident, reflected, and refracted wave, each consisting of vibrations which take place in the plane of incidence. If the media could transmit normal vibrations with velocities com-parable to those with which they transmit tranversal vibrations, the incident wave would occasion two reflected and two refracted waves, one of each consisting of normal, and the other of trans-versal vibrations, provided the angle of incidence were less than the smallest of the three critical angles (when such exist), cor-responding to the refracted transversal vibrations and to the re-flected and refracted normal vibrations respectively. There appear however the strongest reasons for regarding the ether as sensibly incompressible, so that the velocity of propagation of normal vibra-tions is incomparably greater than that of transversal vibrations. On this supposition the two critical angles for the normal vibra-tions vanish, so that there are no normal vibrations transmitted in the regular way whatever be the angle of incidence. Instead of such vibrations there is a sort of superficial undulation in each medium, in which the disturbance is insensible at the distance of a small multiple of λ from the surface: the expressions for these disturbances involve in fact an exponential with a negative index, which contains in its numerator the distance of the point considered from the common surface of the media. It is easy to see that the existence of the superficial undulations above mentioned does not affect the truth of equations (1), (2), (3), (4); for, to obtain these equations, it is sufficient to consider points in the media whose distances from the surface are greater than that for which the superficial undulations are sensible.

No notice has hitherto been taken of a possible motion of the material molecules, which we might conceive to be produced by the vibrations of the ether. If the vibrations of the molecules take place in the same period as those of the ether, and if moreover they are not propagated in the body either regularly, with a velocity of propagation of their own, or in an irregular manner, the material molecules and the ether form a single vibrating system; they are in fact as good as a single medium, and the principle of reversion will apply.

In either of the excepted cases, however, the principle would not apply, for the same reason that it might lead to false results if there were normal vibrations produced as well as transversal, and the normal vibrations were not taken into account. In the case of transparent media, in which there appears to be no sensible loss of light by absorption for the small thicknesses of the media with which we are concerned in considering the laws of reflection and refraction, we are led to suppose, either that the material molecules are not sensibly influenced by the vibrations of the ether, or that they form with the ether a single vibrating system; and consequently the principle of reversion may be applied. In the case of opaque bodies, however, it seems likely that the labouring force brought by the incident luminous vibrations is partly consumed in producing an irregular motion among the molecules themselves.

When a convex lens is merely laid on a piece of glass, the central black spot is not usually seen; the centre is occupied by the colour belonging to a ring of some order. It requires the exertion of a considerable amount of pressure to bring the glasses into sufficiently intimate contact to allow of the perfect formation of the central spot.

Suppose that we deemed the glasses to be in contact when they were really separated by a certain interval Δ, and for simplicity suppose the reflections and refractions unaccompanied by any change of phase, except the loss of half an undulation. It evidently comes to the same thing to suppose the reflections and refractions to take place at the surfaces at which they do actually take place, as to suppose them to take place at a surface midway between the glasses, and to be accompanied by certain changes of phase; and these changes ought to satisfy equations (3) and (4). This may be easily verified. In fact, putting μ, μ' for the refractive indices of the first and second media, i, i'' for the angles of incidence and refraction, we easily find, by calculating the retardations, that

$$\beta = \frac{2\pi\Delta}{\lambda}\mu\cos i, \quad \gamma = \frac{\pi\Delta}{\lambda}\frac{\mu}{\sin i''}\sin(i''-i);$$

from which we get, by interchanging i and i'', μ and μ', and changing the signs, since for the first reflection and refraction the true surface comes before the supposed, but for the second the supposed surface comes before the true,

$$\epsilon = -\frac{2\pi\Delta}{\lambda}\mu'\cos i', \quad \phi = \frac{\pi\Delta}{\lambda}\frac{\mu'}{\sin i}\sin(i'-i);$$

and these values satisfy equations (3) and (4), as was foreseen.

Hitherto the common surface of the media has been spoken of as if the media were separated by a perfectly definite surface, up to which they possessed the same properties respectively as at a distance from the surface. It may be observed, however, that the application of the principle of reversion requires no such restriction. We are at liberty to suppose the nature of the media to change in any manner in approaching the common surface; we may even suppose them to fade insensibly into each other; and these changes may take place within a distance which need not be small in comparison with λ.

It may appear to some to be superfluous to deduce particular results from hypotheses of great generality, when these results may be obtained, along with many others which equally agree with observation, from more refined theories which start with more particular hypotheses. And indeed, if the only object of theories were to group together observed facts, or even to allow us to predict the results of observation in cases not very different from those already observed, and grouped together by the theory, such a view might be correct. But theories have a higher aim than this. A well-established theory is not a mere aid to the memory, but it professes to make us acquainted with the real processes of nature in producing observed phenomena. The evidence in favour of a particular theory may become so strong that the fundamental hypotheses of the theory are hardly less certain than observed facts. The probability of the truth of the hypotheses, however, cannot be greater than the improbability that another set of equally simple hypotheses should be conceivable, which should equally well explain all the phenomena. When the hypotheses are of a general and simple character, the improbability in question may become extremely strong; but it diminishes in proportion as the hypotheses become more particular. In sifting the evidence for the truth of any set of hypotheses, it becomes of great importance to consider whether the phenomena explained, or some of them, are explicable on more simple and general hypotheses, or whether they appear absolutely to require the more particular restrictions adopted. To take an illustration from the case in hand, we may suppose that some theorist, starting with some particular views as to the cause of the diminished velocity of light in refracting media, and supposing that the transition from one medium to another takes place, if not abruptly, at least in a space which is very small compared with λ, has obtained as the result of his analysis M. Arago's law and the loss of half an undulation. We may conceive our theorist pointing triumphantly to these laws as an evidence of the correctness of his particular views. Yet, as we have seen, if these were the only laws obtained, the theorist would have absolutely no solid evidence of the truth of the particular hypotheses with which he started.

This fictitious example leads to the consideration of the experimental evidence for Fresnel's expressions for the intensity of reflected and refracted polarized light.

There are three particular angles of incidence, namely the polarizing angle, the angle of 90°, and the angle 0°, for which special results are deducible from Fresnel's formulæ, which admit of being put, and which have been put, to the test of experiment.

The accordance of the results with theory is sometimes adduced as evidence of the truth of the formulæ: but this point will require consideration.

In the first place, it follows from Fresnel's formula for the intensity of reflected light which is polarized in a plane perpendicular to the plane of incidence, that at a certain angle of incidence the reflected light vanishes; and this angle is precisely that determined by experiment. This result is certainly very remarkable. For Fresnel's expressions are not mere empirical formulæ, chosen so as to satisfy the more remarkable results of experiment. On the contrary, they were obtained by him from dynamical considerations and analogies, which, though occasionally somewhat vague, are sufficient to lead us to regard the formulæ as having a dynamical foundation, as probably true under circumstances which without dynamical absurdity might be conceived to exist; though whether those circumstances agree with the actual state of reflecting transparent media is another question. Consequently we should *a priori* expect the formulæ to be either true or very nearly true, the difference being attributable to some modifying cause left out of consideration, or else to be altogether false: and therefore the verification of the formulæ in a remarkable, though a particular case, may be looked on as no inconsiderable evidence of their general truth. It will be observed that the truth of the formulæ is here spoken of, not the truth of the hypotheses concerned in obtaining them from theory.

Nevertheless, even the complete establishment of the formula for the reflection of light polarized in a plane perpendicular to the plane of incidence would not establish the formula for light polarized in the plane of incidence, although it would no doubt increase the probability of its truth, inasmuch as the two formulæ were obtained in the same sort of way. But, besides this, the simplicity of the law, that the reflected ray vanishes when its direction becomes perpendicular to that of the refracted ray, is such as to lead us to regard it as not improbable that different formulæ, corresponding to different hypotheses, should agree in this point. And in fact the investigation shews that when sound is reflected at the common surface of two gases, the reflected sound vanishes when the angle of incidence becomes equal to what may be called, from the analogy of light, the polarizing angle. It is true that the formula for the intensity of the reflected sound agrees with the

formula for the intensity of reflected light when the light is polarized in a plane perpendicular to the plane of incidence, and that
it is the truth of the formulæ, not that of the hypotheses, which is
under consideration. Nevertheless the formulæ require further
confirmation.

When the angle of incidence becomes 90°, it follows from
Fresnel's expressions that, whether the incident light is polarized
in or perpendicularly to the plane of incidence, the intensity of the
reflected light becomes equal to that of the incident, and consequently the same is true for common light. This result has been
compared with experiment, and the completeness of the reflection
at an incidence of 90° has been established*. The evidence, however, for the truth of Fresnel's formulæ which results from this
experiment is but feeble: for the result follows in theory from the
principle of *vis viva*, provided we suppose none of the labouring
force brought by the incident light to be expended in producing
among the molecules of the reflecting body a disturbance which is
propagated into the interior, as appears to be the case with opaque
bodies. Accordingly a great variety of different particular hypotheses, leading to formulæ differing from one another, and from
Fresnel's, would agree in giving a perfect reflection at an incidence
of 90°. Thus for example the formula which Green has given† for
the intensity of the reflected light, when the incident light is polarized in a plane perpendicular to the plane of incidence, gives the
intensities of the incident and reflected light equal when the angle
of incidence becomes 90°, although the formula in question differs
from Fresnel's, with which it only agrees to a first approximation
when μ is supposed not to differ much from 1. It appeared in
the experiment last mentioned that the sign of the reflected vibration was in accordance with Fresnel's formulæ, and that there was
no change of phase. Still it is probable that a variety of formulæ
would agree in these respects.

When the angle of incidence vanishes, it follows from Fresnel's
expressions, combined with the fundamental hypotheses of the
theory of transversal vibrations, that if the incident light be circularly polarized, the reflected light will be also circularly polarized,
but of the opposite kind, the one being right-handed, and the other

---

* *Transactions of the Royal Irish Academy*, vol. XVII. p. 171.
† *Transactions of the Cambridge Philosophical Society*, vol. VII. p. 22.

left-handed*. The experiment has been performed, at least performed for a small angle of incidence†, from whence the result
which would have been observed at an angle of incidence 0° may
be inferred; and theory has proved to be in complete accordance
with experiment. Yet this experiment, although confirming the.
theory of transversal vibrations, offers absolutely no confirmation of
Fresnel's formulæ. For when the angle of incidence vanishes,
there ceases to be any distinction between light polarized in, and
light polarized perpendicularly to the plane of incidence: be the
intensity of the reflected light what it may, it must be the same in
the two cases; and this is all that is necessary to assume in deducing the result from theory. The result would necessarily be
the same in the case of metallic reflection, although Fresnel's formulæ do not apply to metals.

By the fundamental hypotheses of the theory of transverse
vibrations, are here meant the suppositions, first, that the vibrations,
at least in vacuum and in ordinary media, take place in the front
of the wave; and secondly, that the vibrations in the case of plane
polarized light are, like all the phenomena presented by such light,
symmetrical with respect to the plane of polarization, and consequently are rectilinear, and take place either in, or perpendicularly
to the plane of polarization. From these hypotheses, combined
with the principle of the superposition of vibrations, the nature of
circularly and elliptically polarized light follows. As to the two
suppositions above mentioned respecting the direction of the vibrations in plane polarized light, there appears to be nothing to choose
between them, so far as the geometrical part of the theory is concerned: they represent observed facts equally well. The question
of the direction of the vibrations, it seems, can only be decided, if
decided at all, by a dynamical theory of light. The evidence ac-
cumulated in favour of a particular dynamical theory may be conceived to become so strong as to allow us to regard as decided the
question of the direction of the vibrations of plane polarized light.
It appears, however, that Fresnel's expressions for the intensities,
and the law which gives the velocities of plane waves in different
directions within a crystal, have been deduced, if not exactly, at
least as approximations to the exact result, from different dyna-

* *Philosophical Magazine* (*New Series*), vol. XXII. (1843) p. 92.
† *Ibid.* p. 262.

mical theories, in some of which the vibrations are supposed to be in, and in others perpendicular to the plane of polarization.

It is worthy of remark that, whichever supposition we adopt, the direction of revolution of an ethereal particle in circularly polarized light formed in a given way is the same. Similarly, in elliptically polarized light the direction of revolution is the same on the two suppositions, but the plane which on one supposition contains the major axis of the ellipse described, on the other supposition contains the minor axis. Thus the direction of revolution may be looked on as established, even though it be considered doubtful whether the vibrations of plane polarized light are in, or perpendicular to the plane of polarization.

The verification of Fresnel's formulæ for the three particular angles of incidence above mentioned is, as we have seen, not sufficient: the formulæ however admit of a very searching comparison with experiment in an indirect way, which does not require any photometrical processes. When light, polarized in a plane making a given angle with the plane of incidence, is incident on the surface of a transparent medium, it follows from Fresnel's formulæ that both the reflected and the refracted light are plane polarized, and the azimuths of the planes of polarization are known functions of the angles of incidence and refraction, and of the azimuth of the plane of polarization of the incident light, the same formulæ being obtained whether the vibrations of plane polarized light are supposed to be in, or perpendicular to the plane of polarization. It is found by experiment that the reflected or refracted light is plane polarized, at least if substances of a very high refractive power be excepted, and that the rotation of the plane of polarization produced by reflection or refraction agrees with the rotation determined by theory. This proves that the two formulæ, that is to say the formula for light polarized in, and for light polarized perpendicularly to the plane of incidence, are either both right, within the limits of error of very precise observations, or both wrong in the same ratio, where the ratio in question may be any function of the angles of incidence and refraction. There does not appear to be any reason for suspecting that the two formulæ for reflection are both wrong in the same ratio. As to the formulæ for refraction, the absolute value of the displacement will depend on the particular theory of refraction adopted. Perhaps it would be best, in order to be independent of any particular theory, to speak, not of the absolute displacement within a refracting medium, but of the equivalent displacement in vacuum, of which all that we are concerned to know is, that it is proportional to the absolute displacement. By the *equivalent displacement in vacuum*, is here meant the displacement which would exist if the light were to pass perpendicularly, and therefore without refraction, out of the medium into vacuum, without losing *vis viva* by reflection at the surface. It is easy to prove that Fresnel's formulæ for refraction would be adapted to this mode of estimating the vibrations by multiplying by $\sqrt{\mu}$; indeed, the formulæ for refraction might be thus proved, except as to sign, by means of the principle of *vis viva*, the formulæ for reflection being assumed. It will be sufficient to shew this in the case of light polarized in the plane of incidence.

Let i, i' be the angles of incidence and refraction, A any area taken in the front of an incident wave, l the height of a prism having A for its base and situated in the first medium. Let r be the coefficient of vibration in the reflected wave, that in the incident wave being unity, q the coefficient of the vibration in vacuum equivalent to the refracted vibration. Then the incident light which fills the volume Al will give rise to a quantity of reflected light filling an equal volume Al, and to a quantity of refracted light which, after passing into vacuum in the way supposed, would fill a volume $Al \cos i'/\cos i$. We have therefore, by the principle of *vis viva*,

$$q^2 \frac{\cos i'}{\cos i} = 1 - r^2 = 1 - \frac{\sin^2 (i' - i)}{\sin^2 (i' + i)} = \frac{4 \sin i' \cos i' \sin i \cos i}{\sin^2 (i' + i)} \cdot$$

This equation does not determine the sign of q: but it seems impossible that the vibrations due to the incident light in the ether immediately outside the refracting surface should give rise to vibrations in the opposite direction in the ether immediately inside the surface, so that we may assume q to be positive. We have then

$$q = \frac{2 \cos i \sqrt{(\sin i' \sin i)}}{\sin (i' + i)} = \sqrt{\mu} \cdot \frac{2 \sin i' \cos i}{\sin (i' + i)} \dots (5),$$

as was to be proved. The formula for light polarized perpendicularly to the plane of incidence may be obtained in the same way. The formula (5), as might have been foreseen, applies equally well to the hypothesis that the diminished velocity of propagation within refracting media is due to an increase of density of the ether, which requires us to suppose that the vibrations of plane polarized light are perpendicular to the plane of polarization, and to the hypothesis that the diminution of the velocity of propagation is due to a diminution of elasticity, which requires us to suppose the vibrations to be in the plane of polarization.

If the refraction, instead of taking place out of vacuum into a medium, takes place out of one medium into another, it is easy to shew that we have only got to multiply by $\sqrt{\mu'/\mu}$ instead of $\sqrt{\mu}$; μ, μ' being the refractive indices of the first and second media respectively.

[From the *Cambridge and Dublin Mathematical Journal*, Vol. IV. p. 194
(*May* and *November*, 1849).]

On Attractions, and on Clairaut's Theorem.

CLAIRAUT'S Theorem is usually deduced as a consequence
of the hypothesis of the original fluidity of the earth, and the
near agreement between the numerical values of the earth's ellipticity, deduced independently from measures of arcs of the meridian
and from pendulum experiments, is generally considered as a
strong confirmation of the hypothesis. Although this theorem is
usually studied in connection with the hypothesis just mentioned,
it ought to be observed that Laplace, without making any assumption respecting the constitution of the earth, except that it consists
of nearly spherical strata of equal density, and that its surface
may be regarded as covered by a fluid, has established a connexion
between the form of the surface and the variation of gravity, which
in the particular case of an oblate spheroid gives directly Clairaut's
Theorem*. If, however, we merely assume, as a matter of observation, that the earth's surface is a surface of equilibrium, (the
trifling irregularities of the surface being neglected), that is to say
that it is perpendicular to the direction of gravity, then, independently of any particular hypothesis respecting the state of the
interior, or any theory but that of universal gravitation, there
exists a necessary connexion between the form of the surface and
the variation of gravity along it, so that the one being given the
other follows. In the particular case in which the surface is an

---

* See the *Mécanique Céleste*, Liv. III., or the reference to it in Pratt's *Mechanics*,
Chap. *Figure of the Earth*.

oblate spheroid of small eccentricity, which the measures of arcs
shew to be at least very approximately the form of the earth's
surface, the variation of gravity is expressed by the equation which
is arrived at on the hypothesis of original fluidity.   I am at present
engaged in preparing a paper on this subject for the Cambridge
Philosophical Society: the object of the following pages is to give
a demonstration of Clairaut's Theorem, different from the one
there employed, which will not require a knowledge of the properties of the functions usually known by the name of Laplace's
Functions.   It will be convenient to commence with the demonstration of a few known theorems relating to attractions, the law
of attraction being that of the inverse square of the distance*.

Preliminary Propositions respecting Attractions.

PROP. I.   To express the components of the attraction of any
mass in three rectangular directions by means of a single function.

Let m' be the mass of an attracting particle situated at the
point P', the unit of mass being taken as is usual in central
forces, m the mass of the attracted particle situated at the point
P, x', y', z' the rectangular co-ordinates of P' referred to any origin,
x, y, z those of P; X, Y, Z the components of the attraction of
m' on m, measured as accelerating forces, and considered positive
when they tend to increase x, y, z; then, if $PP' = r'$,

$$X = \frac{m'}{r'^3}(x' - x), \quad Y = \frac{m'}{r'^3}(y' - y), \quad Z = \frac{m'}{r'^3}(z' - z).$$

* My object in giving these demonstrations is simply to enable a reader who
may not have attended particularly to the theory of attractions to follow with
facility the demonstration here given of Clairaut's Theorem.   In speaking of the
theorems as "known" I have, I hope, sufficiently disclaimed any pretence at
originality.   In fact, not one of the "propositions respecting attractions" is new,
although now and then the demonstrations may differ from what have hitherto
been given.   With one or two exceptions, these propositions will all be found in
a paper by Gauss, of which a translation is published in the third volume of
Taylor's *Scientific Memoirs*, p. 153.   The demonstration here given of Prop. IV. is
the same as Gauss's; that of Prop. V., though less elegant than Gauss's, appears to
me more natural.   The ideas on which it depends render it closely allied to a paper
by Professor Thomson, in the third volume of this *Journal* (Old Series), p. 71.
Prop. IX. is given merely for the sake of exemplifying the application of the same
mode of proof to a theorem of Gauss's.

Since $r'^2 = (x' - x)^2 + (y' - y)^2 + (z' - z)^2,$

we have $r' \dfrac{dr'}{dx} = -(x' - x)$; whence $X = -\dfrac{m'}{r'^2} \dfrac{dr'}{dx} = \dfrac{d}{dx} \dfrac{m'}{r'}$;

with similar equations for Y and Z.

If instead of a single particle m' we have any number of attracting particles m', m'' ... situated at the points (x', y', z'), (x'', y'', z'')..., and if we put

$$\frac{m'}{r'} + \frac{m''}{r''} + \ldots = \Sigma \frac{m'}{r'} = V \ldots\ldots\ldots\ldots\ldots (1),$$

we get

$$X = \frac{d}{dx} \left(\frac{m'}{r'} + \frac{m''}{r''} + \ldots \right) = \frac{dV}{dx}; \text{ similarly } Y = \frac{dV}{dy}, \ Z = \frac{dV}{dz} \ldots (2).$$

If instead of a set of distinct particles we have a continuous attracting mass M', and if we denote by dm' a differential element of M', and replace (1) by

$$V = \iiint \frac{dm'}{r'} \ldots\ldots\ldots\ldots\ldots\ldots (3),$$

equations (2) will still remain true, provided at least P be external to M'; for it is only in that case that we are at liberty to consider the continuous mass as the limit of a set of particles which are all situated at finite distances from P. It must be observed that should M' occupy a closed shell, within the inner surface of which P is situated, P must be considered as external to the mass M'. Nevertheless, even when P lies within M', or at its surface, the expressions for V and dV/dx, namely $\iiint \dfrac{dm'}{r'}$ and $\iiint (x' - x) \dfrac{dm'}{r'^3}$, admit of real integration, defined as a limiting summation, as may be seen at once on referring M' to polar co-ordinates originating at P; so that the equations (2) still remain true.

Prop. II. To express the attraction resolved along any line by means of the function V.

Let s be the length of the given line measured from a fixed point up to the point P; λ, μ, ν, the direction-cosines of the tangent to this line at P, F the attraction resolved along this tangent; then

$$F = \lambda X + \mu Y + \nu Z = \lambda \frac{dV}{dx} + \mu \frac{dV}{dy} + \nu \frac{dV}{dz}.$$

Now if we restrict ourselves to points lying in the line s, V will be a function of s alone; or we may regard it as a function of x, y, and z, each of which is a function of s; and we shall have, by Differential Calculus,

$$\frac{dV}{ds} = \frac{dV}{dx}\frac{dx}{ds} + \frac{dV}{dy}\frac{dy}{ds} + \frac{dV}{dz}\frac{dz}{ds};$$

and since $dx/ds = \lambda$, $dy/ds = \mu$, $dz/ds = \nu$, we get

$$F = \frac{dV}{ds} \dotfill (4).$$

PROP. III. To examine the meaning of the function V.

This function is of so much importance that it will be well to dwell a little on its meaning.

In the first place it may be observed that the equation (1) or (3) contains a physical definition of V, which has nothing to do with the system of co-ordinates, rectangular, polar, or any other, which may be used to define algebraically the positions of P and of the attracting particles. Thus V is to be contemplated as a function of the position of P in space, if such an expression may be allowed, rather than as a function of the co-ordinates of P; although, in consequence of its depending upon the position of P, V will be a function of the co-ordinates of P, of whatever kind they may be.

Secondly, it is to be remarked that although an attracted particle has hitherto been conceived as situated at P, yet V has a definite meaning, depending upon the position of the point P, whether any attracted matter exist there or not. Thus V is to be contemplated as having a definite value at each point of space, irrespective of the attracted matter which may exist in some places.

The function V admits of another physical definition which ought to be noticed. Conceive a particle whose mass is m to move along any curve from the point P_0 to P. If F be the attraction of M' resolved along a tangent to m's path, reckoned as an accelerating force, the moving force of the attraction resolved in the same direction will be mF, and therefore the work done by the attraction while m describes the elementary arc ds will be ultimately $mFds$, or by (4) $m \cdot dV/ds \cdot ds$. Hence the whole work done as m moves from P_0 to P is equal to $m(V - V_0)$, V_0 being the value of V at P_0. If P_0 be situated at an infinite distance, V_0 vanishes, and the expression for the work done becomes simply mV. Hence V might be called the *work of the attraction, referred to a unit of mass of the attracted particle;* but besides that such a name would be inconveniently long, a recognized name already exists. The function V is called the *potential* of the attracting mass*.

The first physical definition of V is peculiar to attraction according to the inverse square of the distance. According to the second, V is regarded as a particular case of the more general function whose partial differential coefficients with respect to x, y, z are equal to the components of the accelerating force; a function which exists whenever $Xdx + Ydy + Zdz$ is an exact differential.

PROP. IV.   If S be any closed surface to which all the attracting mass is external, dS an element of S, dn an element of the normal drawn outwards at dS, then

$$\iint \frac{dV}{dn}\, dS = 0 \dots\dots\dots\dots\dots\dots\dots(5),$$

the integral being taken throughout the whole surface S.

Let m' be the mass of any attracting particle which is situated at the point P', P' being by hypothesis external to S. Through P' draw any right line L cutting S, and produce it indefinitely in one direction from P'. The line L will in general cut S in two points; but if the surface S be re-entrant, it may be cut in four, six, or any even number of points. Denote the points of section, taken in order, by P_1, P_2, P_3, &c., P_1 being that which lies nearest to P'. With P' for vertex, describe about the line L a conical surface containing an infinitely small solid angle α, and denote by A_1, A_2... the areas which it cuts out from S about the points P_1, P_2.... Let θ_1, θ_2... be the angles which the normals drawn outwards at P_1, P_2... make with the line L, taken in the direction from P_1 to P'; N_1, N_2... the attractions of m' at P_1, P_2... resolved along the normals; r_1, r_2... the distances of P_1, P_2... from P'. It

---

* [The term "potential," as used in the theory of Electricity, may be defined in the following manner: "The potential at any point P, in the neighbourhood of electrified matter, is the amount of work that would be necessary to remove a small body charged with a unit of negative electricity from that position to an infinite distance."—W. T.]

is evident that the angles θ_1, θ_2... will be alternately acute and obtuse. Then we have

$$N_1 = \frac{m'}{r_1^2} \cos \theta_1, \quad N_2 = -\frac{m'}{r_2^2} \cos (\pi - \theta_2) \text{ \&c.}$$

We have also in the limit

$$A_1 = a r_1^2 \sec \theta_1, \quad A_2 = a r_2^2 \sec (\pi - \theta_2), \text{ \&c.;}$$

and therefore $N_1 A_1 = a m'$, $N_2 A_2 = -a m'$, $N_3 A_3 = a m'$, \&c.;

and therefore, since the number of points P_1, P_2... is even,

$$N_1 A_1 + N_2 A_2 + N_3 A_3 + N_4 A_4 ... = a m' - a m' + a m' - a m' ... = 0.$$

Now the whole solid angle contained within a conical surface described with P' for vertex so as to circumscribe S may be divided into an infinite number of elementary solid angles, to each of which the preceding reasoning will apply; and it is evident that the whole surface S will thus be exhausted. We have therefore

$$\text{limit of } \Sigma N A = 0;$$

or, by the definition of an integral,

$$\iint N dS = 0.$$

The same will be true of each attracting particle m'; and therefore if N refer to the attraction of the whole attracting mass, we shall still have $\iint N dS = 0$. But by (4) $N = dV/dn$, which proves the proposition.

PROP. V. If V be equal to zero at all points of a closed surface S, which does not contain any portion of the attracting mass, it must be equal to zero at all points of the space T contained within S.

For if not, V must be either positive or negative in at least a certain portion of the space T, and therefore must admit of at least one positive or negative maximum value V_1. Call the point, or the assemblage of connected points, at which V has its maximum value V_1, T_1. It is to be observed, first, that T_1 may denote either a space, a surface, a line, or a single point; secondly, that should V happen to have the same value V_1 at other points within T, such points must not be included in T_1. Then, all round T_1, V is decreasing, positively or negatively according as V_1 is positive or negative. Circumscribe a closed surface S_1 around T_1, lying wholly within S, which is evidently possible. Then if S_1 be drawn sufficiently close round T_1, V will be increasing in passing outwards across S_1*; and therefore, if n_1 denote a normal drawn outwards at the element dS_1 of S_1, dV/dn_1 will be negative or positive according as V_1 is positive or negative, and therefore $\iint \dfrac{dV}{dn_1}\, dS_1$, taken throughout the whole surface S_1, will be negative or positive, which is contrary to Prop. IV. Hence V must be equal to zero throughout the space T.

COR. 1. If V be equal to a constant A at all points of the surface S, it must be equal to A at all points within S. For it may be proved just as before that V cannot be either greater or less than A within S.

COR. 2. If V be not constant throughout the surface S, and if A be its greatest, and B its least value in that surface, V cannot anywhere within S be greater than A nor less than B.

COR. 3. All these theorems will be equally true if the space T extend to infinity, provided that instead of the value of V at the bounding surface of T we speak of the value of V at the surface by which T is partially bounded, and its limiting value at an infinite distance in T. This limiting value might be conceived to vary from one direction to another. Thus T might be the infinite space lying within one sheet of a cone, or hyperboloid of one sheet, or the infinite space which lies outside a given closed surface S, which contains within it all the attracting mass. On the latter supposition, if V be equal to zero throughout S, and vanish at an infinite distance, V must be equal to zero everywhere outside S. If V vanish at an infinite distance, and range between the limits A and B at the surface S, V cannot anywhere outside S lie beyond the limits determined by the two extremes of the three quantities A, B, and 0.

---

* It might, of course, be possible to prevent this by drawing S_1 sufficiently puckered, but S_1 is supposed not to be so drawn. Since V is decreasing from T_1 outwards, if we consider the loci of the points where V has the values $V_2, V_3, V_4 \ldots$ decreasing by infinitely small steps from V_1, it is evident that in the immediate neighbourhood of T_1 these loci will be closed surfaces, each lying outside the preceding, the first of which ultimately coincides with T_1 if T_1 be a point, a line, or a surface, or with the surface of T_1 if T_1 be a space. If now we take for S_1 one of these " surfaces of equilibrium," or any surface cutting them at acute angles, what was asserted in the text respecting S_1 will be true.

PROP. VI. At any point (x, y, z) external to the attracting mass, the potential V satisfies the partial differential equation

$$\frac{d^2V}{dx^2} + \frac{d^2V}{dy^2} + \frac{d^2V}{dz^2} = 0 \ldots\ldots\ldots\ldots\ldots\ldots(6).$$

For if V' denote the potential of a single particle m', we have, employing the notation of Prop. I.,

$$V' = \frac{m'}{r'}, \quad \frac{dV'}{dx} = -\frac{m'}{r'^2}\frac{dr'}{dx} = \frac{m'}{r'^3}(x'-x), \quad \frac{d^2V'}{dx^2} = \frac{3m'}{r'^5}(x'-x)^2 - \frac{m'}{r'^3},$$

with similar expressions for d^2V'/dy^2 and d^2V'/dz^2; and therefore V' satisfies (6). This equation will be also satisfied by the potentials V'', V'''... of particles m'', m'''... situated at finite distances from the point (x, y, z), and therefore by the potential V of all the particles, since $V = V' + V'' + V''' + ...$ Now, by supposing the number of particles indefinitely increased, and their masses, as well as the distances between adjacent particles, indefinitely diminished, we pass in the limit to a continuous mass, of which all the points are situated at finite distances from the point (x, y, z). Hence the potential V of a continuous mass satisfies equation (6) at all points of space to which the mass does not reach.

SCHOLIUM to Prop. V. Although the equations (5) and (6) have been proved independently of each other from the definition of a potential, either of these equations is a simple analytical consequence of the other*. Now the only property of a potential

---

* The equation (6) will be proved by means of (5) further on (Prop. VIII.), or rather an equation of which (6) is a particular case, by means of an equation of which (5) is a particular case. Equation (5) may be proved from (6) by a known transformation of the equation $\iiint \nabla V \, dx \, dy \, dz = 0$, where ∇V denotes the first member of (6), and the integration is supposed to extend over the space T. For, taking the first term in ∇V, we get

$$\iiint \frac{d^2V}{dx^2} \, dx \, dy \, dz = \iint \left(\frac{dV}{dx}\right)_{,,} dy \, dz - \iint \left(\frac{dV}{dx}\right)_{,} dy \, dz,$$

where $\left(\dfrac{dV}{dx}\right)_{,,}$, $\left(\dfrac{dV}{dx}\right)_{,}$ denote the values of $\dfrac{dV}{dx}$ at the points where S is cut by a line drawn parallel to the axis of x through the point whose co-ordinates are $0, y, z$. Now if λ be the angle between the normal drawn outwards at the element of surface dS and the axis of x,

$$\iint \left(\frac{dV}{dx}\right)_{,,} dy \, dz = \iint \frac{dV}{dx} \cos \lambda \, dS, \quad \iint \left(\frac{dV}{dx}\right)_{,} dy \, dz = \iint \frac{dV}{dx} \cos (\pi - \lambda) \, dS,$$

where the first integration is to be extended over the portion of S which lies to the

assumed in Prop. v, is, that it is a quantity which varies continuously within the space T, and satisfies the equation (5) for any
closed surface drawn within T. Hence Prop. v, which was enunciated with respect to the potential of a mass lying outside T, is
equally true with respect to any continuously varying quantity
which within the space T satisfies the equation (6). It should be
observed that a quantity like r^{-1} is not to be regarded as such, if r
denote the distance of the point (x, y, z) from a point P, which lies
within T, because r^{-1} becomes infinite at P.

Clairaut's Theorem.

1. Although the earth is really revolving about its axis, so
that all problems relating to the relative equilibrium of the earth
itself and the bodies on its surface are really dynamical problems,
we know that they may be treated statically by introducing, in
addition to the attraction, that fictitious force which we call the
centrifugal force. The force of gravity is the resultant of the
attraction and the centrifugal force; and we know that this force
is perpendicular to the general surface of the earth. In fact, by
far the larger portion of the earth's surface is covered by water,
the equilibrium of which requires, according to the principles of
hydrostatics, that its surface be perpendicular to the direction of
gravity; and the elevation of the land above the level of the sea,
or at least the elevation of large tracts of land, is but trifling compared with the dimensions of the earth. We may therefore regard
the earth's surface as a surface of equilibrium.

positive side of the curve of contact of S and an enveloping cylinder with its generating lines parallel to the axis of x, and the second integration over the remainder
of S. If then we extend the integration over the whole of the surface S, we get

$$\iiint \frac{d^2V}{dx^2}\, dx\, dy\, dz = \iint \frac{dV}{dx} \cos \lambda . dS.$$

Making a similar transformation with respect to the two remaining terms of ∇V,
and observing that if μ, ν be for y, z what λ is for x,

$$\cos \lambda \frac{dV}{dx} + \cos \mu \frac{dV}{dy} + \cos \nu \frac{dV}{dz} = \frac{dV}{dn},$$

we obtain equation (5).

If V be any continuously varying quantity which within the space T satisfies
the equation $\nabla V = 0$, it may be proved that it is always possible to distribute
attracting matter outside T in such a manner as to produce within T a potential
equal to V.

2. Let the earth be referred to rectangular axes, the axis of z coinciding with the axis of rotation. Let V be the potential of the mass, ω the angular velocity, X, Y, Z the components of the whole force at the point (x, y, z); then

$$X = \frac{dV}{dx} + \omega^2 x, \quad Y = \frac{dV}{dy} + \omega^2 y, \quad Z = \frac{dV}{dz}.$$

Now the general equation to surfaces of equilibrium is

$$\int (X dx + Y dy + Z dz) = \text{const.},$$

and therefore we must have at the earth's surface

$$V + \tfrac{1}{2}\omega^2 (x^2 + y^2) = c \dots\dots\dots\dots\dots(7),$$

where c is an unknown constant. Moreover V satisfies the equation (6) at all points external to the earth, and vanishes at an infinite distance. But these conditions are sufficient to determine V at all points of space external to the earth. For if possible let V admit of two different values V_1, V_2 outside the earth, and let $V_1 - V_2 = V'$. Since V_1 and V_2 have the same value

$$c - \tfrac{1}{2}\omega^2 (x^2 + y^2)$$

at the surface, V' vanishes at the surface; and it vanishes likewise at an infinite distance, and therefore by Prop. v. $V' = 0$ at all points outside the earth. Hence if the form of the surface be given, V is determinate at all points of external space, except so far as relates to the single arbitrary constant c which is involved in its complete expression.

3. Now it appears from measures of arcs of the meridian, that the earth's surface is represented, at least very approximately, by an oblate spheroid of small ellipticity, having its axis of figure coinciding with the axis of rotation. It will accordingly be more convenient to refer the earth to polar, than to rectangular coordinates. Let the centre of the surface be taken for origin; let r be the radius vector, θ the angle between this radius and the axis of z, ϕ the angle between the plane passing through these lines and the plane xz. Then if the square of the ellipticity be neglected, the equation to the surface may be put under the form

$$r = a (1 - \epsilon \cos^2\theta) \dots\dots\dots\dots\dots(8);$$

and from (7) we must have at the surface

$$V + \tfrac{1}{2}\omega^2 a^2 \sin^2\theta = c \dots\dots\dots\dots\dots(9).$$

If we denote for shortness the equation (6) by $\nabla V = 0$, we have by transformation to polar co-ordinates*

$$\nabla V = \frac{d^2 V}{dr^2} + \frac{2}{r}\frac{dV}{dr} + \frac{1}{r^2 \sin\theta}\frac{d}{d\theta}\left(\sin\theta \frac{dV}{d\theta}\right) + \frac{1}{r^2 \sin^2\theta}\frac{d^2 V}{d\phi^2}$$
$$= 0 \ldots\ldots(10).$$

4. The form of the equations (8) and (9) suggests the occurrence of terms of the form $\psi(r) + \chi(r)\cos^2\theta$ in the value of V. Assume then

$$V = \psi(r) + \chi(r)\cos^2\theta + w \ldots\ldots\ldots(11).$$

We are evidently at liberty to make this assumption, on account of the indeterminate function w. Now if we observe that

$$\frac{1}{\sin\theta}\frac{d}{d\theta}\left(\sin\theta \frac{d \cdot \cos^2\theta}{d\theta}\right) = 2 - 6\cos^2\theta,$$

we get from (10) and (11)

$$\psi''(r) + \frac{2}{r}\psi'(r) + \frac{2}{r^2}\chi(r) + \left\{\chi''(r) + \frac{2}{r}\chi'(r) - \frac{6}{r^2}\chi(r)\right\}\cos^2\theta$$
$$+ \nabla w = 0 \ldots\ldots(12).$$

If now we determine the functions ψ, χ from the equations

$$\psi''(r) + \frac{2}{r}\psi'(r) + \frac{2}{r^2}\chi(r) = 0 \ldots\ldots\ldots\ldots(13),$$

$$\chi''(r) + \frac{2}{r}\chi'(r) - \frac{6}{r^2}\chi(r) = 0 \ldots\ldots\ldots\ldots(14),$$

we shall have $\nabla w = 0$.

By means of (14), equation (13) may be put under the form

$$\psi''(r) + \frac{2}{r}\psi'(r) = -\tfrac{1}{3}\left\{\chi''(r) + \frac{2}{r}\chi'(r)\right\};$$

and therefore $\psi(r) = -\tfrac{1}{3}\chi(r)$ is a particular integral of (13). The equations (14), and (13) when deprived of its last term, are easily integrated, and we get

$$\psi(r) = \frac{A}{r} + B - \tfrac{1}{3}\chi(r), \quad \chi(r) = \frac{C}{r^3} + Dr^2 \ldots\ldots(15).$$

Now V vanishes at an infinite distance; and the same will be the

* *Cambridge Mathematical Journal*, Vol. I. (Old Series), p. 122, or O'Brien's *Tract on the Figure of the Earth*, p. 12.

case with w provided we take $B = 0$, $D = 0$, when we get from (11) and (15)

$$V = \frac{A}{r} + \frac{C}{r^3}(\cos^2\theta - \tfrac{1}{3}) + w \dots\dots\dots\dots (16).$$

5. It remains to satisfy (9). Now this equation may be satisfied, so far as the large terms are concerned, by means of the constant A, since θ appears only in the small terms. We have a right then to assume C to be a small quantity of the first order. Substituting in (16) the value of r given by (8), putting the resulting value of V in (9), and retaining the first order only of small quantities, we get

$$\frac{A}{a}(1 + \epsilon\cos^2\theta) + \frac{C}{a^3}(\cos^2\theta - \tfrac{1}{3}) + w_, + \frac{\omega^2}{2}a^2(1 - \cos^2\theta) = c \dots (17),$$

$w_,$ being the value of w at the surface of the earth. Now the constants A and C allow us to satisfy this equation without the aid of $w_,$. We get by equating to zero the sum of the constant terms, and the coefficient of $\cos^2\theta$,

$$\left.\begin{aligned} \frac{A}{a} - \frac{C}{3a^3} + \frac{\omega^2 a^2}{2} &= c \\ \frac{A\epsilon}{a} + \frac{C}{a^3} - \frac{\omega^2 a^2}{2} &= 0 \end{aligned}\right\} \dots\dots\dots\dots (18).$$

These equations combined with (17) give $w_1 = 0$. Now we have seen that w satisfies the equation $\nabla w = 0$ at all points exterior to the earth, and that it vanishes at an infinite distance; and since it also vanishes at the surface, it follows from Prop. V. that it is equal to zero every where without the earth.

It is true that $w_,$ is not strictly equal to zero, but only to a small quantity of the second order, since quantities of that order are omitted in (17). But it follows from Prop. V. Cor. 3, that if w', w'' be respectively the greatest and least values of $w_,$, w cannot anywhere outside the earth lie beyond the limits determined by the two extremes of the three quantities w', w'', and 0, and therefore must be a small quantity of the second order; and since we are only considering the potential at external points, we may omit w altogether.

If E be the mass of the earth, the potential at a very great distance r is ultimately equal to E/r. Comparing this with the equation obtained from (16) by leaving out w, we get

$$A = E.$$

The first of equations (18) serves only to determine c in terms of E, and c is not wanted. The second gives

$$C = - Ea^2\epsilon + \tfrac{1}{2}\omega^2 a^5,$$

whence, we get from (16)

$$V = \frac{E}{r} - \left(\frac{E\epsilon}{a} - \tfrac{1}{2}\omega^2 a^2\right)\frac{a^3}{r^3}(\cos^2\theta - \tfrac{1}{3}) \dots\dots\dots (19).$$

6. If g be the force of gravity at any point of the surface, ν the angle between the vertical and the radius vector drawn from the centre, $g\cos\nu$ will be the resolved part of gravity along the radius vector; and we shall have

$$g\cos\nu = -\frac{d}{dr}\left(V + \frac{\omega^2}{2}r^2\sin^2\theta\right)\dots\dots\dots\dots(20),$$

where after differentiation r is to be put equal to the radius vector of the surface. Now ν is a small quantity of the first order, and therefore $\cos\nu$ may be replaced by 1, whence we get from (8), (19), and (20),

$$g = \frac{E}{a^2}(1 + 2\epsilon\cos^2\theta) - 3\left(\frac{E\epsilon}{a^2} - \tfrac{1}{2}\omega^2 a\right)(\cos^2\theta - \tfrac{1}{3}) - \omega^2 a(1 - \cos^2\theta),$$

or $\quad g = (1+\epsilon)\dfrac{E}{a^2} - \tfrac{3}{2}\omega^2 a + \left(\tfrac{5}{2}\omega^2 a - \dfrac{E\epsilon}{a^2}\right)\cos^2\theta \dots\dots\dots(21).$

At the equator $\theta = \tfrac{1}{2}\pi$; and if we put G for gravity at the equator, m for the ratio of the centrifugal force to gravity at the equator, we get $\omega^2 a = mG$, and

$$(1 + \tfrac{3}{2}m)\,G = (1+\epsilon)\frac{E}{a^2},$$

whence $\qquad\qquad E = (1 + \tfrac{3}{2}m - \epsilon)\,Ga^2 \dots\dots\dots\dots(22);$

and (21) becomes $\quad g = G\{1 + (\tfrac{5}{2}m - \epsilon)\cos^2\theta\} \dots\dots\dots(23).$

7. Equation (22) gives the mass of the earth by means of the value of G determined by the pendulum. In the preceding investigation, θ is the complement of the corrected latitude; but since θ occurs only in the small terms, and the squares of small quantities

have been omitted throughout, we may regard θ as the complement of the true latitude, and therefore replace $\cos \theta$ by the sine of the latitude. In the case of the earth, m is about $\frac{1}{289}$ and ϵ about $\frac{1}{300}$, and therefore $\frac{5}{2}m - \epsilon$ is positive. Hence it appears from (23) that the increase of gravity from the equator to the pole varies as the square of the sine of the latitude, and the ratio which the excess of polar over equatorial gravity bears to the latter, added to the ellipticity, is equal to $\frac{5}{2}$ × the ratio of the centrifugal force to gravity at the equator.

8. If instead of the equatorial radius a, and equatorial gravity G, we choose to employ the mean radius a_1, and mean gravity G_1, we have only to remark that the mean value of $\cos^2 \theta$, or

$$\frac{1}{4\pi} \int_0^\pi \int_0^{2\pi} \cos^2 \theta \sin \theta \, d\theta \, d\phi,$$

is $\frac{1}{3}$, which gives

$$a_1 = a \left(1 - \tfrac{1}{3}\epsilon\right), \quad G_1 = G \left(1 + \tfrac{5}{6}m - \tfrac{1}{3}\epsilon\right),$$

which reduces equations (8), (22), and (23) to

$$r = a_1 \left\{1 - \epsilon \left(\cos^2 \theta - \tfrac{1}{3}\right)\right\},$$

$$E = \left(1 + \tfrac{2}{3}m\right) G_1 a_1{}^2,$$

$$g = G_1 \left\{1 + \left(\tfrac{5}{2}m - \epsilon\right) \left(\cos^2 \theta - \tfrac{1}{3}\right)\right\}.$$

9. We get from (19), for the potential at an external point,

$$V = \frac{E}{r} - \left(\epsilon - \tfrac{1}{2}m\right) \frac{Ea^2}{r^3} \left(\cos^2 \theta - \tfrac{1}{3}\right) \quad \ldots \ldots \ldots \ldots \ldots (24).$$

Now the attraction of the moon on any particle of the earth, and consequently the attraction of the whole earth on the moon, will be very nearly the same as if the moon's mass were collected at her centre of gravity. Let r be the distance between the centres of the earth and moon, θ the moon's north polar distance, P the attraction of the earth on the moon, resolved along the radius vector drawn from the earth's centre, Q the attraction perpendicular to the radius vector, a force which will evidently lie in a plane passing through the earth's axis and the centre of the moon. Then, supposing Q measured positive towards the equator, we have from (4),

$$P = -\frac{dV}{dr}, \quad Q = \frac{1}{r} \frac{dV}{d\theta};$$

whence, from (24),

$$\left.\begin{array}{l} P = \dfrac{E}{r^2} - 3 \left(\epsilon - \tfrac{1}{2}m\right) \dfrac{Ea^2}{r^4} \left(\cos^2 \theta - \tfrac{1}{3}\right) \\[2ex] Q = 2 \left(\epsilon - \tfrac{1}{2}m\right) \dfrac{Ea^2}{r^4} \sin \theta \cos \theta \end{array}\right\} \quad \ldots\ldots\ldots(25).$$

The moving force arising from the attraction of the earth on the moon is a force passing through the centre of the moon, and having for components MP along the radius vector, and MQ perpendicular to the radius vector, M being the mass of the moon; and on account of the equality of action and reaction, the moving force arising from the attraction of the moon on the earth is equal and opposite to the former. Hence the latter force is equivalent to a moving force MP passing through the earth's centre in the direction of the radius vector of the moon, a force MQ passing through the earth's centre in a direction perpendicular to the radius vector, and a couple whose moment is MQr tending to turn the earth about an equatorial axis. Since we only want to determine the motion of the moon relatively to the earth, the effect of the moving forces MP, MQ acting on the earth will be fully taken into account by replacing E in equations (25) by $E + M$. If μ be the moment of the couple, we have

$$\mu = 2 \left(\epsilon - \tfrac{1}{2}m\right) \dfrac{MEa^2}{r^3} \sin \theta \cos \theta \ \ldots\ldots\ldots\ldots(26).$$

This formula will of course apply, *mutatis mutandis*, to the moment of the moving force arising from the attraction of the sun.

10.   The force expressed by the second term in the value of P, in equations (25), and the force Q, or rather the forces thence obtained by replacing E by $E + M$, are those which produce the only two sensible inequalities in the moon's motion which depend on the oblateness of the earth. We see that they enable us to determine the ellipticity of the earth independently of any hypothesis respecting the distribution of matter in its interior.

The moment μ, and the corresponding moment for the sun, are the forces which produce the phenomena of precession and nutation. In the observed results, the moments of the forces are divided by the moment of inertia of the earth about an equatorial axis. Call this $Ea^2\kappa$; let $M = E/n$; let b be the annual precession, and f the coefficient in the lunar nutation in obliquity; then we shall have

$$b = \left(A + \frac{B}{n+1}\right) \left(\epsilon - \tfrac{1}{2}m\right) \frac{1}{\kappa}, \quad f = \frac{C}{n+1} \left(\epsilon - \tfrac{1}{2}m\right) \frac{1}{\kappa} *,$$

where A, B, C denote certain known quantities. Hence the observed values of b and f will serve to determine the two unknown quantities n, and the ratio of $\epsilon - \tfrac{1}{2}m$ to κ. If therefore we suppose ϵ to be known otherwise, we shall get the numerical value of κ.

11. In determining the mutual attraction of the moon and earth, the attraction of the moon has been supposed the same as if her mass were collected at her centre, which we know would be strictly true if the moon were composed of concentric spherical strata of equal density, and is very nearly true of any mass, however irregular, provided the distance of the attracted body be very great compared with the dimensions of the attracting mass, and the centre be understood to mean the centre of gravity. It will be desirable to estimate the magnitude of the error which is likely to result from this supposition. For this purpose suppose the moon's surface, or at least a surface of equilibrium drawn immediately outside the moon, to be an oblate spheroid of small ellipticity, having its axis of figure coincident with the axis of rotation. Then the equation (24) will apply to the attraction of the moon on the earth, provided we replace E, a, by M, a', where a' is the moon's radius, take θ to denote the angular distance of the radius vector of the earth from the moon's axis, and suppose ϵ and m to have the values which belong to the moon. Now E is about 80 times as great as M, and a about 4 times as great as a', and therefore Ea^2 is about 1200 times as great as Ma'^2. But m is extremely small in the case of the moon; and there is no reason to think that the value of ϵ for the moon is large in comparison with its value for the earth, but rather the contrary; and therefore the effect of the moon's oblateness on the relative motions of the centres of the earth and moon must be altogether insignificant, especially when we remember that the coefficients of the two sensible inequalities in the moon's motion depending on the earth's

---

* $1/(n+1)$ will appear in these equations rather than $1/n$, because, if S be the mass, and r, the distance of the sun, the ratio of M/r^3 to S/r^3 is equal to $1/(n+1)$ multiplied by that of $(E+M)/r^3$ to S/r^3, and the latter ratio is known by the mean motions of the sun and moon.

oblateness are only about 8″. It is to be observed that the supposition of a spheroidal figure has only been made for the sake of
rendering applicable the equation (24), which had been already
obtained, and has nothing to do with the order of magnitude of the
terms we are considering*.

Although however the effect of the moon's oblateness, or rather
of the possible deviation of her mass from a mass composed of concentric spherical strata, may be neglected in considering the motion
of the moon's centre, it does not therefore follow that it ought to
be neglected in considering the moon's motion about her own axis.
For in the first place, in comparing the effects produced on the
moon and on the earth, the moment of the mutual moving force of
attraction of the moon and earth is divided by the moment of
inertia of the moon, instead of the moment of inertia of the earth,
which is much larger; and in the second place, the effect now considered is not mixed up with any other. In fact, it is well known
that the circumstance that the moon always presents the same face
to us has been accounted for in this manner.

12. In concluding this subject, it may be well to consider the
degree of evidence afforded by the figure of the earth in favour
of the hypothesis of the earth's original fluidity.

In the first place, it is remarkable that the surface of the earth
is so nearly a surface of equilibrium. The elevation of the land
above the level of the sea is extremely trifling compared with the
breadth of the continents. The surface of the sea must of course
necessarily be a surface of equilibrium, but still it is remarkable
that the sea is spread so uniformly over the surface of the earth.
There is reason to think that the depth of the sea does not exceed
a very few miles on the average. Were a roundish solid taken at
random, and a quantity of water poured on it, and allowed to
settle under the action of the gravitation of the solid, the probability is that the depth of the water would present no sort of

---

* If the expression for V be formed directly, and be expanded according to
inverse powers of r, the first term will be M/r. The terms involving r^{-2} will
disappear if the centre of gravity of the moon be taken for origin, those involving
r^{-3} are the terms we are here considering. If the moon's centre of gravity, or
rather its projection on the apparent disk, did not coincide with the centre of the
disk, it is easy to see the nature of the apparent inequality in the moon's motion
which would thence result.

uniformity, and would be in some places very great. Nevertheless the circumstance that the surface of the earth is so nearly a surface of equilibrium might be attributed to the constant degradation of the original elevations during the lapse of ages.

In the second place, it is found that the surface is very nearly an oblate spheroid, having for its axis the axis of rotation. That the surface should *on the whole* be protuberant about the equator is nothing remarkable, because even were the matter of which the earth is composed arranged symmetrically about the centre, a surface of equilibrium would still be protuberant in consequence of the centrifugal force; and were matter to accumulate at the equator by degradation, the ellipticity of the surface of equilibrium would be increased by the attraction of this matter. Nevertheless the ellipticity of the earth is much greater than the ellipticity ($\frac{1}{2}m$) due to the centrifugal force alone, and even greater than the ellipticity which would exist were the earth composed of a sphere touching the surface at the poles, and consisting of concentric spherical strata of equal density and of a spherico-spheroidal shell having the density of the rocks and clay at the surface*. This being the case, the regularity of the surface is no doubt remarkable; and this regularity is accounted for on the hypothesis of original fluidity.

The near coincidence between the numerical values of the ellipticity of the terrestrial spheroid obtained independently from the motion of the moon, from the pendulum, by the aid of Clairaut's theorem, and from direct measures of arcs, affords no additional evidence whatsoever in favour of the hypothesis of original fluidity, being a direct consequence of the law of universal gravitation†.

* It may be proved without difficulty that the value of ϵ corresponding to this supposition is $\frac{1}{470}$ nearly, if we suppose the density of the shell to be to the mean density as 5 to 11.

† With respect to the argument derived from the motion of the moon, this remark has already been made by Professor O'Brien, who has shewn that if the form of the surface and the law of the variation of gravity be given independently, and if we suppose the earth to consist approximately of spherical strata of equal density, without which it seems impossible to account for the observed regularity of gravity at the surface, then the attraction on the moon follows as a necessary consequence, independently of any theory but that of universal gravitation. (*Tract on the Figure of the Earth.*) If the surface be not assumed to be one of equilibrium, nor even nearly spherical, and if the component of gravity in a direction perpen-

If the expression for V given by (24) be compared with the expression which would be obtained by direct integration, it may easily be shewn that the axis of rotation is a principal axis, and that the moments of inertia about the other two principal axes are equal to each other, so that every equatorial axis is a principal axis. These results would follow as a consequence of the hypothesis of original fluidity. Still it should be remembered that we can only affirm them to be accurate to the degree of accuracy to which we are authorized by measures of arcs and by pendulum experiments to affirm the surface to be an oblate spheroid.

The phenomena of precession and nutation introduce a new element to our consideration, namely the moment of inertia of the earth about an equatorial axis. The observation of these phenomena enables us to determine the numerical value of the quantity κ, if we suppose ϵ known otherwise. Now, independently of any hypothesis as to original fluidity, it is probable that the earth consists approximately of spherical strata of equal density. Any material deviation from this arrangement could hardly fail to produce an irregularity in the variation of gravity, and consequently in the form of the surface, since we know that the surface is one of equilibrium. Hence we may assume, when not directly considering the ellipticity, that the density ρ is a function of the distance r from the centre. Now the mean density of the earth as compared with that of water is known from the result of Cavendish's experiment, and the superficial density

dicular to the surface, as well as the form of the surface, be given independently, it may be shewn that the attraction on an external particle follows, independently of any hypothesis respecting the distribution of matter in the interior of the earth. It may be remarked that if the surface be supposed to differ from a surface of equilibrium by a quantity of the order of the ellipticity, the component of gravity in a direction perpendicular to the surface may be considered equal to the whole force of gravity. Since however, as a matter of fact, the surface *is* a surface of equilibrium, if very trifling irregularities be neglected, it seems better to assume it to be such, and then the law of the variation of gravity, as well as the attraction on the moon, follow from the form of the surface.

It must not here be supposed that these irregularities are actually neglected. Such an omission would ill accord with the accuracy of modern measures. In geodetic operations and pendulum experiments, the direct observations are in fact reduced to the level of the sea, and so rendered comparable with a theory in which it is supposed that the earth's surface is accurately a surface of equilibrium. I have considered this subject in detail in the paper referred to at the beginning of this article, which has since been read before the Cambridge Philosophical Society.

may be considered equal to that of ordinary rocks, or about $2\frac{1}{2}$ times that of water; and therefore the ratio of the mean to the superficial density may be considered known. Take for simplicity the earth's radius for the unit of length, and let $\rho = \rho_1$ when $r = 1$. From the mean density and the value of κ we know the ratios of the integrals $\int_0^1 \rho r^2 dr$ and $\int_0^1 \rho r^4 dr$ to ρ_1. Now it is probable that ρ increases, at least on the whole, from the surface to the centre. If we assume this to be the case, and restrict ρ to satisfy the conditions of becoming equal to ρ_1 when $r = 1$, and of giving to the two integrals just written their proper numerical values, it is evident that the law of density cannot range within any very wide limits; and speaking very roughly we may say that the density is *determined*.

Now the preceding results will not be sensibly affected by giving to the nearly spherical strata of equal density one form or another, but the form of the surface will be materially affected. The surface in fact might not be spheroidal at all, or if spheroidal, the ellipticity might range between tolerably wide limits. But according to the hypothesis of original fluidity the surface ought to be spheroidal, and the ellipticity ought to have a certain numerical value depending upon the law of density.

If then there exist a law of density, not in itself improbable *à priori*, which satisfies the required conditions respecting the mean and superficial densities, and which gives to the ellipticity and to the annual precession numerical values nearly agreeing with their observed values, we may regard this law not only as in all probability representing approximately the distribution of matter within the earth, but also as furnishing, by its accordance with observation, a certain degree of evidence in favour of the hypothesis of original fluidity. The law of density usually considered in the theory of the figure of the earth is a law of this kind.

It ought to be observed that the results obtained relative to the attraction of the earth remain just the same whether we suppose the earth to be solid throughout or not; but in founding any argument on the numerical value of κ we are obliged to consider the state of the interior. Thus if the central portions of the earth be, as some suppose, in a state of fusion, the quantity $Ea^2\kappa$ must be taken to mean the moment of inertia of that solid, whatever it may be, which is equivalent to the solid crust together with its fluid or viscous contents. On this supposition it is even conceivable that κ should depend on the period of the disturbing force, so that different numerical values of κ might have to be used in the precession and in the lunar nutation, in which case the mass of the moon deduced from precession and nutation would not be quite correct.

Additional Propositions respecting Attractions.

Although the propositions at the commencement of this paper were given merely for the sake of the applications made of them to the figure of the earth, there are a few additional propositions which are so closely allied to them that they may conveniently be added here.

Prop. VII*. If V be the potential of any mass M_1, and if M_0 be the portion of M_1 contained within a closed surface S,

$$\iint \frac{dV}{dn} \, dS = -4\pi M_0 \quad\ldots\ldots\ldots\ldots (27),$$

n and dS having the same meaning as in Prop. IV., and the integration being extended to the whole surface S.

---

* This and Prop. IV. are expressed respectively by equations (7) and (8) in the article by Professor Thomson already referred to (Vol. III. p. 203), where a demonstration of a theorem comprehending both founded on the equation

$$\frac{d^2v}{dx^2} + \frac{d^2v}{dy^2} + \frac{d^2v}{dz^2} = -4\pi\rho \quad\ldots\ldots\ldots\ldots\ldots\ldots(a)$$

is given. In the present paper a different order of investigation is followed; direct geometrical demonstrations of the equations

$$\iint \frac{dV}{dn} \, dS = 0 \text{ in one case, and } \iint \frac{dV}{dn} \, dS = -4\pi M_0 \text{ in another,}$$

are given in Props. IV. and VII.; and a new proof of the equation (a) is deduced from them in Prop. VIII.

These equations may be obtained as very particular cases of a general theorem originally given by Green (*Essay on Electricity*, p. 12). It will be sufficient to suppose $U=1$ in Green's equation, and to observe that $dw = -dn$, and $\delta V = 0$ or $= -4\pi\rho$, if V be taken to denote the potential of the mass whose attraction is considered.

Let m' be the mass of an attracting particle situated at the point P' inside S. Through P' draw a right line L, and produce it indefinitely in one direction. This line will in general cut S in one point; but if S be a re-entrant* surface it may be cut by L in three, five, or any odd number of points. About L describe a conical surface containing an infinitely small solid angle α, and let the rest of the notation be as in Prop. IV. In this case the angles θ_1, θ_2,......, will be alternately obtuse and acute, and we shall have

$$N_1 = -\frac{m'}{r_1^2}\cos(\pi - \theta_1) = \frac{m'}{r_1^2}\cos\theta_1,$$

$$A_1 = \alpha r_1^2 \sec(\pi - \theta_1) = -\alpha r_1^2 \sec\theta_1,$$

and therefore $\qquad N_1 A_1 = -\alpha m'.$

Should there be more than one point of section, the terms $N_2 A_2$, $N_3 A_3$, &c. will destroy each other two and two, as in Prop. IV. Now all angular space around P' may be divided into an infinite number of solid angles such as α, and it is evident that the whole surface S will thus be exhausted. We get therefore

$$\text{limit of } \Sigma NA = -\Sigma \alpha m' = -m'\Sigma \alpha;$$

or, since $\Sigma \alpha = 4\pi$, $\qquad \iint N dS = -4\pi m'.$

The same formula will apply to any other internal particle, and it has been shewn in Prop. IV. that for an external particle $\iint N dS = 0$. Hence, adding together all the results, and taking N now to refer to the attraction of all the particles, both internal and external, we get $\iint N dS = -4\pi M_0$. But $N = dV/dn$, which proves the proposition.

Prop. VIII. At an internal point (x, y, z) about which the density is ρ, the potential V satisfies the equation

$$\frac{d^2V}{dx^2} + \frac{d^2V}{dy^2} + \frac{d^2V}{dz^2} = -4\pi\rho \ \dotfill(28).$$

Consider the elementary parallelepiped $dx\,dy\,dz$, and apply to it the equation (27). For the face $dy\,dz$ whose abscissa is x, the value of $\iint \frac{dV}{dn} dS$ is ultimately $-dV/dx \cdot dy\,dz$, and for the opposite face it is ultimately $+\left(\frac{dV}{dx} + \frac{d^2V}{dx^2} dx\right) dy\,dz$; and therefore for this

---

* This term is here used, and has been already used in the demonstration of Prop. IV., to denote a closed surface which can be cut by a tangent plane.

pair of faces the value of the integral is ultimately $d^2 V/dx^2 . dx\, dy\, dz$. Treating the two other pairs of faces in the same way, we get ultimately for the value of the first member of equation (27),

$$\left(\frac{d^2 V}{dx^2} + \frac{d^2 V}{dy^2} + \frac{d^2 V}{dz^2}\right) dx\, dy\, dz.$$

But the density being ultimately constant, the value of M_0, which is the mass contained within the parallelepiped, is ultimately $\rho\, dx\, dy\, dz$, whence by passing to the limit we obtain equation (28).

The equation which (28) becomes when the polar co-ordinates r, θ, ϕ are employed in place of rectangular, may readily be obtained by applying equation (27) to the elementary volume $dr . r d\theta . r \sin \theta d\phi$, or else it may be derived from (28) by transformation of co-ordinates. The first member of the transformed equation has already been written down (see equation (10),) ; the second remains $- 4\pi\rho$.

Example of the application of equation (28).—In order to give an example of the practical application of this equation, let us apply it to determine the attraction which a sphere composed of concentric spherical strata of uniform density exerts on an internal particle.

Refer the sphere to polar co-ordinates originating at the centre. Let ρ be the density, which by hypothesis is a function of r, R the external radius, V the potential of the sphere, which will evidently be a function of r only. For a point within the sphere we get from (28)

$$\frac{d^2 V}{dr^2} + \frac{2}{r}\frac{dV}{dr} = - 4\pi\rho \quad\dotfill(29).$$

For a point outside the sphere the equation which V has to satisfy is that which would be obtained from (29) by replacing the second member by zero; but we may evidently apply equation (29) to all space provided we regard ρ as equal to zero outside the sphere. Since the first member of (29) is the same thing as $1/r . d^2 rV/dr^2$, we get

$$V = - \frac{4\pi}{r} \iint \rho r\, dr^2.$$

Now we get by integration by parts,

$$\int (\int \rho r \, dr) \, dr = r \int \rho r \, dr - \int \rho r^2 dr,$$

whence $\qquad V = - 4\pi \int \rho r \, dr + \frac{4\pi}{r} \int \rho r^2 dr,$

where the arbitrary constants are supposed to be included in the signs of integration. Now V vanishes at an infinite distance, and does not become infinite at the centre, and therefore the second integral vanishes when $r = 0$, and the first when $r = \infty$, or, which is the same, when $r = R$, since $\rho = 0$ when $r > R$. We get therefore finally,

$$V = 4\pi \int_r^R \rho r \, dr + \frac{4\pi}{r} \int_0^r \rho r^2 dr.$$

If F be the required force of attraction, we have $F = - dV/dr$; and observing that the two terms arising from the variation of the limits destroy each other, we get

$$F = \frac{4\pi}{r^2} \int_0^r \rho r^2 dr.$$

Now $4\pi \int_0^r \rho r^2 dr$ is the mass contained within a sphere described about the centre with a radius r, and therefore the attraction is the same as if the mass within this sphere were collected at its centre, and the mass outside it were removed.

The attraction of the sphere on an external particle may be considered as a particular case of the preceding, since we may first suppose the sphere to extend beyond the attracted particle, and then make ρ vanish when $r > R$.

Before concluding, one or two more known theorems may be noticed, which admit of being readily proved by the method employed in Prop. v.

Prop. IX. If T be a space which contains none of the attracting matter, the potential V cannot be constant throughout any finite portion of T without having the same constant value throughout the whole of the space T and at its surface. For if possible let V have the constant value A throughout the space T_1, which forms a portion of T, and a greater or less value at the portions of T adjacent to T_1. Let R be a region of T adjacent to T_1 where V is greater than A. By what has been already remarked, V must increase continuously in passing from T_1 into R. Draw a closed surface σ lying partly within T_1 and partly within R, and call the portions lying in T_1 and R, σ_1, σ_2 respectively. Then if ν be a normal to σ, drawn outwards, $dV/d\nu$ will be positive throughout σ_1 if σ_1 be drawn sufficiently close to the space T_1 (see Prop. v. and note), and $dV/d\nu$ is equal to zero throughout the surface σ_2, since V is constant throughout the space T_1; and therefore $\iint \dfrac{dV}{d\nu} d\sigma$, taken throughout the whole surface σ, will be positive, which is contrary to Prop. iv. Hence V cannot be greater than A in any portion of T adjacent to T_1, and similarly it cannot be less, and therefore V must have the constant value A throughout T, and therefore, on account of the continuity of V, at the surface of T.

Combining this with Prop. v. Cor. 1, we see that if V be constant throughout the whole surface of a space T which contains no attracting matter, it will have the same constant value throughout T; but if V be not constant throughout the whole surface, it cannot be constant throughout any finite portion of T, but only throughout a surface. Such a surface cannot be closed, but must abut upon the surface of T, since otherwise V would be constant within it.

Prop. x. The potential V cannot admit of a maximum or minimum value in the space T.

It appears from the demonstration of Prop. v. that V cannot have a maximum or minimum value at a point, or throughout a line, surface, or space, which is isolated in T. But not even can V have the maximum or minimum value V_1 throughout T_1 if T_1 reach up to the surface S of T; though the term maximum or minimum is not strictly applicable to this case. By Prop. ix. V cannot have the value V_1 throughout a space, and therefore T_1 can only be a surface or a line.

If possible, let V have the maximum value V_1 throughout a line L which reaches up to S. Consider the loci of the points where V has the successive values V_2, V_3..., decreasing by infinitely small steps from V_1. In the immediate neighbourhood of L, these loci will evidently be tube-shaped surfaces, each lying outside the preceding, the first of which will ultimately coincide with L. Let s be an element of L not adjacent to S, nor reaching up to the extremity of L, in case L terminate abruptly. At each extremity of s draw an infinite number of *lines of force*, that is, lines traced from point to point in the direction of the force, and therefore perpendicular to the surfaces of equilibrium. The assemblage of these lines will evidently constitute two surfaces cutting the tubes, and perpendicular to s at its extremities. Call the space contained within the two surfaces and one of the tubes T_2, and apply equation (5) to this space. Since V is a maximum at L, dV/dn is negative for the tube surface of T_2, and it vanishes for the other surfaces, as readily follows from equation (4). Hence $\iint \dfrac{dV}{dn} dS$, taken throughout the whole surface T_2, is negative, which is contrary to equation (5). Hence V cannot have a maximum value at the line L; and similarly it cannot have a minimum value.

It may be proved in a similar manner that V cannot have a maximum or minimum value V_1 throughout a surface S_1 which reaches up to S. For this purpose it will be sufficient to draw a line of force through a point in S_1, and make it travel round an elementary area σ which forms part of S_1, and to apply equation (5) to the space contained between the surface generated by this line, and the two portions, one on each side of S_1, of a surface of equilibrium corresponding to a value of V very little different from V_1.

It should be observed that the space T considered in this proposition and in the preceding need not be closed: all that is requisite is that it contain none of the attracting mass. Thus, for instance, T may be the infinite space surrounding an attracting mass or set of masses.

It is to be observed also, that although attractive forces have been spoken of throughout, all that has been proved is equally true of repulsive forces, or of forces partly attractive and partly repulsive. In fact, nothing in the reasoning depends upon the sign of m; and by making m negative we pass to the case of repulsive forces.

Prop. XI. If an isolated particle be in equilibrium under the action of forces varying inversely as the square of the distance, the equilibrium cannot be stable with reference to every possible

130 ON ATTRACTIONS, AND ON CLAIRAUT'S THEOREM.

displacement, nor unstable, but must be stable with reference to some displacements and unstable with reference to others; and therefore the equilibrium of a *free* isolated particle in such circumstances must be unstable*.

For we have seen that V cannot be a maximum or minimum, and therefore either V must be absolutely constant, (as for instance within a uniform spherical shell), in which case the particle may be in equilibrium at any point of the space in which it is situated, or else, if the particle be displaced along any straight line or curve, for some directions of the line or curve V will be increasing and for some decreasing. In the former case the force resolved along a tangent to the particle's path will be directed *from* the position of equilibrium, and will tend to remove the particle still farther from it, while in the latter case the reverse will take place.

---

* This theorem was first given by Mr Earnshaw in his memoir on Molecular Forces read at the Cambridge Philosophical Society, March 18, 1839 (*Trans.* Vol. VII.). See also a paper by Professor Thomson in the first series of this *Journal*, Vol. IV. p. 223.

[From the *Transactions of the Cambridge Philosophical Society*, Vol. VIII. p. 672.]

On the Variation of Gravity at the Surface of the Earth.

[Read *April* 23, 1849.]

On adopting the hypothesis of the earth's original fluidity, it has been shewn that the surface ought to be perpendicular to the direction of gravity, that it ought to be of the form of an oblate spheroid of small ellipticity, having its axis of figure coincident with the axis of rotation, and that gravity ought to vary along the surface according to a simple law, leading to the numerical relation between the ellipticity and the ratio between polar and equatorial gravity which is known by the name of Clairaut's Theorem. Without assuming the earth's original fluidity, but merely supposing that it consists of nearly spherical strata of equal density, and observing that its surface may be regarded as covered by a fluid, inasmuch as all observations relating to the earth's figure are reduced to the level of the sea, Laplace has established a connexion between the form of the surface and the variation of gravity, which in the particular case of an oblate spheroid agrees with the connexion which is found on the hypothesis of original fluidity. The object of the first portion of this paper is to establish this general connexion without making any hypothesis whatsoever respecting the distribution of matter in the interior of the earth, but merely assuming the theory of universal gravitation. It appears that if the form of the surface be given, gravity is determined throughout the whole surface, except so far as regards one arbitrary constant which is contained in its complete expression, and which

9—2

may be determined by the value of gravity at one place. Moreover the attraction of the earth at all external points of space is determined at the same time; so that the earth's attraction on the moon, including that part of it which is due to the earth's oblateness, and the moments of the forces of the sun and moon tending to turn the earth about an equatorial axis, are found quite independently of the distribution of matter within the earth.

The near coincidence between the numerical values of the earth's ellipticity deduced independently from measures of arcs, from the lunar inequalities which depend on the earth's oblateness, and, by means of Clairaut's Theorem, from pendulum experiments, is sometimes regarded as a confirmation of the hypothesis of original fluidity. It appears, however, that the form of the surface (which is supposed to be a surface of equilibrium), suffices to determine both the variation of gravity and the attraction of the earth on an external particle*, and therefore the coincidence in question, being a result of the law of gravitation, is no confirmation of the hypothesis of original fluidity. The evidence in favour of this hypothesis which is derived from the figure and attraction of the earth consists in the perpendicularity of the surface to the direction of gravity, and in the circumstance that the surface is so nearly represented by an oblate spheroid having for its axis the axis of rotation. A certain degree of additional evidence is afforded by the near agreement between the observed ellipticity and that calculated with an assumed law of density which is likely *a priori* to be not far from the truth, and which is confirmed, as to its general correctness, by leading to a value for the annual precession which does not much differ from the observed value.

---

* It has been remarked by Professor O'Brien (*Mathematical Tracts*, p. 56) that if we have given the form of the earth's surface and the variation of gravity, we have data for determining the attraction of the earth on an external particle, the earth being supposed to consist of nearly spherical strata of equal density; so that the motion of the moon furnishes no additional confirmation of the hypothesis of original fluidity.

If we have given the component of the attraction of any mass, however irregular as to its form and interior constitution, in a direction perpendicular to the surface, throughout the whole of the surface, we have data for determining the attraction at every external point, as well as the components of the attraction at the surface in two directions perpendicular to the normal. The corresponding proposition in Fluid Motion is self-evident.

Since the earth's actual surface is not strictly a surface of equilibrium, on account of the elevation of the continents and islands above the sea level, it is necessary to consider in the first instance in what manner observations would have to be reduced in order to render the preceding theory applicable. It is shewn in Art. 13 that the earth may be regarded as bounded by a surface of equilibrium, and therefore the expressions previously investigated may be applied, provided the sea level be regarded as the bounding surface, and observed gravity be reduced to the level of the sea by taking account only of the change of distance from the earth's centre. Gravity reduced in this manner would, however, be liable to vary irregularly from one place to another, in consequence of the attraction of the land between the station and the surface of the sea, supposed to be prolonged underground, since this attraction would be greater or less according to the height of the station above the sea level. In order therefore to render the observations taken at different places comparable with one another, it seems best to correct for this attraction in reducing to the level of the sea; but since this additional correction is introduced in violation of the theory in which the earth's surface is regarded as one of equilibrium, it is necessary to consider what effect the habitual neglect of the small attraction above mentioned produces on the values of mean gravity and of the ellipticity deduced from observations taken at a number of stations. These effects are considered in Arts. 17, 18.

Besides the consideration of the mode of determining the values of mean gravity, and thereby the mass of the earth, and of the ellipticity, and thereby the effect of the earth's oblateness on the motion of the moon, it is an interesting question to consider whether the observed anomalies in the variation of gravity may be attributed wholly or mainly to the irregular distribution of land and sea at the surface of the earth, or whether they must be referred to more deeply seated causes. In Arts. 19, 20, I have considered the effect of the excess of matter in islands and continents, consisting of the matter which is there situated above the actual sea level, and of the defect of matter in the sea, consisting of the difference between the mass of the sea, and the mass of an equal bulk of rock or clay. It appears that besides the attraction of the land lying immediately underneath a continental station,

between it and the level of the sea, the more distant portions of
the continent cause an increase in gravity, since the attraction
which they exert is not wholly horizontal, on account of the curvature of the earth. But besides this direct effect, a continent
produces an indirect effect on the magnitude of apparent gravity.
For the horizontal attraction causes the verticals to point more
inwards, that is, the zeniths to be situated further outwards, than
if the continent did not exist; and since a level surface is everywhere perpendicular to the vertical, it follows that the sea level
on a continent is higher than it would be at the same place if the
continent did not exist.   Hence, in reducing an observation taken
at a continental station to the level of the sea, we reduce it to
a point more distant from the centre of the earth than if the
continent were away ; and therefore, on this account alone, gravity
is less on the continent than on an island. It appears that this
latter effect more than counterbalances the former, so that on the
whole, gravity is less on a continent than on an island, especially
if the island be situated in the middle of an ocean. This circumstance has already been noticed as the result of observation. In
consequence of the inequality to which gravity is subject, depending on the character of the station, it is probable that the
value of the ellipticity which Mr Airy has deduced from his discussion of pendulum observations is a little too great, on account
of the decided preponderance of oceanic stations in low latitudes
among the group of stations where the observations were taken.

The alteration of attraction produced by the excess and defect
of matter mentioned in the preceding paragraph does not constitute the whole effect of the irregular distribution of land and
sea, since if the continents were cut off at the actual sea level,
and the sea were replaced by rock and clay, the surface so formed
would no longer be a surface of equilibrium, in consequence of
the change produced in the attraction. In Arts 25—27, I have
investigated an expression for the reduction of observed gravity to
what would be observed if the elevated solid portions of the earth
were to become fluid, and to run down, so as to form a level bottom
for the sea, which in that case would cover the whole earth.   The
expressions would be very laborious to work out numerically, and
besides, they require data, such as the depth of the sea in a great
many places, &c., which we do not at present possess; but from a

consideration of the general character of the correction, and from the estimation given in Art. 21 of the magnitude which such corrections are likely to attain, it appears probable that the observed anomalies in the variation of gravity are mainly due to the irregular distribution of land and sea at the surface of the earth.

---

1. Conceive a mass whose particles attract each other according to the law of gravitation, and are besides acted on by a given force f, which is such that if X, Y, Z be its components along three rectangular axes, $Xdx + Ydy + Zdz$ is the exact differential of a function U of the co-ordinates. Call the surface of the mass S, and let V be the potential of the attraction, that is to say, the function obtained by dividing the mass of each attracting particle by its distance from the point of space considered, and taking the sum of all such quotients. Suppose S to be a surface of equilibrium. The general equation to such surfaces is

$$V + U = c \quad \dots\dots\dots\dots\dots\dots\dots (1),$$

where c is an arbitrary constant; and since S is included among these surfaces, equation (1) must be satisfied at all points of the surface S, when some one particular value is assigned to c. For any point external to S, the potential V satisfies, as is well known, the partial differential equation

$$\frac{d^2V}{dx^2} + \frac{d^2V}{dy^2} + \frac{d^2V}{dz^2} = 0 \quad \dots\dots\dots\dots (2);$$

and evidently V cannot become infinite at any such point, and must vanish at an infinite distance from S. Now these conditions are sufficient for the complete determination of the value of V for every point external to S, the quantities U and c being supposed known. The mathematical problem is exactly the same as that of determining the permanent temperature in a homogeneous solid, which extends infinitely around a closed space S, on the conditions, (1) that the temperature at the surface S shall be equal to $c - U$, (2) that it shall vanish at an infinite distance. This problem is evidently possible and determinate. The possibility has moreover been demonstrated mathematically.

If U alone be given, and not c, the general value of V will contain one arbitrary constant, which may be determined if we know the value of V, or of one of its differential coefficients, at one point situated either in the surface S or outside it. When V is known, the components of the force of attraction will be obtained by mere differentiation.

Nevertheless, although we know that the problem is always determinate, it is only for a very limited number of forms of the surface S that the solution has hitherto been effected. The most important of these forms is the sphere. When S has very nearly one of these forms the problem may be solved by approximation.

2. Let us pass now to the particular case of the earth. Although the earth is really revolving about its axis, so that the bodies on its surface are really describing circular orbits about the axis of rotation, we know that the relative equilibrium of the earth itself, or at least its crust, and the bodies on its surface, would not be affected by supposing the crust at rest, provided that we introduce, in addition to the attraction, that fictitious force which we call the centrifugal force. The vertical at any place is determined by the plumb-line, or by the surface of standing fluid, and its determination is therefore strictly a question of relative equilibrium. The intensity of gravity is determined by the pendulum; but although the result is not mathematically the same as if the earth were at rest and acted on by the centrifugal force, the difference is altogether insensible. It is only in consequence of its influence on the direction and magnitude of the force of gravity that the earth's actual motion need be considered at all in this investigation: the mere question of attraction has nothing to do with motion; and the results arrived at will be equally true whether the earth be solid throughout or fluid towards the centre, even though, on the latter supposition, the fluid portions should be in motion relatively to the crust.

We know, as a matter of observation, that the earth's surface is a surface of equilibrium, if the elevation of islands and continents above the level of the sea be neglected. Consequently the law of the variation of gravity along the surface is determinate, if the form of the surface be given, the force f of Art. 1 being in this case the centrifugal force. The nearly spherical form of the surface renders the determination of the variation easy.

3. Let the earth be referred to polar co-ordinates, the origin being situated in the axis of rotation, and coinciding with the centre of a sphere which nearly represents the external surface. Let r be the radius vector of any point, θ the angle between the radius vector and the northern direction of the axis, ϕ the angle which the plane passing through these two lines makes with a plane fixed in the earth and passing through the axis. Then the equation (2) which V has to satisfy at any external point becomes by a common transformation

$$r\frac{d^2 . rV}{dr^2} + \frac{1}{\sin\theta}\frac{d}{d\theta}\left(\sin\theta\frac{dV}{d\theta}\right) + \frac{1}{\sin^2\theta}\frac{d^2V}{d\phi^2} = 0 \dots (3).$$

Let ω be the angular velocity of the earth; then

$$U = \tfrac{1}{2}\omega^2 r^2 \sin^2\theta,$$

and equation (1) becomes

$$V + \tfrac{1}{2}\omega^2 r^2 \sin^2\theta = c \dots\dots (4),$$

which has to be satisfied at the surface of the earth.

For a given value of r, greater than the radius of the least sphere which can be described about the origin as centre so as to lie wholly without the earth, V can be expanded in a series of Laplace's functions

$$V_0 + V_1 + V_2 + \dots;$$

and therefore in general, provided r be greater than the radius of the sphere above mentioned, V can be expanded in such a series, but the general term V_n will be a function of r, as well as of θ and ϕ. Substituting the above series in equation (3), and observing that from the nature of Laplace's functions

$$\frac{1}{\sin\theta}\frac{d}{d\theta}\left(\sin\theta\frac{dV_n}{d\theta}\right) + \frac{1}{\sin^2\theta}\frac{d^2V_n}{d\phi^2} = -n(n+1)V_n \dots\dots(5),$$

we get

$$\Sigma\left\{r\frac{d^2 . rV_n}{dr^2} - n(n+1)V_n\right\} = 0,$$

where all integral values of n from 0 to ∞ are to be taken.

Now the differential coefficients of V_n with respect to r are Laplace's functions of the nth order as well as V_n itself; and since a series of Laplace's functions cannot be equal to zero unless the Laplace's functions of the same order are separately equal to zero, we must have

$$r \frac{d^2 \cdot rV_n}{dr^2} - n\,(n+1)\,V_n = 0 \quad \dotfill (6).$$

The integral of this equation is

$$V_n = \frac{Y_n}{r^{n+1}} + Z_n r^n,$$

where Y_n and Z_n are arbitrary constants so far as r is concerned, but contain θ and ϕ. Since these functions are multiplied by different powers of r, V_n cannot be a Laplace's function of the nth order unless the same be true of Y_n and Z_n. We have for the complete value of V

$$\frac{Y_0}{r} + \frac{Y_1}{r^2} + \frac{Y_2}{r^3} + \dots + Z_0 + Z_1 r + \dots\dots.$$

Now V vanishes when $r = \infty$, which requires that $Z_0 = 0$, $Z_1 = 0$, &c.; and therefore

$$V = \frac{Y_0}{r} + \frac{Y_1}{r^2} + \frac{Y_2}{r^3} + \quad \dotfill (7).$$

4. The preceding equation will not give the value of the potential throughout the surface of a sphere which lies partly within the earth, because although V, as well as any arbitrary but finite function of θ and ϕ, can be expanded in a series of Laplace's functions, the second member of equation (3) is not equal to zero in the case of an internal particle, but to $-4\pi\rho r^2$, where ρ is the density. Nevertheless we may employ equation (7) for values of r corresponding to spheres which lie partly within the earth, provided that in speaking of an internal particle we slightly change the signification of V, and interpret it to mean, not the actual potential, but what would be the potential if the protuberant matter were distributed within the least sphere which cuts the surface, in such a manner as to leave the potential unchanged throughout the actual surface. The possibility of such a distribution will be justified by the result, provided the series to which we are led prove convergent. Indeed, it might easily be shewn that the potential at any internal point near the surface differs from what would be given by (7) by a small quantity of the second order only; but its differential coefficient with respect to r, which

gives the component of the attraction along the radius vector, differs by a small quantity of the first order. We do not, however, want the potential at any point of the interior, and in fact it cannot be found without making some hypothesis as to the distribution of the matter within the earth.

5.   It remains now to satisfy equation (4).   Let $r = a\,(1+u)$ be the equation to the earth's surface, where u is a small quantity of the first order, a function of θ and ϕ.   Let u be expanded in a series of Laplace's functions $u_0 + u_1 + \ldots$ The term u_0 will vanish provided we take for a the mean radius, or the radius of a sphere of equal volume.   We may, therefore, take for the equation to the surface

$$r = a\,(1+u_1+u_2+\ldots)\ldots\ldots\ldots\ldots\ldots\ldots(8).$$

If the surface were spherical, and the earth had no motion of rotation, V would be independent of θ and ϕ, and the second member of equation (7) would be reduced to its first term.   Hence, since the centrifugal force is a small quantity of the first order, as well as u, the succeeding terms must be small quantities of the first order; so that in substituting in (7) the value of r given by (8) it will be sufficient to put $r = a$ in these terms.   Since the second term in equation (4) is a small quantity of the first order, it will be sufficient in that term likewise to put $r = a$.   We thus get from (4), (7), and (8), omitting the squares of small quantities,

$$\frac{Y_0}{a}\,(1-u_1-u_2-u_3\ldots) + \frac{Y_1}{a^2} + \frac{Y_2}{a^3} + \ldots + \frac{\omega^2 a^2}{2}\sin^2\theta = c \ \ldots\ldots (9).$$

The most general Laplace's function of the order 0 is a constant; and we have

$$\sin^2\theta = \tfrac{2}{3} + (\tfrac{1}{3} - \cos^2\theta),$$

of which expression the two parts are Laplace's functions of the orders 0, 2, respectively.   We thus get from (9), by equating to zero Laplace's functions of the same order,

$$Y_0 = ac - \tfrac{1}{3}\omega^2 a^3,$$

$$Y_1 = aY_0 u_1,$$

$$Y_2 = a^2 Y_0 u_2 - \tfrac{1}{2}\omega^2 a^5\,(\tfrac{1}{3} - \cos^2\theta),$$

$$Y_3 = a^3 Y_0 u_3,\ \&c.$$

The first of these equations merely gives a relation between the arbitrary constants Y_0 and c; the others determine Y_1, Y_2, &c.; and we get by substituting in (7)

$$V = Y_0 \left(\frac{1}{r} + \frac{a}{r^2} u_1 + \frac{a^2}{r^3} u_2 + \ldots \right) - \frac{\omega^2 a^5}{2r^3} (\tfrac{1}{3} - \cos^2 \theta) \ldots\ldots(10).$$

6.   Let g be the force of gravity at any point of the surface of the earth, dn an element of the normal drawn outwards at that point; then $g = -d(V + U)/dn$. Let ψ be the angle between the normal and the radius vector; then $g \cos \psi$ is the resolved part of gravity along the radius vector, and this resolved part is equal to $-d(V + U)/dr$. Now ψ is a small quantity of the first order, and therefore we may put $\cos \psi = 1$, which gives

$$g = -\frac{d}{dr}(V + U),$$

where, after differentiation, r is to be replaced by the radius vector of the surface, which is given by (8). We thus get

$$g = \frac{Y_0}{a^2}(1 - 2u_1 - 2u_2 - 2u_3 \ldots) + \frac{Y_0}{a^2}(2u_1 + 3u_2 + 4u_3 \ldots)$$
$$- \tfrac{3}{2} \omega^2 a (\tfrac{1}{3} - \cos^2 \theta) - \omega^2 a (\tfrac{2}{3} + \tfrac{1}{3} - \cos^2 \theta),$$

which gives, on putting

$$\frac{Y_0}{a^2} - \tfrac{2}{3} \omega^2 a = G, \qquad \frac{\omega^2 a}{G} = m \quad \ldots\ldots\ldots\ldots (11),$$

and neglecting squares of small quantities,

$$g = G \{1 - \tfrac{5}{2} m (\tfrac{1}{3} - \cos^2 \theta) + u_2 + 2u_3 + 3u_4 \ldots\ldots\} \ldots\ldots(12).$$

In this equation G is the mean value of g taken throughout the whole surface, since we know that $\int_0^\pi \int_0^{2\pi} u_n \sin \theta \, d\theta \, d\phi = 0$, if n be different from zero. The second of equations (11) shews that m is the ratio of the centrifugal force at a distance from the axis equal to the mean distance to mean gravity, or, which is the same, since the squares of small quantities are neglected, the ratio of the centrifugal force to gravity at the equator. Equation (12) makes known the variation of gravity when the form of the surface is given, the surface being supposed to be one of equilibrium; and, conversely, equation (8) gives the form of the surface if the variation of gravity be known. It may be observed that on the latter supposition there is nothing to determine u_1. The most general form of u_1 is

$$\alpha \sin \theta \cos \phi + \beta \sin \theta \sin \phi + \gamma \cos \theta,$$

where α, β, γ are arbitrary constants; and it is very easy to prove that the co-ordinates of the centre of gravity of the volume are equal to $a\alpha$, $a\beta$, $a\gamma$ respectively, the line from which θ is measured being taken for the axis of z, and the plane from which ϕ is measured for the plane of xz. Hence the term u_1 in (8) may be made to disappear by taking for origin the centre of gravity of the volume. It is allowable to do this even should the centre of gravity fall a little out of the axis of rotation, because the term involving the centrifugal force, being already a small quantity of the first order, would not be affected by supposing the origin to be situated a little out of the axis.

Since the variation of gravity from one point of the surface to another is a small quantity of the first order, its expression will remain the same whether the earth be referred to one origin or another nearly coinciding with the centre, and therefore a knowledge of the variation will not inform us what point has been taken for the origin to which the surface has been referred.

7. Since the angle between the vertical at any point and the radius vector drawn from the origin is a small quantity of the first order, and the angles θ, ϕ occur in the small terms only of equations (8), (10), and (12), these angles may be taken to refer to the direction of the vertical, instead of the radius vector.

8. If E be the mass of the earth, the potential of its attraction at a very great distance r is ultimately equal to E/r. Comparing this with (10), we get $Y_0 = E$, and therefore, from the first of equations (11),

$$E = Ga^2 + \tfrac{2}{3}\,\omega^2 a^3 = Ga^2\,(1 + \tfrac{2}{3}\,m) \ldots\ldots\ldots\ldots\ldots(13),$$

which determines the mass of the earth from the value of G determined by pendulum experiments.

9. If we suppose that the surface of the earth may be represented with sufficient accuracy by an oblate spheroid of small ellipticity, having its axis of figure coincident with the axis of rotation, equation (8) becomes

$$r = a\,\{1 + \epsilon\,(\tfrac{1}{3} - \cos^2 \theta)\}\ldots\ldots\ldots\ldots\ldots\ldots\ldots(14),$$

where ϵ is a constant which may be considered equal to the ellipticity. We have therefore in this case $u_1 = 0$, $u_2 = \frac{1}{3} - \cos^2\theta$, $u_n = 0$ when $n > 2$; so that (12) becomes

$$g = G\left\{1 - \left(\tfrac{5}{2}m - \epsilon\right)\left(\tfrac{1}{3} - \cos^2\theta\right)\right\} \quad\ldots\ldots\ldots\ldots\ldots(15),$$

which equation contains Clairaut's Theorem. It appears also from this equation that the value of G which must be employed in (13) is equal to gravity at a place the square of the sine of whose latitude is $\frac{1}{3}$.

10. Retaining the same supposition as to the form of the surface, we get from (10), on replacing Y_0 by E, and putting in the small term at the end $\omega^2 a^5 = m\,Ga^4 = mEa^2$,

$$V = \frac{E}{r} + \left(\epsilon - \tfrac{1}{2}m\right)\frac{Ea^2}{r^3}\left(\tfrac{1}{3} - \cos^2\theta\right)\ldots\ldots\ldots\ldots(16).$$

Consider now the effect of the earth's attraction on the moon. The attraction of any particle of the earth on the moon, and therefore the resultant attraction of the whole earth, will be very nearly the same as if the moon were collected at her centre. Let therefore r be the distance of the centre of the moon from that of the earth, θ the moon's North Polar Distance, P the accelerating force of the earth on the moon resolved along the radius vector, Q the force perpendicular to the radius vector, which acts evidently in a plane passing through the earth's axis; then

$$P = -\frac{dV}{dr}, \quad Q = \frac{dV}{rd\theta},$$

whence we get from (16)

$$P = \frac{E}{r^2} + 3\left(\epsilon - \tfrac{1}{2}m\right)\frac{Ea^2}{r^4}\left(\tfrac{1}{3} - \cos^2\theta\right),$$

$$Q = 2\left(\epsilon - \tfrac{1}{2}m\right)\frac{Ea^2}{r^4}\sin\theta\cos\theta \ldots\ldots\ldots\ldots\ldots(17).$$

The moving forces arising from the attraction of the earth on the moon will be obtained by multiplying by M, where M denotes the mass of the moon; and these are equal and opposite to the moving forces arising from the attraction of the moon on the earth. The component MQ of the whole moving force is equivalent to an equal and parallel force acting at the centre of the earth and a couple. The accelerating forces acting on the earth will be obtained by dividing by E; and since we only want to determine the relative motions of the moon and earth, we may conceive equal and opposite accelerating forces applied both to the earth and to the moon, which comes to the same thing as replacing E by $E + M$ in (17). If K be the moment of the couple arising from the attraction of the moon, which tends to turn the earth about an equatorial axis, $K = MQr$, whence

$$K = 2 \left(\epsilon - \tfrac{1}{2}m\right) \frac{MEa^2}{r^3} \sin\theta \cos\theta \dots\dots\dots(18).$$

The same formula will of course apply, *mutatis mutandis*, to the attraction of the sun.

11. The spheroidal form of the earth's surface, and the circumstance of its being a surface of equilibrium, will afford us some information respecting the distribution of matter in the interior. Denoting by x', y', z' the co-ordinates of an internal particle whose density is ρ', and by x, y, z those of the external point of space to which V refers, we have

$$V = \iiint \frac{\rho'\, dx'\, dy'\, dz'}{\{(x-x')^2 + (y-y')^2 + (z-z')^2\}^{\frac{1}{2}}},$$

the integrals extending throughout the interior of the earth. Writing dm' for $\rho'\, dx'\, dy'\, dz'$, putting λ, μ, ν for the direction-cosines of the radius vector drawn to the point (x, y, z), so that $x = \lambda r$, $y = \mu r$, $z = \nu r$, and expanding the radical according to inverse powers of r, we get

$$V = \frac{1}{r}\iiint dm' + \Sigma \frac{\lambda}{r^2}\iiint x'\, dm' + \frac{1}{2r^3}\Sigma (3\lambda^2 - 1)\iiint x'^2\, dm'$$

$$+ \frac{3}{r^3}\Sigma\lambda\mu \iiint x'\, y'\, dm' + \dots\dots(19),$$

Σ denoting the sum of the three expressions necessary to form a symmetrical function. Comparing this expression for V with that given by (10), which in the present case reduces itself to (16), we get $Y_0 = \iiint dm' = E$, as before remarked, and

$$\iiint x'\, dm' = 0, \quad \iiint y'\, dm' = 0, \quad \iiint z'\, dm' = 0 \dots\dots\dots(20),$$

$$\tfrac{1}{2}\Sigma (3\lambda^2 - 1)\iiint x'^2\, dm' + 3\Sigma\lambda\mu \iiint x'\, y'\, dm'$$
$$= (\epsilon - \tfrac{1}{2}m)\, Ea^2 (\tfrac{1}{3} - \cos^2\theta) \dots\dots(21);$$

together with other equations, not written down, obtained by equating to zero the coefficients of $1/r^4$, $1/r^5$ &c. in (19).

Equations (20) shew that the centre of gravity of the mass coincides with the centre of gravity of the volume. In treating equation (21), it is to be remarked that λ, μ, ν are not independent, but connected by the equation $\lambda^2 + \mu^2 + \nu^2 = 1$. If now we insert $\lambda^2 + \mu^2 + \nu^2$ as a coefficient in each term of (21) which does not contain λ, μ, or ν, the equation will become homogeneous with respect to λ, μ, ν, and will therefore only involve the two independent ratios which exist between these three quantities, and consequently we shall have to equate to zero the coefficients of corresponding powers of λ, μ, ν. By the transformation just mentioned, equation (21) becomes, since $\cos\theta = \nu$,

$$\Sigma \left(\lambda^2 - \tfrac{1}{2}\mu^2 - \tfrac{1}{2}\nu^2\right) \iiint x'^2\, dm' + 3\Sigma\lambda\mu \iiint x'y'\, dm'$$
$$= (\epsilon - \tfrac{1}{2}m)\, Ea^2 \left(\tfrac{1}{3}\lambda^2 + \tfrac{1}{3}\mu^2 - \tfrac{2}{3}\nu^2\right);$$

and we get

$$\iiint x'y'\, dm' = 0, \quad \iiint y'z'\, dm' = 0, \quad \iiint z'x'\, dm' = 0 \ldots\ldots(22),$$

$$\left.\begin{aligned}
\iiint x'^2 dm' &- \tfrac{1}{2} \iiint y'^2 dm' - \tfrac{1}{2} \iiint z'^2 dm' \\
&= \iiint y'^2 dm' - \tfrac{1}{2} \iiint z'^2 dm' - \tfrac{1}{2} \iiint x'^2 dm' \\
&= -\tfrac{1}{2} \iiint z'^2 dm' + \tfrac{1}{4} \iiint x'^2 dm' + \tfrac{1}{4} \iiint y'^2 dm' \\
&= \tfrac{1}{3} (\epsilon - \tfrac{1}{2}m)\, Ea^2
\end{aligned}\right\} \ldots\ldots(23).$$

Equations (22) shew that the co-ordinate axes are principal axes. Equations (23) give in the first place

$$\iiint x'^2 dm' = \iiint y'^2 dm',$$

which shews that the moments of inertia about the axes of x and y are equal to each other, as might have been seen at once from (22), since the principal axes of x and y are any two rectangular axes in the plane of the equator. The two remaining equations of the system (23) reduce themselves to one, which is

$$\iiint x'^2 dm' - \iiint z'^2 dm' = \tfrac{2}{3}(\epsilon - \tfrac{1}{2}m)\, Ea^2.$$

If we denote the principal moments of inertia by A, A, C, this equation becomes

$$C - A = \tfrac{2}{3}(\epsilon - \tfrac{1}{2}m)\, Ea^2 \ldots\ldots\ldots\ldots\ldots (24),$$

which reconciles the expression for the couple K given by (18) with the expression usually given, which involves moments of inertia, and which, like (18), is independent of any hypothesis as to the distribution of the matter within the earth.

It should be observed that in case the earth be not solid to the centre, the quantities A, C must be taken to mean what would be the moments of inertia if the several particles of which the earth is composed were rigidly connected.

12. In the preceding article the surface has been supposed spheroidal. In the general case of an arbitrary form we should have to compare the expressions for V given by (10) and (19). In the first place it may be observed that the term u_1 can always be got rid of by taking for origin the centre of gravity of the volume. Equations (20) shew that in the general case, as well as in the particular case considered in the last article, the centre of gravity of the mass coincides with the centre of gravity of the volume.

Now suppress the term u_1 in u, and let $u = u' + u''$, where $u'' = \frac{1}{2}m\left(\frac{1}{3} - \cos^2\theta\right)$. Then u' may be expanded in a series of Laplace's functions $u'_2 + u'_3 + \dots$; and since $Y_0 = E$, equation (10) will be reduced to

$$V = E\left(\frac{1}{r} + \frac{a^2}{r^3}u'_2 + \frac{a^3}{r^4}u'_3 \dots\right) \quad \dots\dots\dots\dots\dots (25).$$

If the mass were collected at the centre of gravity, the second member of this equation would be reduced to its first term, which requires that $u'_2 = 0$, $u'_3 = 0$, &c. Hence (8) would be reduced to $r = a(1 + u'')$, and therefore au'' is the alteration of the surface due to the centrifugal force, and au' the alteration due to the difference between the actual attraction and the attraction of a sphere composed of spherical strata. Consider at present only the term u'_2 of u'. From the general form of Laplace's functions it follows that au'_2 is the excess of the radius vector of an ellipsoid not much differing from a sphere over that of a sphere having a radius equal to the mean radius of the ellipsoid. If we take the principal axes of this ellipsoid for the axes of co-ordinates, we shall have

$$u'_2 = \epsilon'\left(\tfrac{1}{3} - \sin^2\theta\cos^2\phi\right) + \epsilon''\left(\tfrac{1}{3} - \sin^2\theta\sin^2\phi\right) + \epsilon'''\left(\tfrac{1}{3} - \cos^2\theta\right),$$

ϵ', ϵ'', ϵ''' being three arbitrary constants, and θ, ϕ denoting angles related to the new axes of x, y, z in the same way that the angles before denoted by θ, ϕ were related to the old axes. Substituting the preceding expression for u'_2 in (25), and comparing the result with (19), we shall again obtain equations (22).

Consequently the principal axes of the mass passing through the centre of gravity coincide with the principal axes of the ellipsoid. It will be found that the three equations which replace (23) are equivalent to but two, which are

$$A - \tfrac{2}{3}\epsilon' E a^2 = B - \tfrac{2}{3}\epsilon'' E a^2 = C - \tfrac{2}{3}\epsilon''' E a^2,$$

where A, B, C denote the principal moments.

The permanence of the earth's axis of rotation shews however that one of the principal axes of the ellipsoid coincides, at least very nearly, with the axis of rotation; although, strictly speaking, this conclusion cannot be drawn without further consideration except on the supposition that the earth is solid to the centre. If we assume this coincidence, the term $\epsilon''' (\tfrac{1}{3} - \cos^2 \theta)$ will unite with the term u'' due to the centrifugal force. Thus the most general value of u is that which belongs to an ellipsoid having one of its principal axes coincident with the axis of rotation, added to a quantity which, if expanded in a series of Laplace's functions, would furnish no terms of the order 0, 1, or 2.

It appears from this and the preceding article that the coincidence of the centres of gravity of the mass and volume, and that of the axis of rotation and one of the principal axes of the ellipsoid whose equation is $r = a (1 + u_2)$, which was established by Laplace on the supposition that the earth consists of nearly spherical strata of equal density, holds good whatever be the distribution of matter in the interior.

13. Hitherto the surface of the earth has been regarded as a surface of equilibrium. This we know is not strictly true, on account of the elevation of the land above the level of the sea. The question now arises, By what imaginary alteration shall we reduce the surface to one of equilibrium ?

Now with respect to the greater portion of the earth's surface, which is covered with water, we have a surface of equilibrium ready formed. The expression *level of the sea* has a perfectly definite meaning as applied to a place in the middle of a continent, if it be defined to mean the level at which the sea-water would stand if introduced by a canal. The surface of the sea, supposed to be prolonged in the manner just considered, forms indeed a surface of equilibrium, but the preceding investigation does not apply directly to this surface, inasmuch as a portion of the attracting matter lies outside it. Conceive however the land which lies above the level of the sea to be depressed till it gets below it, or, which is the same, conceive the land cut off at the level of the sea produced, and suppose the density of the earth or rock which lies immediately below the sea-level to be increased, till the increase of mass immediately below each superficial element is equal to the mass which has been removed from above it. The whole of the attracting matter will thus be brought inside the original sea-level; and it is easy to see that the attraction at a point of space external to the earth, even though it be close to the surface, will not be sensibly affected. Neither will the sea-level be sensibly changed, even in the middle of a continent. For, suppose the sea-water introduced by a pipe, and conceive the land lying above the sea-level condensed into an infinitely thin layer coinciding with the sea-level. The attraction of an infinite plane on an external particle does not depend on the distance of the particle from the plane; and if a line be drawn through the particle inclined at an angle α to the perpendicular let fall on the plane, and be then made to revolve around the perpendicular, the resultant attraction of the portion of the plane contained within the cone thus formed will be to that of the whole plane as versin α to 1. Hence the attraction of a piece of table-land on a particle close to it will be sensibly the same as that of a solid of equal thickness and density comprised between two parallel infinite planes, and that, even though the lateral extent of the table-land be inconsiderable, only equal, suppose, to a small multiple of the length of a perpendicular let fall from the attracted particle on the further bounding plane. Hence the attraction of the land on the water in the tube will not be sensibly altered by the condensation we have supposed, and therefore we are fully justified in regarding the level of the sea as unchanged.

The surface of equilibrium which by the imaginary displacement of matter just considered has also become the bounding surface, is that surface which at the same time coincides with the surface of the actual sea, where the earth is covered by water, and belongs to the system of surfaces of equilibrium which lie wholly outside the earth. To reduce observed gravity to what would have been observed just above this imaginary surface, we must evidently increase it in the inverse ratio of the square of the distance from the centre of the earth, without taking account of the attraction of the table-land which lies between the level of the station and the level of the sea. The question now arises, How shall we best determine the numerical value of the earth's ellipticity, and how best compare the form which results from observation with the spheroid which results from theory on the hypothesis of original fluidity?

14. Before we consider how the numerical value of the earth's ellipticity is to be determined, it is absolutely necessary that we define what we mean by ellipticity; for, when the irregularities of the surface are taken into account, the term must be to a certain extent conventional.

Now the attraction of the earth on an external body, such as the moon, is determined by the function V, which is given by (10). In this equation, the term containing r^{-2} will disappear if r be measured from the centre of gravity; the term containing r^{-4}, and the succeeding terms, will be insensible in the case of the moon, or a more distant body. The only terms, therefore, after the first, which need be considered, are those which contain r^{-3}. Now the most general value of u_2 contains five terms, multiplied by as many arbitrary constants, and of these terms one is $\frac{1}{3} - \cos^2 \theta$, and the others contain as a factor the sine or cosine of ϕ or of 2ϕ. The terms containing $\sin \phi$ or $\cos \phi$ will disappear for the reason mentioned in Art. 12; but even if they did not disappear their effect would be wholly insensible, inasmuch as the corresponding forces go through their period in a day, a lunar day if the moon be the body considered. These terms therefore, even if they existed, need not be considered; and for the same reason the terms containing $\sin 2\phi$ or $\cos 2\phi$ may be neglected; so that nothing remains but a term which unites with the last term in equation (10). Let ϵ be the coefficient of the term $\frac{1}{3} - \cos^2 \theta$ in the expansion of u: then ϵ is the constant which determines the effect of the earth's oblateness on the motion of the moon, and which enters into the expression for the moment of the attractions of the sun and moon on the earth; and in the particular case in which the earth's surface is an oblate spheroid, having its axis coincident with the axis of rotation, ϵ is the ellipticity. Hence the constant ϵ seems of sufficient dignity to deserve a name, and it may be called in any case the *ellipticity*.

Let r be the radius vector of the earth's surface, regarded as coincident with the level of the sea; and take for shortness $\mathfrak{m}\{f(\theta,\phi)\}$ to denote the mean value of the function $f(\theta,\phi)$ throughout all angular space, or

$$\frac{1}{4\pi} \int_0^\pi \int_0^{2\pi} f(\theta,\phi) \sin\theta d\theta d\phi.$$

Then it follows from the theory of Laplace's functions that

$$\epsilon = \frac{45}{4a} \mathfrak{m} \{(\tfrac{1}{3} - \sin^2 l)\, r\} \quad\ldots\ldots\ldots\ldots\ldots\ldots\ldots(26),$$

l being the latitude, or the complement of θ. To obtain this equation it is sufficient to multiply both sides of (8) by $1/4\pi \times (\tfrac{1}{3} - \cos^2\theta) \sin\theta d\theta d\phi$, and to integrate from $\theta = 0$ to $\theta = \pi$, and from $\phi = 0$ to $\phi = 2\pi$. Since $\tfrac{1}{3} - \cos^2\theta$ is a Laplace's function of the second order, none of the terms at the second side of (8) will furnish any result except u_2, and even in the case of u_2 the terms involving the sine or cosine of ϕ or of 2ϕ will disappear.

15. Let g be gravity reduced to the level of the sea by taking account only of the height of the station. Then this is the quantity to which equation (12) is applicable; and putting for u_2 its value we get by means of the properties of Laplace's functions

$$G = \mathfrak{m}(g), \quad G(\tfrac{5}{2} m - \epsilon) = -\tfrac{45}{4} \mathfrak{m} \{(\tfrac{1}{3} - \sin^2 l)\, g\} \ldots\ldots(27).$$

If we were possessed of the values of g at an immense number of stations scattered over the surface of the whole earth, we might by combining the results of observation in the manner indicated by equations (27) obtain the numerical values of G and ϵ. We cannot, however, obtain by observation the values of g at the surface of the sea, and the stations on land where the observations have been made from which the results are to be obtained are not very numerous. We must consider therefore in what way the variations of gravity due to merely local causes are to be got rid of, when we know the causes of disturbance; for otherwise a local irregularity, which would be lost in the mean of an immense number of observations, would acquire undue importance in the result.

16. Now the most obvious cause of irregularity consists in the attraction of the land lying between the level of the station and the level of the sea. This attraction would render the values of g

sensibly different, which would be obtained at two stations only a
mile or two apart, but situated at different elevations. To render
our observations comparable with one another, it seems best to
correct for the attraction of the land which lies underneath the
pendulum; but then we must consider whether the habitual
neglect of this attraction may not affect the mean values from
which G and ϵ are to be found.

Let $g = g_1 + g'$, where g' is the attraction just mentioned, so
that g_1 is the result obtained by reducing the observed value of
gravity to the level of the sea by means of Dr Young's formula*.
Let h be the height of the station above the level of the sea, σ the
superficial density of the earth where not covered by water; then
by the formula for the attraction of an infinite plane we have
$g' = 2\pi\sigma h$. To make an observation, conceived to be taken at the
surface of the sea, comparable with one taken on land, the correction for local attraction would be additive, instead of subtractive;
we should have in fact to add the excess of the attraction of a
layer of earth or rock, of a thickness equal to the depth of the sea
at that place, over the attraction of so much water. The formula
$g' = 2\pi\sigma h$ will evidently apply to the surface of the sea, provided
we regard h as a negative quantity, equal to the depth of the sea,
and replace σ by $\sigma - 1$, the density of water being taken for the
unit of density; or we may retain σ as the coefficient, and diminish
the depth in the ratio of σ to $\sigma - 1$.

Let ρ be the mean density of the earth, then

$$g' = 2\pi\sigma h = G\,\frac{2\pi\sigma h}{\frac{4}{3}\pi\rho a} = G\,\frac{3\sigma h}{2\rho a}\,.$$

If we suppose $\sigma = 2\frac{1}{2}$, $\rho = 5\frac{1}{2}$, $a = 4000$ miles, and suppose h
expressed in miles, with the understanding that in the case of the
sea h is a negative quantity equal to $\frac{3}{5}$ths of the actual depth, we
have $g' = \cdot00017\,Gh$ nearly.

---

* *Phil. Trans.* for 1819. Dr Young's formula is based on the principle of taking
into account the attraction of the table-land existing between the station and the
level of the sea, in reducing the observation to the sea level. On account of this
attraction, the multiplier $2h/a$ which gives the correction for elevation alone must
be reduced in the ratio of 1 to $1 - 3\sigma/4\rho$, or 1 to $\cdot66$ nearly, if $\sigma = 2\frac{1}{2}$, $\rho = 5\frac{1}{4}$. Mr
Airy, observing that the value $\sigma = 2\frac{1}{2}$ is a little too small, and $\rho = 5\frac{1}{2}$ a little too
great, has employed the factor $\cdot6$, instead of $\cdot66$.

17. Consider first the value of G. We have by the preceding formula, and the first of equations (27),

$$G = \mathfrak{m}\,(g_1) + G \times \cdot 00017\,\mathfrak{m}\,(h).$$

According to Professor Rigaud's determination, the quantity of land on the surface of the earth is to that of water as 100 to 276[*]. If we suppose the mean elevation of the land $\frac{1}{5}$th of a mile, and the mean depth of the sea $3\frac{1}{2}$ miles, we shall have

$$\mathfrak{m}\,(h) = -\frac{\frac{3}{5} \times 3\frac{1}{2} \times 276 - \frac{1}{5} \times 100}{376} = -1\cdot49 \text{ nearly};$$

so that the value of G determined by g_1 would be too great by about $\cdot000253$ of the whole. Hence the mass of the earth determined by the pendulum would be too great by about the one four-thousandth of the whole; and therefore the mass of the moon, obtained by subtracting from the sum of the masses of the earth and moon, as determined by means of the coefficient of lunar parallax, the mass of the earth alone, as determined by means of the pendulum, would be too small by about the one four-thousandth of the mass of the earth, or about the one fiftieth of the whole.

18. Consider next the value of ϵ. Let ϵ_1 be the value which would be determined by substituting g_1 for g in (27), and let

$$\tfrac{4}{4}\mathfrak{m}\,\{(\sin^2 l - \tfrac{1}{3})g'\} = Gq.$$

In considering the value of q we may attend only to the land, provided we transfer the defect of density of the sea with an opposite sign to the land, because if g' were constant, q would vanish. This of course proceeds on the supposition that the depth of the sea is constant. Since $\epsilon = \epsilon_1 - q$, if q were positive, the ellipticity determined by the pendulum would appear too great in consequence of the omission of the force g'. I have made a sort of rough integration by means of a map of the world, by counting the quadrilaterals of land bounded each by two meridians distant 10°, and by two parallels of latitude distant 10°, estimating the fraction of a broken quadrilateral which was partly occupied by sea. The number of quadrilaterals of land between two consecutive parallels, as for example 50° and 60°, was multiplied by $12\,(\frac{1}{3} - \sin^2 l)\cos l$, or $3\cos 3l + \cos l$, where for l was taken the mean latitude, (55° in the example,) the sum of the results was taken for the whole surface,

---

[*] *Cambridge Philosophical Transactions*, Vol. VI. p. 297.

and multiplied by the proper coefficient. The north pole was supposed to be surrounded by water, and the south pole by land, as far as latitude 80°. It appeared that the land lying beyond the parallels for which $\sin^2 l = \frac{1}{3}$, that is, beyond the parallels 35° N. and 35° S. nearly, was almost exactly neutralized by that which lay within those parallels. On the whole, q appeared to have a very small positive value, which on the same suppositions as before respecting the height of the land and the depth of the sea, was ·0000012. It appears, therefore, that the omission of the force g' will produce no sensible increase in the value of ϵ, unless the land be on the whole higher, or the sea shallower, in high latitudes than in low. If the land had been collected in a great circular continent around one pole, the value of q would have been ·000268; if it had been collected in a belt about the equator, we should have had $q = -\cdot000362$. The difference between these values of q is about one fifth of the whole ellipticity.

19. The attraction g' is not the only irregularity in the magnitude of the force of gravity which arises from the irregularity in the distribution of land and sea, and in the height of the land and depth of the sea, although it is the only irregularity, arising from that cause, which is liable to vary suddenly from one point at the surface to another not far off. The irregular coating of the earth will produce an irregular attraction besides that produced by the part of this coating which lies under and in the immediate neighbourhood of the station considered, and it will moreover cause an irregular elevation or depression in the level of the sea, and thereby cause a diminution or increase in the value of g_1.

Consider the attraction arising from the land which lies above the level of the sea, and from the defect of attracting matter in the sea. Call this excess or defect of matter the *coating* of the earth : conceive the coating condensed into a surface coinciding with the level of the sea, and let $A\delta$ be the mass contained in a small element A of this surface. Then $\delta = \sigma h$ in the case of the land, and $\delta = -(\sigma - 1) h$ in the case of the sea, h being in that case the depth of the sea. Let V_e be the potential of the coating, V', V'' the values of V_e outside and inside the surface respectively. Conceive δ expanded in a series of Laplace's functions $\delta_0 + \delta_1 + ...$, then it is easily proved that

$$V' = 4\pi a^2 \left(\frac{1}{r} \delta_0 + \frac{a}{3r^2} \delta_1 + \frac{a^2}{5r^3} \delta_2 + \ldots \right)$$

$$V'' = 4\pi a^2 \left(\frac{1}{a} \delta_0 + \frac{r}{3a^2} \delta_1 + \ldots \right)$$

$$\left. \right\} \ldots\ldots\ldots(28),$$

r being the distance of the point considered from the centre. These equations give

$$\frac{dV'}{dr} = - 4\pi \Sigma \frac{i+1}{2i+1} \left(\frac{a}{r} \right)^{i+2} \delta_i$$

$$\frac{dV''}{dr} = 4\pi \Sigma \frac{i}{2i+1} \left(\frac{r}{a} \right)^{i-1} \delta_i$$

$$\left. \right\} \ldots\ldots\ldots(29).$$

Consider two points, one external, and the other internal, situated along the same radius vector very close to the surface. Let E be an element of this surface lying around the radius vector, an element which for clear ideas we may suppose to be a small circle of radius s, and let s be at the same time infinitely small compared with a, and infinitely great compared with the distance between the points. Then the limiting values of dV'/dr and dV''/dr will differ by the attraction of the element E, an attraction which, as follows from what was observed in Art. 13, will be ultimately the same as that of an infinite plane of the same density, or $2\pi\delta$*. The mean of the values of dV'/dr and dV''/dr will express the attraction of the general coating in the direction of the radius vector, the general coating being understood to mean the whole coating, with the exception of a superficial element lying adjacent to the points where the attraction is considered. Denoting this mean by dV_c/dr, we get, on putting $r = a$,

$$\frac{dV_c}{dr} = - 2\pi \Sigma \frac{\delta_i}{2i+1}.$$

This equation becomes by virtue of either of the equations (28)

$$\frac{dV_c}{dr} = - \frac{V_c}{2a} \ldots\ldots\ldots\ldots\ldots\ldots(30),$$

---

* This result readily follows from equations (29), which give, on putting $r = a$,

$$\frac{dV''}{dr} - \frac{dV'}{dr} = 4\pi \Sigma \delta_i = 4\pi \delta.$$

This difference of attraction at points infinitely close can evidently only arise from the attraction of the interposed element of surface, which, being ultimately plane, will act equally at both points; and, therefore, the attraction will be in each case $2\pi\delta$, and will act outwards in the first case, and inwards in the second.

which is a known equation. Let either member of this equation
be denoted by $-g''$. Then gravity will be increased by g'', in
consequence of the attraction of the general coating.

20.   But besides its direct effect, the attraction of the coating
will produce an indirect effect by altering the sea-level. Since the
potential at any place is increased by V_c in consequence of the
coating, in passing from what would be a surface of equilibrium if
the coating were removed, to the actual surface of equilibrium
corresponding to the same parameter, {that is, the same value of
the constant c in equation (1),} we must ascend till the labouring
force expended in raising a unit of mass is equal to V_c, that is, we
must ascend through a space V_c/g, or V_c/G nearly. In consequence
of this ascent, gravity will be diminished by the quantity corresponding to the height V_c/G, or h' suppose. If we take account
only of the alteration of the distance from the centre of the earth,
this diminution will be equal to $G . 2h'/a$, or $2V_c/a$, or $4g''$, and
therefore the combined direct and indirect effects of the general
coating will be to diminish gravity by $3g''$.

But the attraction of that portion of the stratum whose thickness is h', which lies immediately about the station considered,
will be a quantity which involves h' as a factor, and to include this
attraction we must correct for the change of distance h' by Dr
Young's rule, instead of correcting merely according to the square
of the distance. In this way we shall get for the diminution of
gravity due to the general coating, not $3g''$, but only $4 (1 - 3\sigma/4\rho)$
$g'' - g''$, or kg'' suppose. If $\sigma : \rho :: 5 : 11$, we have $k = 16\cdot4$
nearly.

If we cared to leave the mean value of gravity unaltered, we
should have to use, instead of δ, its excess over its mean value δ_0.
In considering however, only the variation of gravity from one place
to another, this is a point of no consequence.

21.   In order to estimate the magnitude which the quantity
$3g''$ is likely to attain, conceive two stations, of which the first is
surrounded by land, and the second by sea, to the distance of 1000
miles, the distribution of land and sea beyond that distance being
on the average the same at the two stations. Then, by hypothesis,
the potential due to the land and sea at a distance greater than

1000 miles is the same at the two stations; and as we only care for the difference between the values of the potential of the earth's coating at the two stations, we may transfer the potential due to the defect of density at the second station with an opposite sign to the first station. We shall thus have around the first station, taking h' for the depth of the sea around the second station, $\delta = \sigma h + (\sigma - 1) h'$. In finding the difference V of the potentials of the coating, it will be amply sufficient to regard the attracting matter as spread over a plane disk, with a radius s equal to 1000 miles. On this supposition we get

$$V = \int_0^s s^{-1} \cdot 2\pi \delta s \, ds = 2\pi \delta s.$$

Now $G = \frac{4}{3} \pi \rho a$, and therefore

$$3g'' = \frac{3V}{2a} = \frac{9 \delta s}{4 \rho a^2} \, G = \frac{9}{4} \cdot \frac{\sigma h + (\sigma - 1) h'}{\rho a} \cdot \frac{s}{a} \, G.$$

Making the same suppositions as before with regard to the numerical values of σ, ρ, h, h', and a, we get $3g'' = \cdot 000147 \, G$. This corresponds to a difference of $6\cdot35$ vibrations a day in a seconds' pendulum. Now a circle with a radius of 1000 miles looks but small on a map of the world, so that we may readily conceive that the difference depending on this cause between the number of vibrations observed at two stations might amount to 15 or 20, that is 7·5 or 10 on each side of the mean, or even more if the height of the land or the depth of the sea be under-estimated. This difference will however be much reduced by using kg'' in place of $3g''$*.

22. The value of V_c at any station is expressed by a double integral, which is known if δ be known, and which may be calculated numerically with sufficient accuracy by dividing the surface into small portions and performing a summation. Theoretically speaking, V_c could be expressed for the whole surface at once by means of a series of Laplace's functions; the constants in this series could be determined by integration, or at least the approximate integration obtained by summation, and then the value of V_c could be obtained by substituting in the series the

---

* The effect of the irregularity of the earth's surface is greater than what is represented by kg'', for a reason which will be explained further on (Art. 25).

latitude and longitude of the given station for the general latitude
and longitude. But the number of terms which would have to be
retained in order to represent with tolerable accuracy the actual
state of the earth's surface would be so great that the method, I
apprehend, would be practically useless; although the leading
terms of the series would represent the effect of the actual
distribution of land and sea in its broad features. It seems
better to form directly the expression for V_c at any station. This
expression may be calculated numerically for each station by
using the value of δ most likely to be correct, if the result be
thought worth the trouble; but even if it be not calculated
numerically, it will enable us to form a good estimate of the
variation of the quantity $3g''$ or kg'' from one place to another.

Let the surface be referred to polar co-ordinates originating at
the centre, and let the angles ψ, χ be with reference to the station
considered what θ, ϕ were with reference to the north pole. The
mass of a superficial element is equal to $\delta a^2 \sin \psi d\psi d\chi$, and its
distance from the station is $2a \sin \frac{1}{2}\psi$. Hence we have

$$V_c = a \iint \delta \cos \tfrac{1}{2}\psi \; d\psi d\chi. \;\; \ldots\ldots (31)$$

Let δ_m be the mean value of δ throughout a circle with an
angular radius ψ, then the part of V_c which is due to an annulus
having a given infinitely small angular breadth $d\psi$ is proportional
to $\delta_m \cos \frac{1}{2}\psi$, or to δ_m nearly when ψ is not large. If we regard
the depth of the sea as uniform, we may suppose $\delta = 0$ for the
sea, and transfer the defect of density of the sea with an opposite
sign to the land. We have seen that if we set a circle of land
$\frac{1}{5}$ mile high of 1000 miles radius surrounding one station against
a circle of sea $3\frac{1}{2}$ miles deep, and of the same radius, surround-
ing another, we get a difference of about $\frac{1}{3} \times 1\cdot 64 \times 6\cdot 35$, or $3\frac{1}{2}$
nearly, in the number of vibrations performed in one day by a
seconds' pendulum. It is hardly necessary to remark that high
table-land will produce considerably more effect than land only
just raised above the level of the sea, but it should be observed
that the principal part of the correction is due to the depth of the
sea. Thus it would require a uniform elevation of about $2\cdot 1$
miles, in order that the land elevated above the level of the sea
should produce as much effect as is produced by the difference
between a stratum of land $3\frac{1}{2}$ miles thick and an equal stratum of
water.

23. These considerations seem sufficient to account, at least in a great measure, for the apparent anomalies which Mr Airy has noticed in his discussion of pendulum experiments*. The first table at p. 230 contains a comparison between the observations which Mr Airy considers first-rate and theory. The column headed "Error in Vibrations" gives the number of vibrations *per diem* in a seconds' pendulum corresponding to the excess of observed gravity over calculated gravity. With respect to the errors Mr Airy expressly remarks "upon scrutinizing the errors of the first-rate observations, it would seem that, *cæteris paribus*, gravity is greater on islands than on continents." This circumstance appears to be fully accounted for by the preceding theory. The greatest positive errors appear to belong to oceanic stations, which is just what might be expected. Thus the only errors with the sign + which amount to 5 are, Isle of France + 7·0; Marian Islands + 6·8; Sandwich Islands + 5·2; Pulo Gaunsah Lout (a small island near New Guinea and almost on the equator), + 5·0. The largest negative errors are, California − 6·0; Maranham − 5·6; Trinidad − 5·2. These stations are to be regarded as continental, because generally speaking the stations which are the most continental in character are but on the coasts of continents, and Trinidad may be regarded as a coast station. That the negative errors just quoted are larger than those that stand opposite to more truly continental stations such as Clermont, Milan, &c. is no objection, because the errors in such different latitudes cannot be compared except on the supposition that the value of the ellipticity used in the comparison is correct.

Now if we divide the 49 stations compared into two groups, an equatorial group containing the stations lying between latitudes 35° N. and 35° S., and a polar group containing the rest, it will be found that most if not all of the oceanic stations are contained in the former group, while the stations belonging to the latter are of a more continental character. Hence the observations will make gravity appear too great about the equator and too small towards the poles, that is, they will on the whole make gravity vary too little from the equator to the poles; and since the variation depends upon $\frac{5}{2}m - \epsilon$, the observations will be best satisfied by a value of ϵ which is too great. This is in fact pre-

---

* *Encyclopædia Metropolitana.* Art. Figure of the Earth.

cisely the result of the discussion, the value of ϵ which Mr Airy
has obtained from the pendulum experiments ('003535) being
greater than that which resulted from the discussion of geodetic
measures ('003352), or than any of the values ('003370, '003360,
and '003407), obtained from the two lunar inequalities which
depend upon the earth's oblateness.

Mr Airy has remarked that in the high north latitudes the
greater number of errors have the sign +, and that those about
the latitude 45° have the sign −; those about the equator being
nearly balanced. To destroy the errors in high and mean latitudes
without altering the others, he has proposed to add a term
$- A \sin^2\lambda \cos^2\lambda$, where λ is the latitude. But a consideration of the
character of the stations seems sufficient, with the aid of the
previous theory, to account for the apparent anomaly. About
latitude 45° the stations are all continental; in fact, ten consecutive stations including this latitude are Paris, Clermont, Milan,
Padua, Fiume, Bordeaux, Figeac, Toulon, Barcelona, New York.
These stations *ought*, as a group, to appear with considerable negative errors. Mr Airy remarks " If we increased the multiplier of
$\sin^2\lambda$," and consequently diminished the ellipticity, " we might
make the errors at high latitudes as nearly balanced as those at
the equator: but then those about latitude 45° would be still
greater than at present."

The largeness of the ellipticity used in the comparison accounts
for the circumstance that the stations California, Maranham,
Trinidad, appear with larger negative errors than any of the
stations about latitude 45°, although some of the latter appear
more truly continental than the former. On the whole it would
seem that the best value of the ellipticity is one which, supposing
it left the errors in high latitudes nearly balanced, would give a
decided preponderance to the negative errors about latitude 45° N.
and a certain preponderance to the positive errors about the
equator, on account of the number of oceanic stations which occur
in low latitudes.

If we follow a chain of stations from the sea inland, or from the
interior to the coast, it is remarkable how the errors decrease
algebraically from the sea inwards. The chain should not extend
over too large a portion of the earth's surface, as otherwise a small
error in the assumed ellipticity might effect the result. Thus for

example, Spitzbergen + 4·3, Hammerfest − 0·4, Drontheim − 2·7. In comparing Hammerfest with Drontheim, we may regard the former as situated at the vertex of a slightly obtuse angle, and the latter as situated at the edge of a straight coast. Again, Dunkirk − 0·1, Paris − 1·9, Clermont − 3·9, Figeac − 3·8, Toulon − 0·1, Barcelona 0·0, Fomentera + 0·2. Again, Padua + 0·7, Milan − 2·8. Again, Jamaica − 0·8, Trinidad − 5·2.

24. Conceive the correction kg'' calculated, and suppose it applied, as well as the correction $-g'$, to observed gravity reduced to the level of the sea, or to g, and let the result be $g_{,,}$ Let $\epsilon_{,,}$ be the ellipticity which would be determined by means of $g_{,,}$, $\epsilon_{,,} + \Delta\epsilon_{,,}$ the true ellipticity. Since $g_{,,} = g - g' + kg''$, and therefore $g = g_{,,} + g' - kg''$, we get by (27)

$$\Delta\epsilon_{,,} = \frac{45}{4G} \mathfrak{m} \left\{ (\tfrac{1}{3} - \sin^2 l)\, (g' - kg'') \right\} \ldots\ldots\ldots\ldots(32).$$

Now $g' = 2\pi\sigma h = 2\pi\delta = 2\pi\Sigma\delta_i$; and we get from (30) and (28)

$$kg'' = -k\frac{dV_e}{dr} = \frac{kV_e}{2a} = 2k\pi\Sigma\, \frac{\delta_i}{2i+1}.$$

All the terms δ_i will disappear from the second side of (32) except δ_2, and we therefore get

$$\Delta\epsilon_{,,} = \frac{45}{4G} \mathfrak{m} \left\{ (\tfrac{1}{3} - \sin^2 l)(1 - \tfrac{1}{5}k)\, 2\pi\delta_2 \right\}.$$

Hence the correction $\Delta\epsilon_{,,}$ is less than that considered in Art. 18, in the ratio of $5 - k$ to 5, and is therefore probably insensible on account of the actual distribution of land and water at the surface of the earth.

25. Conceive the islands and continents cut off at the level of the sea, and the water of the sea replaced by matter having the same density as the land. Suppose gravity to be observed at the surface which would be thus formed, and to be reduced by Dr Young's rule to the level of what would in the altered state of the earth be a surface of equilibrium. It is evident that $g_{,,}$ expresses the gravity which would be thus obtained.

The irregularities of the earth's coating would still not be wholly allowed for, because the surface which would be formed in the manner just explained would no longer be a surface of equilibrium, in consequence of the fresh distribution of attracting matter. The surface would thus preserve traces of its original irregularity. A repetition of the same process would give a surface still more regular, and so on indefinitely. It is easy to see the general nature of the correction which still remains. Where a small island was cut off, there was previously no material elevation of the sea-level, and therefore the surface obtained by cutting off the island will be very nearly a surface of equilibrium, except in so far as that may be prevented by alterations which take place on a large scale. But where a continent is cut off there was a considerable elevation in the sea-level, and therefore the surface which is left will be materially raised above the surface of equilibrium which most nearly represents the earth's surface in its altered state. Hence the general effect of the additional correction will be to increase that part of g'' which is due to causes which act on a larger scale, and to leave nearly unaffected that part which is due to causes which are more local.

The form of the surface of equilibrium which would be finally obtained depends on the new distribution of matter, and conversely, the necessary distribution of matter depends on the form of the final surface. The determination of this surface is however easy by means of Laplace's analysis.

26. Conceive the sea replaced by solid matter, of density σ, having a height from the bottom upwards which is to the depth of the sea as 1 to σ. Let h be the height of the land above the actual sea-level, h being negative in the case of the sea, and equal to the depth of the sea multiplied by $1 - 1/\sigma$. Let x be the unknown thickness of the stratum which must be removed in order to leave the surface a surface of equilibrium, and suppose the mean value of x to be zero, so that on the whole matter is neither added nor taken away. The surface of equilibrium which would be thus obtained is evidently the same as that which would be formed if the elevated portions of the irregular surface were to become fluid and to run down.

Let V be the potential of the whole mass in its first state, V_x the potential of the stratum removed. The removal of this stratum will depress the surface of equilibrium by the space $G^{-1}V_x$; and the condition to be satisfied is, that this new surface of equilibrium, or else a surface of equilibrium belonging to the same system, and therefore derived from the former by further diminishing the radius vector by the small quantity c', shall coincide with the actual surface. We must therefore have

$$G^{-1}V_x + c' = x - h \dots\dots\dots\dots\dots (33).$$

Let h and x be expanded in series of Laplace's functions $h_0 + h_1 + \dots$ and $x_0 + x_1 + \dots$ Then the value of V_x at the surface will be obtained from either of equations (28) by replacing δ by σx and putting $r = a$. We have therefore

$$V_x = 4\pi\sigma a \left(x_0 + \tfrac{1}{3}x_1 + \tfrac{1}{5}x_2 + \dots\right)\dots\dots\dots\dots(34).$$

After substituting in (33) the preceding expressions for V_x, h, and x, we must equate to zero Laplace's functions of the same order. The condition that $x_0 = 0$ may be satisfied by means of the constant c', and we shall have

$$G^{-1} \cdot 4\pi\sigma a \frac{x_i}{2i+1} = x_i - h_i,$$

which gives, on replacing $G^{-1} \cdot 4\pi\sigma a$ by its equivalent $3\sigma/\rho$,

$$x_i = \frac{(2i+1)\rho}{(2i+1)\rho - 3\sigma} h_i = \left\{1 + \frac{3\sigma}{(2i+1)\rho - 3\sigma}\right\} h_i \dots\dots(35).$$

We see that for terms of a high order x_i is very nearly equal to h_i, but for terms of a low order, whereby the distribution of land and sea would be expressed as to its broad features, x_i is sensibly greater than h_i.

27. Let it be required to reduce gravity g to the gravity which would be observed, in the altered state of the surface, along what would then be a surface of equilibrium. Let the correction be denoted by $g' - 3g'''$, where g' is the same as before. The correction due to the alteration of the coating in the manner considered in Art. 20 has been shewn to be equal to

$$2\pi\delta - 6\pi\Sigma \frac{\delta_i}{2i+1},$$

and the required correction will evidently be obtained by replacing δ by σx. Putting for x_i its value got from (35) we have

$$g' - 3g''' = 2\pi\sigma\Sigma \frac{(2i-2)\rho}{(2i+1)\rho - 3\sigma} h_i = 2\pi\sigma\Sigma \left\{1 - \frac{3\rho - 3\sigma}{(2i+1)\rho - 3\sigma}\right\} h_i,$$

which gives, since $2\pi\sigma\Sigma h_i = 2\pi\sigma h = g'$ and $G = \frac{4}{3}\pi\rho a$,

$$3g''' = G\,\frac{3\sigma}{2\rho}\,\Sigma\,\frac{3\rho - 3\sigma}{(2i+1)\,\rho - 3\sigma}\,\frac{h_i}{a} \quad\dots\dots\dots\dots (36).$$

If we put $\sigma = 2\frac{1}{2}$, $\rho = 5\frac{1}{2}$, $a = 4000$, and suppose h expressed in miles, we get

$$3g''' = G\,.\,\frac{15}{88000}\,\Sigma\,\frac{9h_i}{11i-2} = G \times \cdot00017 \times$$

$$(-4.5h_0 + h_1 + .45h_2 + .290h_3 + .214h_4 + \dots)\dots\dots(37).$$

Had we treated the approximate correction $3g''$ in the same manner we should have had

$$3g'' = G\,\frac{3\sigma}{2\rho a}\,\Sigma\,\frac{3h_i}{2i+1} = G \times .00017 \times$$

$$(3h_0 + h_1 + .6h_2 + .429h_3 + .333h_4 + \dots)$$

whereas, since $k = 3\,(1 - \sigma/\rho)$, we get

$$kg'' = G\,\frac{3\sigma}{2\rho a}\,\Sigma\,\frac{(3\rho - 3\sigma)\,h_i}{(2i+1)\,\rho} = G \times .00017 \times$$

$$(1.636h_0 + .545h_1 + .327h_2 + .234h_3 + .182h_4 + \dots)\dots\dots(38).$$

The general expressions for $3g'''$, $3g''$, and kg'' shew that the approximate correction kg'' agrees with the true correction $3g'''$ so far as regards terms of a high order, whereas the leading terms, beginning with the first variable term, are decidedly too small; so that, as far as regards these terms, $3g'''$ is better represented by $3g''$ than by kg''. This agrees with what has been already remarked in Art. 25.

If we put $g - g' + 3g''' = g_{,,,}$, and suppose G and ϵ determined by means of $g_{,,,}$, small corrections similar to those already investigated will have to be applied in consequence of the omission of the quantity $g' - 3g'''$ in the value of g. The correction to ϵ would probably be insensible for the reason mentioned in Art. 18. If we are considering only the variation of gravity, we may of course leave out the term h_0.

The series (37) would probably be too slowly convergent to be of much use. A more convergent series may be obtained by subtracting kg'' from $3g'''$, since the terms of a high order in $3g'''$ are ultimately equal to those in kg''. We thus get

$$3g''' = kg'' + G \times .00017 \times$$

$$(-6.136h_0 + .455h_1 + .123h_2 + .056h_3 + .032h_4 + \dots)\,\dots\,(39),$$

which gives g''' if g'' be known by quadratures for the station considered.

Although for facility of calculation it has been supposed that the sea was first replaced by a stratum of rock or earth of less thickness, and then that the elevated portions of the earth's surface became fluid and ran down, it may be readily seen that it would come to the same thing if we supposed the water to remain as it is, and the land to become fluid and run down, so as to form for the bottom of the sea a surface of equilibrium. The gravity $g_{,,,}$ would apply to the earth so altered.

28. Let us return to the quantity V_e of Art. 19, and consider how the attraction of the earth's irregular coating affects the direction of the vertical. Let l be the latitude of the station, which for the sake of clear ideas may be supposed to be situated in the northern hemisphere, ϖ its longitude west of a given place, ξ the displacement of the zenith towards the south produced by the attraction of the coating, η its displacement towards the east. Then

$$\xi = \frac{1}{Ga} \frac{dV_e}{dl}, \qquad \eta = \frac{\sec l}{Ga} \frac{dV_e}{d\varpi},$$

because $a^{-1} dV_e/dl$ and $\sec l \,.\, a^{-1} dV_e/d\varpi$ are the horizontal components of the attraction towards the north and towards the west respectively, and G may be put for g on account of the smallness of the displacements.

Suppose the angle χ of Art. 22 measured from the meridian, so as to represent the north azimuth of the elementary mass $\delta a^2 \sin \psi d\psi d\chi$. On passing to a place on the same meridian whose latitude is $l + dl$, the angular distance of the elementary mass is shortened by $\cos \chi \,.\, dl$, and therefore its linear distance, which was a chord ψ, or $2a \sin \frac{1}{2}\psi$, becomes

$$2a \sin \tfrac{1}{2}\psi - a \cos \tfrac{1}{2}\psi \cos \chi \,.\, dl.$$

Hence the reciprocal of the linear distance is increased by

$$1/4a \,.\, \cos \tfrac{1}{2}\psi \operatorname{cosec}^2 \tfrac{1}{2}\psi \cos \chi \,.\, dl,$$

and therefore the part of V_e due to this element is increased by

$$\tfrac{1}{2}\delta a \cos^2 \tfrac{1}{2}\psi \operatorname{cosec} \tfrac{1}{2}\psi \cos \chi \,.\, d\psi d\chi dl.$$

Hence we have

$$\frac{dV_e}{dl} = \frac{a}{2} \iint \frac{\cos^2 \frac{1}{2}\psi \cos \chi}{\sin \frac{1}{2}\psi} \delta d\psi d\chi \quad \dots\dots\dots\dots (40).$$

11—2

Although the quantity under the integral sign in this expression becomes infinite when ψ vanishes, the integral itself has a finite value, at least if we suppose δ to vary continuously in the immediate neighbourhood of the station. For if δ becomes δ' when χ becomes $\chi + \pi$, we may replace δ under the integral sign by $\delta - \delta'$, and integrate from $\chi = 0$ to $\chi = \pi$, instead of integrating from $\chi = 0$ to $\chi = 2\pi$, and the limiting value of $(\delta - \delta') / \sin \frac{1}{2}\psi$ when ψ vanishes is $4 d\delta / d\psi$, which is finite.

To get the easterly displacement of the zenith, we have only to measure χ from the west instead of from the north, or, which comes to the same, to write $\chi + \frac{1}{2}\pi$ for χ, and continue to measure χ from the north. We get

$$\sec l \, \frac{dV_e}{d\varpi} = -\frac{a}{2} \iint \cos^2 \tfrac{1}{2}\psi \, \operatorname{cosec} \tfrac{1}{2}\psi \, \sin \chi \, . \, \delta d\psi \, d\chi \, ...(41).$$

29. The expressions (40) and (41) are not to be applied to points very near the station if δ vary abruptly, or even very rapidly, about such points. Recourse must in such a case be had to direct triple integration, because it is not allowable to consider the attracting matter as condensed into a surface. If however δ vary gradually in the neighbourhood of the station, the expression (40) or (41) may be used without further change. For if we modify (40) in the way explained in the preceding article, or else by putting the integral under the form

$$\int_0^\pi \int_0^{2\pi} \cos^2 \tfrac{1}{2}\psi \, \operatorname{cosec} \tfrac{1}{2}\psi \, \cos \chi \, (\delta - \delta_1) \, d\psi \, d\chi,$$

where δ_1 denotes the value of δ at the station, we see that the part of the integral due to a very small area surrounding the station is very small. If δ vary abruptly, in consequence suppose of the occurrence of a cliff, we may employ the expressions (40), (41), provided the distance of the cliff from the station be as much as three or four times its height.

These expressions shew that the vertical is liable to very irregular deviations depending on attractions which are quite local. For it is only in consequence of the opposition of attractions in opposite quarters that the value of the integral is not considerable, and it is of course larger in proportion as that opposition is less complete. Since $\sin \frac{1}{2}\psi$ is but small even at the distance of two or three hundred miles, a distant coast, or on the other hand a distant tract of high land of considerable extent, may produce a sensible effect; although of course in measuring an arc of the meridian those attractions may be neglected which arise from masses which are so distant as to affect both extremities of the arc in nearly the same way.

If we compare (40) or (41) with the expression for g'' or g''', we shall see that the direction of the vertical is liable to far more irregular fluctuations on account of the inequalities in the earth's coating than the force of gravity, except that part of the force which has been denoted by g', and which is easily allowed for. It has been supposed by some that the force of gravity alters irregularly along the earth's surface; and so it does, if we compare only distant stations. But it has been already remarked with what apparent regularity gravity when corrected for the inequality g' appears to alter, in the direction in which we should expect, in passing from one station to another in a chain of neighbouring stations.

30. There is one case in which the deviation of the vertical may become unusually large, which seems worthy of special consideration.

For simplicity, suppose δ to be constant for the land, and equal to zero for the sea, which comes to regarding the land as of constant height, the sea as of uniform depth, and transferring the defect of density of the sea with an opposite sign to the land. Apply the integral (40) to those parts only of the earth's surface which are at no great distance from the station considered, so that we may put $\cos \frac{1}{2}\psi = 1$, $\sin \frac{1}{2}\psi = \frac{1}{2}\psi = s/2a$, if s be the distance of the element, measured along a great circle. In going from the station in the direction determined by the angle χ, suppose that we pass from land to sea at distances s_1, s_3, s_5,... and from sea to land at the intermediate distances s_2, s_4... On going in the opposite direction suppose that we pass from land to sea at the distances s_{-1}, s_{-3}, s_{-5}, ... and from sea to land at the distances s_{-2}, s_{-4}.... Then we get from (40),

$$\frac{dV_c}{dl} = a\delta \int \{\log s_1 - \log s_{-1} - (\log s_2 - \log s_{-2}) + \log s_3 - \log s_{-3} \\ - ...\} \cos \chi . d\chi.$$

If the station be near the coast, one of the terms $\log s_1$, $\log s_{-1}$ will be large, and the zenith will be sensibly displaced towards the

sea by the irregular attraction.   On account of the shelving of the
coast, the preceding expression, which has been formed on the
supposition that δ vanished suddenly, would give too great a
displacement; but the object of this article is not to perform any
precise calculation, but merely to shew how the analysis indicates
a case in which there would be unusual disturbance.   A cliff
bounding a tract of table-land would have the same sort of effect
as a coast, and indeed the effect might be greater, on account of
the more sudden variation of δ.   The effect would be nearly the
same at equal horizontal distances from the edge above and
below, that distance being supposed as great as a small multiple
of the height of the cliff, in order to render the expression (40)
applicable without modification.

31.   Let us return now to the force of gravity, and leaving the
consideration of the connexion between the irregularities of gravity
and the irregularities of the earth's coating, and of the possibility
of destroying the former by making allowance for the latter, let us
take the earth such as we find it, and consider further the connexion between the variations of gravity and the irregularities of
the surface of equilibrium which constitutes the sea-level.

Equation (12) gives the variation of gravity if the form of the
surface be known, and conversely, (8) gives the form of the surface
if the variation of gravity be known.   Suppose the variation of
gravity known by means of pendulum-experiments performed at a
great many stations scattered over the surface of the earth; and
let it be required from the result of the observations to deduce
the form of the surface.   According to what has been already
remarked, a series of Laplace's functions would most likely be
practically useless for this purpose, unless we are content with
merely the leading terms in the expression for the radius vector;
and the leading character of those terms depends, not necessarily
upon their magnitude, but only on the wide extent of the inequalities which they represent.   We must endeavour therefore
to reduce the determination of the radius vector to quadratures.

For the sake of having to deal with small terms, let g be
represented, as well as may be, by the formula which applies to an
oblate spheroid, and let the variable term in the radius vector be
calculated by Clairaut's Theorem.   Let g_c be calculated gravity,

r_c the calculated radius vector, and put $g = g_c + \Delta g$, $r = r_c + a\Delta u$. Suppose Δg and Δu expanded in series of Laplace's functions. It follows from (12) that Δg will have no term of the order 1; indeed, if this were not the case, it might be shewn that the mutual forces of attraction of the earth's particles would have a resultant. Moreover the constant term in Δg may be got rid of by using a different value of G. No constant term need be taken in the expansion of Δu, because such a term might be got rid of by using a different value of a, and a of course cannot be determined by pendulum-experiments. The term of the first order will disappear if r be measured from the common centre of gravity of the mass and volume. The remaining terms in the expansion of Δu will be determined from those in the expansion of Δg by means of equations (8) and (12).

Let
$$\Delta g = G (v_2 + v_3 + v_4 + \ldots) \ldots\ldots\ldots\ldots (42),$$
and we shall have
$$\Delta u = v_2 + \tfrac{1}{2}v_3 + \tfrac{1}{3}v_4 + \ldots\ldots\ldots\ldots\ldots (43).$$

Suppose $\Delta g = GF(\theta, \phi)$. Let ψ be the angle between the directions determined by the angular co-ordinates θ, ϕ and θ', ϕ'. Let $(1 - 2\zeta \cos \psi + \zeta^2)^{\frac{1}{2}}$ be denoted by R, and let Q_i be the coefficient of ζ^i in the expansion of R^{-1} in a series according to ascending powers of ζ. Then

$$v_i = \frac{2i+1}{4\pi} \int_0^\pi \int_0^{2\pi} F(\theta', \phi') \, Q_i \sin \theta' d\theta' d\phi',$$

and therefore if ζ be supposed to be less than 1, and to become 1 in the limit, we shall have $4\pi \Delta u = $ limit of

$$\int_0^\pi \int_0^{2\pi} F(\theta', \phi') \left(5\zeta Q_2 + \tfrac{7}{2}\zeta^2 Q_3 \ldots + \frac{2i+1}{i-1} \zeta^{i-1} Q_i + \ldots\right) \sin \theta' d\theta' d\phi' \ldots (44).$$

Now assume

$$\gamma = 5\zeta Q_2 + \tfrac{7}{2}\zeta^2 Q_3 \ldots + \frac{2i+1}{i-1} \zeta^{i-1} Q_i + \ldots,$$

and we shall have

$$\frac{d\gamma}{d\zeta} = 5Q_2 + 7\zeta Q_3 \ldots + (2i+1) \zeta^{i-2} Q_i + \ldots;$$

$$\int_0^{\sqrt{\zeta}} \zeta^2 \frac{d\gamma}{d\zeta} d . \zeta^{\frac{1}{2}} = \zeta^{\frac{5}{2}} Q_2 + \zeta^{\frac{7}{2}} Q_3 \ldots + \zeta^{i+\frac{1}{2}} Q_i + \ldots = \zeta^{\frac{1}{2}} (R^{-1} - Q_0 - \zeta Q_1);$$

whence we get, putting Z for $R^{-1} - Q_0 - \zeta Q_1$, $\gamma = 2\int \zeta^{-\frac{3}{2}} d . \zeta^{\frac{1}{2}} Z$.

Integrating by parts, and observing that γ vanishes with ζ, we get

$$\gamma = 2\zeta^{-1}Z + 3\int_0^\zeta \zeta^{-2}Zd\zeta.$$

The last integral may be obtained by rationalization. If we assume $R = w - \zeta$, and observe that $Q_0 = 1$, $Q_1 = \cos\psi$, and that $w = 1$ when ζ vanishes, we shall find

$$\int_0^\zeta \zeta^{-2}Zd\zeta = \cos\psi . \log\frac{w - \cos\psi}{1 - \cos\psi} - (1 + \cos\psi)\frac{w - 1}{w + 1} - 2\cos\psi . \log\frac{w + 1}{2}.$$

When $\zeta = 1$ we have $Z = (2 - 2\cos\psi)^{-\frac{1}{2}} - (1 + \cos\psi)$, $w = 1 + 2\sin\frac{1}{2}\psi$, and

$$\int_0^\zeta \zeta^{-2}Zd\zeta = -2\sin\tfrac{1}{2}\psi\,(1 - \sin\tfrac{1}{2}\psi) - \cos\psi \log\{\sin\tfrac{1}{2}\psi\,(1 + \sin\tfrac{1}{2}\psi)\}.$$

Putting $f(\psi)$ for the value of γ when $\zeta = 1$, we have

$$f(\psi) = \operatorname{cosec}\tfrac{1}{2}\psi + 1 - 6\sin\tfrac{1}{2}\psi$$
$$- 5\cos\psi - 3\cos\psi \log\{\sin\tfrac{1}{2}\psi\,(1 + \sin\tfrac{1}{2}\psi)\}\ldots\ldots\ldots(45).$$

In the expression for Δu, we may suppose the line from which θ' is measured to be the radius vector of the station considered. We thus get, on replacing $F(\theta', \phi')$ by $G^{-1}\Delta g$, and employing the notation of Art. 22,

$$\Delta u = \frac{1}{4\pi G}\int_0^\pi\int_0^{2\pi}\Delta g . f(\psi)\sin\psi d\psi d\chi\ldots\ldots\ldots\ldots(46).$$

32. Let $\Delta g = g' + \Delta'g$. Then $\Delta'g$ is the excess of observed gravity reduced to the level of the sea by Dr Young's rule over calculated gravity; and of the two parts g' and $\Delta'g$ of which Δg consists, the former is liable to vary irregularly and abruptly from one place to another, the latter varies gradually. Hence, for the sake of interpolating between the observations taken at different stations, it will be proper to separate Δg into these two parts, or, which comes to the same, to separate the whole integral into two parts, involving g' and $\Delta'g$ respectively, so as to get the part of Δu which is due to g' by our knowledge of the height of the land and the depth of the sea, and the part which depends on $\Delta'g$ by the result of pendulum-experiments. It may be observed that a constant error, or a slowly varying error, in the height of the land would be of no consequence, because it would enter with opposite signs into g' and $\Delta'g$.

It appears, then, that the results of pendulum-experiments furnish sufficient data for the determination of the variable part of the radius vector of the earth's surface, and consequently for the determination of the particular value which is to be employed at any observatory in correcting for the lunar parallax, subject however to a constant error depending on an error in the assumed value of a.

33. The expression for g''' in Art. 27 might be reduced to quadratures by the method of Art. 31, but in this case the integration with respect to ζ could not be performed in finite terms, and it would be necessary in the first instance to tabulate, once for all, an integral of the form $\int_0^1 f(\zeta, \cos \psi)\, d\zeta$ for values of ψ, which need not be numerous, from 0 to π. This table being made, the tabulated function would take the place of $f(\psi)$ in (46), and the rest of the process would be of the same degree of difficulty as the quadratures expressed by the equations (31) and (46).

34. Suppose Δu known approximately, either as to its general features, by means of the leading terms of the series (43), or in more detail from the formula (46), applied in succession to a great many points on the earth's surface. By interpolating between neighbouring places for which Δu has been calculated, find a number of points where Δu has one of the constant values $- 2\beta$, $- \beta, 0, \beta, 2\beta \ldots$, mark these points on a map of the world, and join by a curve those which belong to the same value of Δu. We shall thus have a series of contour lines representing the elevation or depression of the actual sea-level above or below the surface of the oblate spheroid, which has been employed as most nearly representing it. If we suppose these lines traced on a globe, the reciprocal of the perpendicular distance between two consecutive contour lines will represent in magnitude, and the perpendicular itself in direction, the deviation of the vertical from the normal to the surface of the spheroid, or rather that part of the deviation which takes place on an extended scale : for sensible deviations may be produced by attractions which are merely local, and which would not produce a sensible elevation or depression of the sea-level ; although of course, as to the merely mathematical question, if the contour lines could be drawn sufficiently close and exact, even local deviations of the vertical would be represented.

Similarly, by joining points at which the quantity denoted in Art. 19 by V_c has a constant value, contour lines would be formed representing the elevation of the actual sea-level above what would be a surface of equilibrium if the earth's irregular coating were removed. By treating V_x in the same way, contour lines would be formed corresponding to the elevation of the actual sea-level above what would be the sea-level if the solid portions of the earth's crust which are elevated were to become fluid and to run down, so as to form a level bottom for the sea, which would in that case cover the whole earth.

These points of the theory are noticed more for the sake of the ideas than on account of any application which is likely to be made of them; for the calculations indicated, though possible with a sufficient collection of data, would be very laborious, at least if we wished to get the results with any detail.

35. The squares of the ellipticity, and of quantities of the same order, have been neglected in the investigation. Mr Airy, in the Treatise already quoted, has examined the consequence, on the hypothesis of fluidity, of retaining the square of the ellipticity, in the two extreme cases of a uniform density, and of a density infinitely great at the centre and evanescent elsewhere, and has found the correction to the form of the surface and the variation of gravity to be insensible, or all but insensible. As the connexion between the form of the surface and the variation of gravity follows independently of the hypothesis of fluidity, we may infer that the terms depending on the square of the ellipticity which would appear in the equations which express that connexion would be insensible. It may be worth while, however, just to indicate the mode of proceeding when the square of the ellipticity is retained.

By the result of the first approximation, equation (1) is satisfied at the surface of the earth, as far as regards quantities of the first order, but not necessarily further, so that the value of $V + U$ at the surface is not strictly constant, but only of the form $c + H$, where H is a small variable quantity of the second order. It is to be observed that V satisfies equation (3) exactly, not approximately only. Hence we have merely to add to V a potential V' which satisfies equation (3) outside the earth, vanishes at an infinite distance, and is equal to H at the surface. Now if we suppose V' to have the value H at the surface of a sphere whose radius is a, instead of the actual surface of the earth, we shall only commit an error which is a small quantity of the first order compared with H, and H is itself of the second order, and therefore the error will be only of the third order. But by this modification of one of the conditions which V' is to satisfy, we are enabled to find V' just as V was found, and we shall thus have a solution which is correct to the second order of approximation. A repetition of the same process would give a solution which would be correct to the third order, and so on. It need hardly be remarked that in going beyond the first order of approximation, we must distinguish in the small terms between the direction of the vertical, and that of the radius vector.

[From the *Report of the British Association* for 1849.   Part II. p. 10.]

ON A MODE OF MEASURING THE ASTIGMATISM OF A DEFECTIVE EYE.

BESIDES the common defects of long sight and short sight, there exists a defect, not very uncommon, which consists in the eye's refracting the rays of light with different power in different planes, so that the eye, regarded as an optical instrument, is not symmetrical about its axis.   This defect was first noticed by the present Astronomer Royal in a paper published about 20 years ago in the Transactions of the Cambridge Philosophical Society. It may be detected by making a small pin-hole in a card, which is to be moved from close to the eye to arm's length, the eye meanwhile being directed to the sky, or any bright object of sufficient size.   With ordinary eyes the indistinct image of the hole remains circular at all distances; but to an eye having this peculiar defect it becomes elongated, and, when the card is at a certain distance, passes into a straight line.   On further removing the card, the image becomes elongated in a perpendicular direction, and finally, if the eye be not too long-sighted, passes into a straight line perpendicular to the former.   Mr Airy has corrected the defect in his own case by means of a spherico-cylindrical lens, in which the required curvature of the cylindrical surface was calculated by means of the distances of the card from the eye when the two focal lines were formed.   Others however have found a difficulty in preventing the eye from altering its state of adaptation during the measurement of the distances.   The author has constructed an instrument for determining the nature of the required lens, which is based on the following proposition :—

Conceive a lens ground with two cylindrical surfaces of equal radius, one concave and the other convex, with their axes crossed at right angles; call such a lens an *astigmatic lens;* let the reciprocal of its focal length in one of the principal planes be called its *power*, and a line parallel to the axis of the convex surface its *astigmatic axis.* Then if two thin astigmatic lenses be combined with their astigmatic axes inclined at any angle, they will be equivalent to a third astigmatic lens, determined by the following construction:—In a plane perpendicular to the common axis of the lenses, or axis of vision, draw through any point two straight lines, representing in magnitude the powers of the respective lenses, and inclined to a fixed line drawn arbitrarily in a direction perpendicular to the axis of vision at angles equal to twice the inclinations of their astigmatic axes, and complete the parallelogram. Then the two lenses will be equivalent to a single astigmatic lens, represented by the diagonal of the parallelogram in the same way in which the single lenses are represented by the sides. A plano-cylindrical or spherico-cylindrical lens is equivalent to a common lens, the power of which is equal to the semi-sum of the reciprocals of the focal lengths in the two principal planes, combined with an astigmatic lens, the power of which is equal to their semi-difference.

If two plano-cylindrical lenses of equal radius, one concave and the other convex, be fixed, one in the lid and the other in the body of a small round wooden box, with a hole in the top and bottom, so as to be as nearly as possible in contact, the lenses will neutralize each other when the axes of the surfaces are parallel; and, by merely turning the lid round, an astigmatic lens may be formed of a power varying continuously from zero to twice the astigmatic power of either lens. When a person who has the defect in question has turned the lid till the power suits his eye, an extremely simple numerical calculation, the data for which are furnished by the chord of double the angle through which the lid has been turned, enables him to calculate the curvature of the cylindrical surface of a lens for a pair of spectacles which will correct the defect of his eye.

[The proposition here employed is easily demonstrated by a method founded on the notions of the theory of undulations, though of course, depending as it does simply on the laws of reflection and refraction, it does not involve the adoption of any particular theory of light.

Consider a thin lens bounded by cylindrical surfaces, the axes of the cylinders being crossed at right angles. Refer points in the neighbourhood of the lens to the rectangular axes of x, y, z, the axis of z being the axis of the lens, and those of x and y parallel to the axes of the two cylindrical surfaces respectively, the origin being in or near the lens, suppose in its middle point. Let r, s, measured positive when the surfaces are convex, be the radii of curvature in the planes of xz, yz respectively. Then if T be the central thickness of the lens, the thickness near the point (x, y) will be

$$T - \tfrac{1}{2} \left(\frac{x^2}{r} + \frac{y^2}{s} \right)$$

very nearly. As T is constant, and is supposed very small, we may neglect it, and regard the thickness as negative, and expressed by the second term in the above formula. The incident pencil being supposed to be direct, or only slightly oblique, and likewise slender, the retardation of the ray which passes through the point (x, y) may be calculated as if it were incident perpendicularly on a parallel plate of thickness

$$- \tfrac{1}{2} \left(\frac{x^2}{r} + \frac{y^2}{s} \right),$$

so that if R be the retardation, measured by equivalent space in air, and μ be the index of refraction

$$- R = (\mu - 1) . \tfrac{1}{2} \left(\frac{x^2}{r} + \frac{y^2}{s} \right)$$

$$= (\mu - 1) . \tfrac{1}{4} \left(\frac{1}{r} + \frac{1}{s} \right) (x^2 + y^2) + (\mu - 1) . \tfrac{1}{4} \left(\frac{1}{r} - \frac{1}{s} \right) (x^2 - y^2).$$

The effect therefore of our lens, to the lowest order of approximation, which gives the geometrical foci in the principal planes, is the same as that of two thin lenses placed in contact, one an ordinary lens, and the other an astigmatic lens. If r' be the radius of curvature of the plano-spherical lens equivalent to the ordinary lens, and r'' that of the astigmatic lens, we have

$$\frac{1}{r'} = \tfrac{1}{2} \left(\frac{1}{r} + \frac{1}{s} \right), \quad \frac{1}{r''} = \tfrac{1}{2} \left(\frac{1}{r} - \frac{1}{s} \right),$$

as above enunciated. If p be the power of the astigmatic lens,

$$p = (\mu - 1) . \tfrac{1}{2} \left(\frac{1}{r} - \frac{1}{s} \right),$$

and for the retardation produced by this lens alone

$$- R = \tfrac{1}{2} p \left(x^2 - y^2 \right) = \tfrac{1}{2} p\rho^2 \cos 2\,\theta,$$

where ρ, θ are polar co-ordinates in the plane of xy.

If two thin astigmatic lenses of powers p, p' and with their astigmatic axes inclined at azimuths α, α' to the axis of y be combined, we shall have for the combination

$$- R = \tfrac{1}{2} p\rho^2 \cos 2 \left(\theta - \alpha \right) + \tfrac{1}{2} p'\rho^2 \cos 2 \left(\theta - \alpha' \right),$$

which is the same as would be given by a single astigmatic lens of power p_1 at an azimuth α_1, provided

$$p\rho^2 \cos 2 \left(\theta - \alpha \right) + p'\rho^2 \cos 2 \left(\theta - \alpha' \right) = p_1\rho^2 \cos 2 \left(\theta - \alpha_1 \right),$$

which will be satisfied for all values of θ provided

$$p \cos 2\alpha + p' \cos 2\alpha' = p_1 \cos 2\alpha_1,$$
$$p \sin 2\alpha + p' \sin 2\alpha' = p_1 \sin 2\alpha_1.$$

These two equations geometrically interpreted give the proposition enunciated above for the combination of astigmatic lenses.]

[From the *Report of the British Association* for 1849. Part II. p. 11.]

On the Determination of the Wave Length corresponding
with any point of the spectrum.

Mr Stokes said it was well known to all engaged in optical researches that Fraunhofer had most accurately measured the wave lengths of seven of the principal fixed lines of the spectrum. Now he found that by a very simple species of interpolation, which he described, he could find the wave length for any point intermediate between the two of them. He then exemplified the accuracy to be obtained by his method by applying it to the actually known points, and shewed that in these far larger intervals than he ever required to apply the method to the error was only in the eighth, and in one case in the seventh, place of decimals. By introducing a term depending on the square into the interpolation still greater accuracy was attainable. The mode of interpolation depended on the known fact that, if substances of extremely high refractive power be excepted, the increment $\Delta\mu$ of the refractive index in passing from one point of the spectrum to another is nearly proportional to the increment $\Delta\lambda^{-2}$ of the squared reciprocal of the wave length. Even in the case of flint glass, the substance usually employed in the prismatic analysis of light, this law is nearly true for the whole spectrum, and will be all but exact if restricted to the interval between two consecutive fixed lines. Hence we have only to consider μ as a function, not of λ, but of λ^{-2}, and then take proportional parts.

On examining in this way Fraunhofer's indices for flint glass, it appeared that the wave length $B\lambda$ of the fixed line B was too great by about 4 in the last, or eighth, place of decimals. It is remarkable that the line B was not included in Fraunhofer's second and more accurate determination of the wave lengths, and that the proposed correction to $B\lambda$ is about the same, both as to sign and magnitude, as one would have guessed from Fraunhofer's own corrections of the other wave lengths, obtained from his second series of observations.

[A map of the spectrum laid down according to the values of λ^{-2} instead of λ refers equally to a natural standard, that is, one independent of the material of any prism, and is much more convenient for comparison with spectra obtained by dispersion, not diffraction.]

[From the *Transactions of the Cambridge Philosophical Society*, Vol. VIII. p. 707.]

DISCUSSION OF A DIFFERENTIAL EQUATION RELATING TO THE BREAKING OF RAILWAY BRIDGES.

[Read *May* 21, 1849.]

To explain the object of the following paper, it will be best to relate the circumstance which gave rise to it. Some time ago Professor Willis requested my consideration of a certain differential equation in which he was interested, at the same time explaining its object, and the mode of obtaining it. The equation will be found in the first article of this paper, which contains the substance of what he communicated to me. It relates to some experiments which have been performed by a Royal Commission, of which Professor Willis is a member, appointed on the 27th of August, 1847, "for the purpose of inquiring into the conditions to be observed by engineers in the application of iron in structures exposed to violent concussions and vibration." The object of the experiments was to examine the effect of the velocity of a train in increasing or decreasing the tendency of a girder bridge over which the train is passing to break under its weight. In order to increase the observed effect, the bridge was purposely made as slight as possible: it consisted in fact merely of a pair of cast or wrought iron bars, nine feet long, over which a carriage, variously loaded in different sets of experiments, was made to pass with different velocities. The remarkable result was obtained that the deflection of the bridge increased with the velocity of the carriage, at least up to a certain point, and that it amounted in some cases to two or three times the central statical deflection, or that which would be produced by the carriage placed at rest on the middle of the bridge. It seemed highly desirable to investigate the motion mathematically, more especially as the maximum deflection of the bridge, considered as depending on the velocity of the carriage, had not been reached in the experiments*, in some cases because it corresponded to a velocity greater than any at command, in others because the bridge gave way by the fracture of the bars on increasing the velocity of the carriage. The exact calculation of the motion, or rather a calculation in which none but really insignificant quantities should be omitted, would however be extremely difficult, and would require the solution of a partial differential equation with an ordinary differential equation for one of the equations of condition by which the arbitrary functions would have to be determined. In fact, the forces acting on the body and on any element of the bridge depend upon the positions and motions, or rather changes of motion, both of the body itself and of every other element of the bridge, so that the exact solution of the problem, even when the deflection is supposed to be small, as it is in fact, appears almost hopeless.

In order to render the problem more manageable, Professor Willis neglected the inertia of the bridge, and at the same time regarded the moving body as a heavy particle. Of course the masses of bridges such as are actually used must be considerable; but the mass of the bars in the experiments was small compared with that of the carriage, and it was reasonable to expect a near accordance between the theory so simplified and experiment. This simplification of the problem reduces the calculation to an ordinary differential equation, which is that which has been already mentioned; and it is to the discussion of this equation that the present paper is mainly devoted.

This equation cannot apparently be integrated in finite terms†, except for an infinite number of particular values of a certain constant involved in it; but I have investigated rapidly convergent series whereby numerical results may be obtained. By merely altering the scale of the abscissæ and ordinates, the differential equation is reduced to one containing a single constant β, which is defined by equation (5). The meaning of the letters which appear in this equation will be seen on referring to the beginning of Art. 1. For the present it will be sufficient to observe that β varies inversely as the square of the horizontal velocity of the

---

* The details of the experiments will be found in the Report of the Commission, to which the reader is referred.

† [The integral can be expressed by definite integrals. See Art. 7, and last paragraph but one in the paper.]

body, so that a small value of β corresponds to a high velocity, and a large value to a small velocity.

It appears from the solution of the differential equation that the trajectory of the body is unsymmetrical with respect to the centre of the bridge, the maximum depression of the body occurring beyond the centre. The character of the motion depends materially on the numerical value of β. When β is not greater than $\frac{1}{4}$, the tangent to the trajectory becomes more and more inclined to the horizontal beyond the maximum ordinate, till the body gets to the second extremity of the bridge, when the tangent becomes vertical. At the same time the expressions for the central deflection and for the tendency of the bridge to break become infinite. When β is greater than $\frac{1}{4}$, the analytical expression for the ordinate of the body at last becomes negative, and afterwards changes an infinite number of times from negative to positive, and from positive to negative. The expression for the reaction becomes negative at the same time with the ordinate, so that in fact the body leaps.

The occurrence of these infinite quantities indicates one of two things : either the deflection really becomes very large, after which of course we are no longer at liberty to neglect its square; or else the effect of the inertia of the bridge is really important. Since the deflection does not really become very great, as appears from experiment, we are led to conclude that the effect of the inertia is not insignificant, and in fact I have shewn that the value of the expression for the *vis viva* neglected at last becomes infinite. Hence, however light be the bridge, the mode of approximation adopted ceases to be legitimate before the body reaches the second extremity of the bridge, although it may be sufficiently accurate for the greater part of the body's course.

In consequence of the neglect of the inertia of the bridge, the differential equation here discussed fails to give the velocity for which T, the tendency to break, is a maximum. When β is a good deal greater than $\frac{1}{4}$, T is a maximum at a point not very near the second extremity of the bridge, so that we may apply the result obtained to a light bridge without very material error. Let T_1 be this maximum value. Since it is only the inertia of the bridge that keeps the tendency to break from becoming extremely great, it appears that the *general* effect of that inertia is to preserve the bridge, so that we cannot be far wrong in regarding

T_1 as a superior limit to the actual tendency to break.   When β is very large, T_1 may be calculated to a sufficient degree of accuracy with very little trouble.

Experiments of the nature of those which have been mentioned may be made with two distinct objects; the one, to analyse experimentally the laws of some particular phenomenon, the other, to apply practically on a large scale results obtained from experiments made on a small scale.   With the former object in view, the experiments would naturally be made so as to render as conspicuous as possible, and isolate as far as might be, the effect which it was desired to investigate; with the latter, there are certain relations to be observed between the variations of the different quantities which are in any way concerned in the result.   These relations, in the case of the particular problem to which the present paper refers, are considered at the end of the paper.

---

1.   It is required to determine, in a form adapted to numerical computation, the value of y' in terms of x', where y' is a function of x' defined by satisfying the differential equation

$$\frac{d^2y'}{dx'^2} = a - \frac{by'}{(2cx' - x'^2)^2} \quad \dots\dots\dots\dots\dots\dots\dots(1),$$

with the particular conditions

$$y' = 0, \quad \frac{dy'}{dx'} = 0, \quad \text{when } x' = 0 \dots\dots\dots\dots\dots\dots(2),$$

the value of y' not being wanted beyond the limits 0 and $2c$ of x'. It will appear in the course of the solution that the first of the conditions (2) is satisfied by the complete integral of (1), while the second serves of itself to determine the two arbitrary constants which appear in that integral.

The equation (1) relates to the problem which has been explained in the introduction.   It was obtained by Professor Willis in the following manner.   In order to simplify to the very utmost the mathematical calculation of the motion, regard the carriage as a heavy particle, neglect the inertia of the bridge, and suppose the deflection very small.   Let x', y' be the co-ordinates of the moving body, x' being measured horizontally from the beginning of the bridge, and y' vertically downwards.   Let M be the mass of the body, V its velocity on entering the bridge, $2c$ the length of the bridge, g the force of gravity, S the deflection produced by the body placed at rest on the centre of the bridge, R the reaction between the moving body and the bridge. Since the deflection is very small, this reaction may be supposed to act vertically, so that the horizontal velocity of the body will remain constant, and therefore equal to V. The bridge being regarded as an elastic bar or plate, propped at the extremities, and supported by its own stiffness, the depth to which a weight will sink when placed in succession at different points of the bridge will vary as the weight multiplied by $(2cx' - x'^2)^2$, as may be proved by integration, on assuming that the curvature is proportional to the moment of the bending force. Now, since the inertia of the bridge is neglected, the relation between the depth y' to which the moving body has sunk at any instant and the reaction R will be the same as if R were a weight resting at a distance x' from the extremity of the bridge; and we shall therefore have

$$y' = CR\,(2cx' - x'^2)^2,$$

C being a constant, which may be determined by observing that we must have $y' = S$ when $R = Mg$ and $x' = c$; whence

$$C = \frac{S}{Mgc^4}.$$

We get therefore for the equation of motion of the body

$$\frac{d^2y'}{dt^2} = g - \frac{gc^4 y'}{S\,(2cx' - x'^2)^2},$$

which becomes on observing that $\dfrac{dx'}{dt} = V$

$$\frac{d^2y'}{dx'^2} = \frac{g}{V^2} - \frac{gc^4}{V^2 S}\frac{y'}{(2cx' - x'^2)^2},$$

which is the same as equation (1), a and b being defined by the equations

$$a = \frac{g}{V^2}, \qquad b = \frac{gc^4}{V^2 S} \dotfill (3).$$

2. To simplify equation (1) put

$$x' = 2cx, \qquad y' = 16c^4 ab^{-1}y, \qquad b = 4c^2\beta,$$

which gives

$$\frac{d^2y}{dx^2} = \beta - \frac{\beta y}{(x - x^2)^2} \dotfill (4).$$

It is to be observed that x denotes the ratio of the distance of the body from the beginning of the bridge to the length of the bridge; y denotes a quantity from which the depth of the body below the horizontal plane in which it was at first moving may be obtained by multiplying by $16c^4ab^{-1}$ or $16S$; and β, on the value of which depends the form of the body's path, is a constant defined by the equation

$$\beta = \frac{gc^2}{4V^2S} \quad\quad\quad (5).$$

3.   In order to lead to the required integral of (4), let us first suppose that x is very small.   Then the equation reduces itself to

$$\frac{d^2y}{dx^2} = \beta - \frac{\beta y}{x^2} \quad\quad\quad (6),$$

of which the complete integral is

$$y = \frac{\beta x^2}{2+\beta} + Ax^{\frac{1}{2}+\sqrt{\frac{1}{4}-\beta}} + Bx^{\frac{1}{2}-\sqrt{\frac{1}{4}-\beta}} \quad\quad\quad (7),$$

and (7) is the approximate integral of (4) for very small values of x.   Now the second of equations (2) requires that $A = 0, B = 0$, so that the first term in the second member of equation (7) is the leading term in the required solution of (4).

4.   Assuming in equation (4) $y = (x - x^2)^2 z$, we get

$$\frac{d^2}{dx^2}\{(x - x^2)^2 z\} + \beta z = \beta \quad\quad\quad (8).$$

Since (4) gives $y = (x - x^2)^2$ when $\beta = \infty$, and (5) gives $\beta = \infty$ when $V = 0$, it follows that z is the ratio of the depression of the body to the equilibrium depression.   It appears also from Art. 3, that for the particular integral of (8) which we are seeking, z is ultimately constant when x is very small.

---

* When $\beta > \frac{1}{4}$, the last two terms in (7) take the form $x^{\frac{1}{2}} \{C \cos (q \log x) + D \sin (q \log x)\}$; and if y_1 denote this quantity we cannot in strictness speak of the limiting value of dy_1/dx when $x=0$. If we give x a small positive value, which we then suppose to decrease indefinitely, dy_1/dx will fluctuate between the constantly increasing limits $\pm x^{-\frac{3}{2}}\sqrt{\{(\frac{1}{2}C+qD)^2+(\frac{1}{2}D-qC)^2\}}$, or $\pm x^{-\frac{3}{2}}\sqrt{\{\beta(C^2+D^2)\}}$, since $q=\sqrt{(\beta-\frac{1}{4})}$. But the body is supposed to enter the bridge horizontally, that is, in the direction of a tangent, since the bridge is supposed to be horizontal, so that we must clearly have $C^2+D^2=0$, and therefore $C=0, D=0$. When $\beta=\frac{1}{4}$ the last two terms in (7) take the form $x^{\frac{1}{2}} (E + F \log x)$, and we must evidently have $E=0, F=0$.

To integrate (8) assume then

$$z = A_0 + A_1 x + A_2 x^2 + \ldots = \Sigma A_i x^i \ldots\ldots\ldots(9),$$

and we get

$$\Sigma (i+2)(i+1) A_i x^i - 2\Sigma (i+3)(i+2) A_i x^{i+1}$$
$$+ \Sigma (i+4)(i+3) A_i x^{i+2} + \beta \Sigma A_i x^i = \beta,$$

or

$$\Sigma \{[(i+1)(i+2) + \beta] A_i - 2(i+1)(i+2) A_{i-1}$$
$$+ (i+1)(i+2) A_{i-2}\} x^i = \beta \ldots\ldots\ldots(10)$$

where it is to be observed that no coefficients A_i with negative suffixes are to be taken.

Equating to zero the coefficients of the powers 0, 1, 2... of x in (10), we get

$$(2+\beta) A_0 = \beta,$$
$$(6+\beta) A_1 - 12 A_0 = 0, \ \&c.$$

and generally

$$\{(i+1)(i+2) + \beta\} A_i - 2(i+1)(i+2) A_{i-1}$$
$$+ (i+1)(i+2) A_{i-2} = 0 \ldots\ldots\ldots (11).$$

The first of these equations gives for A_0 the same value which would have been got from (7). The general equation (11), which holds good from $i = 1$ to $i = \infty$, if we conventionally regard A_{-1} as equal to zero, determines the constants A_1, A_2, A_3... one after another by a simple and uniform arithmetical process. It will be rendered more convenient for numerical computation by putting it under the form

$$A_i = \{A_{i-1} + \Delta A_{i-2}\} \left\{1 - \frac{\beta}{(i+1)(i+2)+\beta}\right\} \ldots\ldots(12);$$

for it is easy to form a table of differences as we go along; and when i becomes considerable, the quantity to be subtracted from $A_{i-1} + \Delta A_{i-2}$ will consist of only a few figures.

5. When i becomes indefinitely great, it follows from (11) or (12) that the relation between the coefficients A_i is given by the equation

$$A_i - 2A_{i-1} + A_{i-2} = 0 \ldots\ldots\ldots\ldots (13),$$

of which the integral is

$$A_i = C + C'i \ldots\ldots\ldots\ldots\ldots(14).$$

Hence the ratio of consecutive coefficients is ultimately a ratio of equality, and therefore the ratio of the $(i+1)$th term of the series (9) to the ith is ultimately equal to x. Hence the series is convergent when x lies between the limits -1 and $+1$; and it is only between the limits 0 and 1 of x that the integral of (8) is wanted. The degree of convergency of the series will be ultimately the same as in a geometric series whose ratio is x.

6. When x is moderately small, the series (9) converges so rapidly as to give z with little trouble, the coefficients A_1, A_2... being supposed to have been already calculated, as far as may be necessary, from the formula (12). For larger values, however, it would be necessary to keep in a good many terms, and the labour of calculation might be abridged in the following manner.

When i is very large, we have seen that equation (12) reduces itself to (13), or to $\Delta^2 A_{i-2} = 0$, or, which is the same, $\Delta^2 A_i = 0$. When i is large, $\Delta^2 A_i$ will be small; in fact, on substituting in the small term of (12) the value of A_i given by (14), we see that $\Delta^2 A_i$ is of the order i^{-1}. Hence $\Delta^3 A_i$, $\Delta^4 A_i$... will be of the orders i^{-2}, i^{-3}..., so that the successive differences of A_i will rapidly decrease. Suppose i terms of the series (9) to have been calculated directly, and let it be required to find the remainder. We get by finite integration by parts

$$\Sigma A_i x^i = \text{const.} + A_i \frac{x^i}{x-1} - \Delta A_i \frac{x^{i+1}}{(x-1)^2} + \Delta^2 A_i \frac{x^{i+2}}{(x-1)^3} - ...,$$

and taking the sum between the limits i and ∞ we get
$A_i x^i + A_{i+1} x^{i+1} + ...$ to inf.

$$= x^{i-1} \left\{ A_i \frac{x}{1-x} + \Delta A_i \left(\frac{x}{1-x} \right)^2 + \Delta^2 A_i \left(\frac{x}{1-x} \right)^3 + ... \right\} ...... (15);$$

z will however presently be made to depend on series so rapidly convergent that it will hardly be worth while to employ the series (15), except in calculating the series (9) for the particular value $\frac{1}{2}$ of x, which will be found necessary in order to determine a certain arbitrary constant*.

---

* A mode of calculating the value of z for $x = \frac{1}{2}$ will presently be given, which is easier than that here mentioned, unless β be very large. See equation (42) at the end of this paper.

7. If the constant term in equation (4) be omitted, the equation reduces itself to

$$\frac{d^2y}{dx^2} + \frac{\beta y}{(x - x^2)^2} = 0 \dots\dots\dots\dots\dots (16).$$

The form of this equation suggests that there may be an integral of the form $y = x^m (1 - x)^n$. Assuming this expression for trial, we get

$$(x - x^2)^2 \frac{d^2y}{dx^2} = x^m (1-x)^n \{m(m-1)(1-x)^2 - 2mnx(1-x) + n(n-1)x^2\}$$

$$= y\{m(m-1) - 2m(m+n-1)x + (m+n)(m+n-1)x^2\}.$$

The second member of this equation will be proportional to y, if

$$m + n - 1 = 0 \dots\dots\dots\dots\dots\dots (17),$$

and will be moreover equal to $-\beta y$, if

$$m^2 - m + \beta = 0 \dots\dots\dots\dots\dots (18).$$

It appears from (17) that m, n are the two roots of the quadratic (18). We have for the complete integral of (16)

$$y = A x^m (1 - x)^n + B x^n (1 - x)^m \dots\dots\dots\dots (19).$$

The complete integral of (4) may now be obtained by replacing the constants A, B by functions R, S of x, and employing the method of the variation of parameters. Putting for shortness

$$x^m (1 - x)^n = u, \quad x^n (1 - x)^m = v,$$

we get to determine R and S the equations

$$u \frac{dR}{dx} + v \frac{dS}{dx} = 0,$$

$$\frac{du}{dx}\frac{dR}{dx} + \frac{dv}{dx}\frac{dS}{dx} = \beta.$$

Since $v\frac{du}{dx} - u\frac{dv}{dx} = m - n$, we get from the above equations

$$\frac{dR}{dx} = \frac{\beta v}{m - n}, \quad \frac{dS}{dx} = -\frac{\beta u}{m - n},$$

whence we obtain for a particular integral of (4)

$$y = \frac{\beta}{m-n}\left\{x^m(1-x)^n \int_0^x x^n(1-x)^m dx - x^n(1-x)^m \int_0^x x^m(1-x)^n dx\right\} \dots (20);$$

and the complete integral will be got by adding together the second members of equations (19), (20). Now the second member of equation (20) varies ultimately as x^2, when x is very small, and therefore, as shewn in Art. 3, we must have $A = 0$, $B = 0$, so that (20) is the integral we want.

When the roots of the quadratic (18) are real and commensurable, the integrals in (20) satisfy the criterion of integrability, so that the integral of (4) can be expressed in finite terms without the aid of definite integrals. The form of the integral will, however, be complicated, and y may be readily calculated by the method which applies to general values of β.

8.   Since $\int_0^x F(x)\,dx = \int_0^1 F(x)\,dx - \int_0^{1-x} F(1-x)\,dx$, we have from (20)

$$y = \frac{\beta}{m-n}\left\{ x^m(1-x)^n \int_0^1 x^n(1-x)^m\,dx - x^n(1-x)^m \int_0^1 x^m(1-x)^n\,dx \right\}$$

$$+ \frac{\beta}{m-n}\left\{ x^n(1-x)^m \int_0^{1-x}(1-x)^m x^n\,dx - x^m(1-x)^n \int_0^{1-x}(1-x)^n x^m\,dx \right\}.$$

If we put $f(x)$ for the second member of equation (20), the equation just written is equivalent to

$$f(x) = f(1-x) + \phi(x) \dots\dots\dots\dots\dots\dots(21),$$

where

$$\phi(x) = \frac{\beta}{m-n}\left\{ x^m(1-x)^n \int_0^1 x^n(1-x)^m\,dx - x^n(1-x)^m \int_0^1 x^m(1-x)^n\,dx \right\} \dots (22).$$

Now since $m + n = 1$,

$$\int x^n(1-x)^m\,dx = \int x(x^{-1}-1)^m\,dx = -\int w^{-1}(w-1)^m w^{-2}\,dw = -\int \frac{s^m\,ds}{(1+s)^3}.$$

At the limits $x = 0$ and $x = 1$, we have $w = \infty$ and $w = 1$, $s = \infty$ and $s = 0$, whence if I denote the definite integral,

$$I = \int_0^1 x^n(1-x)^m\,dx = \int_0^\infty \frac{s^m\,ds}{(1+s)^3}.$$

We get by integration by parts

$$\int \frac{s^m\,ds}{(1+s)^3} = -\frac{s^m}{2(1+s)^2} + \frac{m}{2}\int \frac{s^{m-1}\,ds}{(1+s)^2},$$

and again by a formula of reduction

$$\int \frac{s^{m-1}\, ds}{(1+s)^2} = \frac{s^m}{1+s} + (1-m)\int \frac{s^{m-1}\, ds}{1+s}.$$

Now β being essentially positive, the roots of the quadratic (18) are either real, and comprised between 0 and 1, or else imaginary with a real part equal to $\frac{1}{2}$. In either case the expressions which are free from the integral sign vanish at the limits $s = 0$ and $s = \infty$, and we have therefore, on replacing $m\,(1-m)$ by its value β,

$$I = \frac{\beta}{2}\int_0^\infty \frac{s^{m-1}\, ds}{1+s}.$$

The function $\phi\,(x)$ will have different forms according as the roots of (18) are real or imaginary. First suppose the roots real, and let $m = \frac{1}{2}+r$, $n = \frac{1}{2}-r$, so that

$$r = \sqrt{\tfrac{1}{4} - \beta}.............................(23).$$

In this case m is a real quantity lying between 0 and 1, and we have therefore by a known formula

$$\int_0^\infty \frac{s^{m-1}\, ds}{1+s} = \frac{\pi}{\sin m\pi} = \frac{\pi}{\cos r\pi}.....................(24),$$

whence we get from (22), observing that the two definite integrals in this equation are equal to each other,

$$\phi\,(x) = \frac{\beta^2 \pi}{4r \cos r\pi}\; \sqrt{x-x^2}\left\{\left(\frac{x}{1-x}\right)^r - \left(\frac{x}{1-x}\right)^{-r}\right\}.....(25).$$

This result might have been obtained somewhat more readily by means of the properties of the first and second Eulerian integrals.

When β becomes equal to $\frac{1}{4}$, r vanishes, the expression for $\phi\,(x)$ takes the form $\frac{0}{0}$, and we easily find

$$\phi\,(x) = \frac{\pi}{32}\; \sqrt{x-x^2}\log \frac{x}{1-x}\; ....(26).$$

When $\beta > \frac{1}{4}$, the roots of (18) become imaginary, and r becomes $\rho \sqrt{-1}$, where

$$\rho = \sqrt{\beta - \tfrac{1}{4}}.........................(27).$$

The formula (25) becomes

$$\phi\,(x) = \frac{\beta^2 \pi}{\rho\,(\epsilon^{\rho\pi} + \epsilon^{-\rho\pi})}\; \sqrt{x-x^2}\sin\left(\rho \log \frac{x}{1-x}\right).......(28).$$

If $f(x)$ be calculated from $x = 0$ to $x = \frac{1}{2}$, equation (21) will enable us to calculate it readily from $x = \frac{1}{2}$ to $x = 1$, since it is easy to calculate $\phi(x)$.

9. A series of a simple form, which is more rapidly convergent than (9) when x approaches the value $\frac{1}{2}$, may readily be investigated.

Let $x = \frac{1}{2}(1 + w)$; then substituting in equation (8) we get

$$\frac{1}{4}\frac{d^2}{dw^2}\{(1 - w^2)^2 z\} + \beta z = \beta \quad\ldots\ldots\ldots\ldots(29).$$

Assume

$$z = B_0 + B_1 w^2 + B_2 w^4 \ldots = \Sigma B_i w^{2i}\ldots\ldots\ldots\ldots(30),$$

then substituting in (29) we get

$$\Sigma B_i\{2i(2i - 1) w^{2i-2} - 2(2i + 2)(2i + 1) w^{2i}$$
$$+ (2i + 4)(2i + 3) w^{2i+2} + 4\beta w^{2i}\} = 4\beta,$$

or,

$$\Sigma\{i(2i - 1) B_i - 2[i(2i - 1) - \beta] B_{i-1} + i(2i - 1) B_{i-2}\} w^{2i-2} = 2\beta.$$

This equation leaves B_0 arbitrary, and gives on dividing by $i(2i - 1)$, and putting in succession $i = 1$, $i = 2$, &c.,

$$B_1 - 2\left(1 - \frac{\beta}{1 \cdot 1}\right) B_0 = 2\beta\ldots\ldots\ldots\ldots\ldots(31),$$

$$B_2 - 2\left(1 - \frac{\beta}{2 \cdot 3}\right) B_1 + B_0 = 0, \text{ &c. ;}$$

and generally when $i > 1$,

$$B_i = B_{i-1} + \Delta B_{i-2} - \frac{2\beta}{i(2i - 1)} B_{i-1} \ldots\ldots\ldots\ldots(32).$$

The constants B_1, B_2,... being thus determined, the series (30) will be an integral of equation (29), containing one arbitrary constant. An integral of the equation derived from (29) by replacing the second member by zero may be obtained in just the same way by assuming $z = C_0 w + C_1 w^3 + \ldots$ when C_1, C_2... will be determined in terms of C_1, which remains arbitrary. The series will both be convergent between the limits $w = -1$ and $w = 1$, that is, between the limits $x = 0$ and $x = 1$. The sum of the two series will be the complete integral of (29), and will be equal to $(x - x^2)^{-2} f(x)$ if the constants B_0, C_0 be properly determined. Denoting the sums of the two series by $F_e(w)$, $F_0(w)$ respectively, and writing $\sigma(x)$ for $(x - x^2)^{-2} f(x)$, so that $z = \sigma(x)$, we get

$$\sigma(x) = F_e(w) + F_0(w), \quad \sigma(1 - x) = F_e(w) - F_0(w);$$

and since $2F_0(w) = \sigma(x) - \sigma(1 - x) = (x - x^2)^{-2} \phi(x)$ by (21), we get

$$\left. \begin{aligned} \sigma(x) &= F_e(w) + \tfrac{1}{2}(x - x^2)^{-2}\phi(x), \\ \sigma(1 - x) &= F_e(w) - \tfrac{1}{2}(x - x^2)^{-2}\phi(x) \end{aligned} \right\} \dots\dots (33).$$

To determine B_0 we have

$$B_0 = \sigma(\tfrac{1}{2}) \dots\dots\dots\dots\dots\dots\dots (34),$$

which may be calculated by the series (9).

10.   The series (9), (30) will ultimately be geometric series with ratios x, w^2, or x, $(2x - 1)^2$, respectively. Equating these ratios, and taking the smaller root of the resulting quadratic, we get $x = \tfrac{1}{4}$. Hence if we use the series (9) for the calculation of $\sigma(x)$ from $x = 0$ to $x = \tfrac{1}{4}$, and (30) for the calculation of $\sigma(x)$ from $x = \tfrac{1}{4}$ to $x = \tfrac{1}{2}$, we shall have to calculate series which are ultimately geometric series with ratios ranging from 0 to $\tfrac{1}{4}$.

Suppose that we wish to calculate $\sigma(x)$ or z for values of x increasing by ·02. The process of calculation will be as follows. From the equation $(2 + \beta) A_0 = \beta$ and the general formula (12), calculate the coefficients A_0, A_1, A_2,... as far as may be necessary. From the series (9), or else from the series (9) combined with the formula (15), calculate $\sigma(\tfrac{1}{2})$ or B_0, and then calculate B_1, B_2... from equations (31), (32). Next calculate $\sigma(x)$ from the series (9) for the values ·02, ·04,... ·26 of x, and $F_e(w)$ from (30) for the values ·04, ·08..., ·44 of w, and lastly $(x - x^2)^{-2}\phi(x)$ for the values ·52, ·54..., ·98 of x. Then we have $\sigma(x)$ calculated directly from $x = 0$ to $x = ·26$; equations (33) will give $\sigma(x)$ from $x = ·28$ to $x = ·72$, and lastly the equation $\sigma(x) = \sigma(1 - x) + (x - x^2)^{-2}\phi(x)$ will give $\sigma(x)$ from $x = ·74$ to $x = 1$.

11.   The equation (21) will enable us to express in finite terms the vertical velocity of the body at the centre of the bridge. For according to the notation of Art. 2, the horizontal and vertical co-ordinates of the body are respectively $2cx$ and $16Sy$, and we have also $d.2cx/dt = V$, whence, if v be the vertical velocity, we get

$$v = \frac{d.16Sy}{dx}\frac{dx}{dt} = \frac{8SV}{c} f'(x).$$

But (21) gives $f'\left(\tfrac{1}{2}\right) = \tfrac{1}{2}\,\phi'\left(\tfrac{1}{2}\right)$, whence if v_c be the value of v at the centre, we get from (25) or (28)

$$v_c = \frac{4\pi S V \beta^2}{c\,\cos r\pi}, \text{ or } = \frac{8\pi S V \beta^2}{c\,(\epsilon^{\rho\pi} + \epsilon^{-\rho\pi})} \quad\ldots\ldots\ldots\ldots (35),$$

according as $\beta < > \tfrac{1}{4}$.

In the extreme cases in which V is infinitely great and infinitely small respectively, it is evident that v_c must vanish, and therefore for some intermediate value of V, v_c must be a maximum. Since $V \propto \beta^{-\frac{1}{2}}$ when the same body is made to traverse the same bridge with different velocities, v_c will be a maximum when p or q is a minimum, where

$$p = 2\beta^{-\frac{3}{2}}\cos r\pi, \quad q = \beta^{-\frac{3}{2}}(\epsilon^{\rho\pi} + \epsilon^{-\rho\pi}).$$

Putting for $\cos r\pi$ its expression in a continued product, and replacing r by its expression (23) in terms of β, we get

$$p = 8\beta^{-\frac{1}{2}}\left(1 - \frac{1-4\beta}{3^2}\right)\left(1 - \frac{1-4\beta}{5^2}\right)\ldots\ldots$$

whence

$$\frac{d\log p}{d\beta} = -\frac{1}{2\beta} + \frac{1}{1\,.\,2+\beta} + \frac{1}{2\,.\,3+\beta} + \ldots\ldots (36).$$

The same expression would have been obtained for $d\log q/d\beta$. Call the second member of equation (36) $F(\beta)$, and let $-N$, P be the negative and positive parts respectively of $F(\beta)$. When $\beta = 0$, $N = \infty$, and $P = \dfrac{1}{1\,.\,2} + \dfrac{1}{2\,.\,3}\ldots = 1$, and therefore $F(\beta)$ is negative. When β becomes infinite, the ratio of P to N becomes infinite, and therefore $F(\beta)$ is positive when β is sufficiently large; and $F(\beta)$ alters continuously with β. Hence the equation $F(\beta) = 0$ must have at least one positive root. But it cannot have more than one; for the rates of proportionate decrease of the quantities N, P, or $-1/N\,.\,dN/d\beta$, $-1/P\,.\,dP/d\beta$, are respectively

$$\frac{1}{\beta}, \quad \frac{(1\,.\,2+\beta)^{-2} + (2\,.\,3+\beta)^{-2} + \ldots}{(1\,.\,2+\beta)^{-1} + (2\,.\,3+\beta)^{-1} + \ldots},$$

and the several terms of the denominator of the second of these expressions are equal to those of the numerator multiplied by $1\,.\,2+\beta$, $2\,.\,3+\beta,\ldots$ respectively, and therefore the denominator is equal to the numerator multiplied by a quantity greater than $2 + \beta$, and therefore greater than β; so that the value of the expression is less than $1/\beta$. Hence for a given infinitely small increment of β the change $- dN$ in N bears to N a greater ratio than $- dP$ bears to P, so that when N is greater than or equal to P it is decreasing more rapidly than P, and therefore after having once become equal to P it must remain always less than P. Hence v_c admits of but one maximum or minimum value, and this must evidently be a maximum.

When $\beta = \frac{1}{4}$, $N = 2$, and $P < \dfrac{1}{1.2} + \dfrac{1}{2.3} + \ldots$ or < 1, and therefore $F(\beta)$ has the same sign as when β is indefinitely small. Hence it is q and not p which becomes a minimum. Equating $dq/d\beta$ to zero, employing (27), and putting $2\pi\rho = \log_e \zeta$, we find

$$\frac{6(\zeta + 1)}{\zeta - 1} = \log_e \zeta + \pi^2 (\log_e \zeta)^{-1}.$$

The real positive root of this equation will be found by trial to be $36 \cdot 3$ nearly, which gives $\rho = \cdot 5717$, $\beta = \frac{1}{4} + \rho^2 = \cdot 5768$. If V_1 be the velocity which gives v_c a maximum, v_1 the maximum value of v_o, U the velocity due to the height S, we get

$$V_1 = \sqrt{\frac{gc^2}{4\beta S}} = \frac{c}{S} \frac{U}{\sqrt{8\beta}}, \text{ and } v_1 = \frac{8\pi\beta^2}{\zeta^{\frac{1}{2}} + \zeta^{-\frac{1}{2}}} \frac{S}{c} V_1, \text{ whence}$$

$$V_1 = \cdot 4655 \frac{c}{S} U, \ v_1 = \cdot 6288 U.$$

12. Conceive a weight W placed at rest on a point of the bridge whose distance from the first extremity is to the whole length as x to 1. The reaction at this extremity produced by W will be equal to $(1 - x) W$, and the moment of this reaction about a point of the bridge whose abscissa $2cx_1$ is less than $2cx$ will be $2c(1 - x)x_1 W$. This moment measures the tendency of the bridge to break at the point considered, and it is evidently greatest when $x_1 = x$, in which case it becomes $2c(1 - x)xW$. Now, if the inertia of the bridge be neglected, the pressure R produced by the moving body will be proportional to $(x - x^2)^{-2}y$, and the tendency to break under the action of a weight equal to R placed at rest on the bridge will be proportional to $(1 - x)x \times (x - x^2)^{-2}y$, or to $(x - x^2)z$.

Call this tendency T, and let T be so measured that it may be equal to 1 when the moving body is placed at rest on the centre of the bridge.   Then $T = C\,(x - x^2)\,z$, and $1 = C\,(\tfrac{1}{2} - \tfrac{1}{4})$, whence

$$T = 4\,(x - x^2)\,z.$$

The tendency to break is actually liable to be somewhat greater than T, in consequence of the state of vibration into which the bridge is thrown, in consequence of which the curvature is alternately greater and less than the statical curvature due to the same pressure applied at the same point.   In considering the motion of the body, the vibrations of the bridge were properly neglected, in conformity with the supposition that the inertia of the bridge is infinitely small compared with that of the body.

The quantities of which it will be most interesting to calculate the numerical values are z, which expresses the ratio of the depression of the moving body at any point to the statical depression, T, the meaning of which has just been explained, and y', the actual depression.   When z has been calculated in the way explained in Art. 10, T will be obtained by multiplying by $4\,(x - x^2)$, and then y'/S will be got by multiplying T by $4\,(x - x^2)$.

13.   The following Table gives the values of these three quantities for each of four values of β, namely $\tfrac{5}{36}$, $\tfrac{1}{4}$, $\tfrac{1}{2}$, and $\tfrac{5}{4}$, to which correspond $r = \tfrac{1}{3}$, $r = 0$, $\rho = \tfrac{1}{2}$, $\rho = 1$, respectively.   In performing the calculations I have retained five decimal places in calculating the coefficients A_0, A_1, A_2... and B_0, B_1, B_2... and four in calculating the series (9) and (30).   In calculating $\phi\,(x)$ I have used four-figure logarithms, and I have retained three figures in the result. The calculations have not been re-examined, except occasionally, when an irregularity in the numbers indicated an error.

14.   Let us first examine the progress of the numbers.   For the first two values of β, z increases from a small positive quantity up to ∞ as x increases from 0 to 1.   As far as the table goes, z is decidedly greater for the second of the two values of β than for the first.   It is easily proved however that before x attains the value 1, z becomes greater for the first value of β than for the second.   For if we suppose x very little less than 1, $f(1 - x)$ will be extremely small compared with $\phi\,(x)$, or, in case $\phi\,(x)$ contain a sine, compared with the coefficient of the sine.   Writing x_1 for

TABLE I.*

x	z				T				y'/S			
	$\beta=\tfrac{5}{36}$	$\beta=\tfrac{1}{4}$	$\beta=\tfrac{1}{2}$	$\beta=\tfrac{5}{4}$	$\beta=\tfrac{5}{36}$	$\beta=\tfrac{1}{4}$	$\beta=\tfrac{1}{2}$	$\beta=\tfrac{5}{4}$	$\beta=\tfrac{5}{36}$	$\beta=\tfrac{1}{4}$	$\beta=\tfrac{1}{2}$	$\beta=\tfrac{5}{4}$
·00	·065	·111	·200	·385	·000	·000	·000	·000	·000	·000	·000	·000
·05	·071	·122	·220	·419	·014	·023	·042	·079	·003	·004	·008	·016
·10	·080	·136	·243	·457	·029	·049	·087	·165	·010	·018	·032	·059
·15	·089	·151	·270	·502	·045	·077	·137	·256	·023	·039	·070	·131
·20	·100	·170	·302	·553	·064	·109	·193	·354	·041	·070	·124	·227
·25	·114	·193	·339	·614	·085	·144	·254	·460	·064	·108	·191	·345
·30	·130	·220	·386	·676	·109	·185	·324	·568	·092	·155	·272	·477
·35	·150	·253	·441	·767	·137	·231	·402	·697	·125	·210	·366	·635
·40	·176	·295	·509	·870	·169	·283	·489	·835	·162	·272	·470	·802
·45	·208	·348	·595	·993	·206	·344	·589	·980	·204	·336	·583	·974
·50	·250	·416	·705	1·14	·250	·416	·705	1·14	·250	·416	·705	1·14
·55	·307	·508	·850	1·33	·304	·503	·841	1·32	·300	·497	·833	1·31
·60	·385	·633	1·05	1·58	·370	·608	1·00	1·52	·355	·584	·964	1·46
·65	·499	·814	1·32	1·90	·454	·740	1·20	1·73	·413	·673	1·09	1·57
·70	·671	1·08	1·72	2·32	·563	·910	1·45	1·95	·473	·765	1·22	1·64
·75	·956	1·52	2·36	2·94	·715	1·14	1·76	2·20	·535	·852	1·32	1·65
·80	1·46	2·30	3·43	3·81	·935	1·46	2·19	2·44	·598	·933	1·40	1·56
·85	2·56	3·93	5·56	5·10	1·29	1·99	2·82	2·59	·658	1·00	1·43	1·32
·90	5·47	8·12	10·6	6·81	1·97	2·92	3·80	2·45	·708	1·05	1·37	·883
·92	8·34	12·1	15·0	7·27	2·45	3·57	4·41	2·14	·723	1·05	1·30	·630
·94	14·3	20·3	23·2	6·44	3·25	4·58	5·23	1·45	·730	1·00	1·18	·318
·96	29·6	43·5	41·8	− ·600	4·55	6·69	6·42	− ·09	·699	1·00	1·00	− ·014
·98	112	139	106	− 65·8	8·80	10·9	8·32	− 5·16	·690	·857	·987	− ·404
1·00	∞	∞	±∞	±∞	∞	∞	±∞	±∞	·690	·000	·652	·000

* [Abridged from the original, which is given to intervals ·02 of x.]

$1-x$, and retaining only the most important term in $f(x)$, we get from (21), (25), (26), and (28)

$$f(x) = \frac{\beta^2 \pi}{4r \cos r\pi} x_1^{\frac{1}{2}-r}, \ = \frac{\pi}{32} x_1^{\frac{1}{2}} \log \frac{1}{x_1},$$

or
$$= \frac{\beta^2 \pi}{\rho (\epsilon^{\rho\pi} + \epsilon^{-\rho\pi})} x_1^{\frac{1}{2}} \sin \left(\rho \log \frac{1}{x_1} \right) \ \ldots\ldots\ldots\ldots (37)$$

according as $\beta < \frac{1}{4}$, $\beta = \frac{1}{4}$, or $\beta > \frac{1}{4}$; and z will be obtained by dividing $f(x)$ by x_1^2 nearly. Hence if $\frac{1}{4} > \beta_2 > \beta_1 > 0$, z is ultimately incomparably greater when $\beta = \beta_1$ than when $\beta = \beta_2$, and when $\beta = \beta_2$ than when $\beta = \frac{1}{4}$. Since $f(0) = A_0 = \beta (2 + \beta)^{-1} = (2\beta^{-1} + 1)^{-1}$, $f(0)$ increases with β, so that $f(x)$ is at first larger when $\beta = \beta_2$ than when $\beta = \beta_1$, and afterwards smaller.

When $\beta > \frac{1}{4}$, z vanishes for a certain value of x, after which it becomes negative, then vanishes again and becomes positive, and so on an infinite number of times. The same will be true of T. If ρ be small, $f(x)$ will not greatly differ, except when x is nearly equal to 1, from what it would be if ρ were equal to zero, and therefore $f(x)$ will not vanish till x is nearly equal to 1. On the other hand, if ρ be extremely large, which corresponds to a very slow velocity, z will be sensibly equal to 1 except when x is nearly equal to 1, so that in this case also $f(x)$ will not vanish till x is nearly equal to 1. The table shews that when $\beta = \frac{1}{2}$, $f(x)$ first vanishes between $x = {\cdot}98$ and $x = 1$, and when $\beta = \frac{5}{4}$ between $x = {\cdot}94$ and $x = {\cdot}96$. The first value of x for which $f(x)$ vanishes is probably never much less than 1, because as β increases from $\frac{5}{4}$ the denominator $\rho (\epsilon^{\rho\pi} + \epsilon^{-\rho\pi})$ in the expression for $\phi(x)$ becomes rapidly large.

15. Since when $\beta > \frac{1}{4}$, T vanishes when $x = 0$, and again for a value of x less than 1, it must be a maximum for some intermediate value. When $\beta = \frac{1}{2}$ the table appears to indicate a maximum beyond $x = {\cdot}98$. When $\beta = \frac{5}{4}$, the maximum value of T is about $2{\cdot}61$, and occurs when $x = {\cdot}86$ nearly. As β increases indefinitely, the first maximum value of T approaches indefinitely to 1, and the corresponding value of x to $\frac{1}{2}$. Besides the first maximum, there are an infinite number of alternately negative and positive maxima; but these do not correspond to the problem, for a reason which will be considered presently.

16. The following curves represent the trajectory of the body for the four values of β contained in the preceding table. These curves, it must be remembered, correspond to the ideal limiting case in which the inertia of the bridge is infinitely small.

In this figure the right line AB represents the bridge in its position of equilibrium, and at the same time represents the trajectory of the body in the ideal limiting case in which $\beta = 0$ or $V = \infty$. $AeeeB$ represents what may be called the *equilibrium trajectory*, or the curve the body would describe if it moved along the bridge with an infinitely small velocity. The trajectories corresponding to the four values of β contained in the above table are marked by $1, 1, 1, 1$; $2, 2, 2$; $3, 3, 3$; $4, 4, 4, 4$ respectively. The dotted curve near B is meant to represent the parabolic arc which the body really describes after it rises above the horizontal line AB*. C is the centre of the right line AB: the curve $AeeeB$ is symmetrical with respect to an ordinate drawn through C.

17. The inertia of the bridge being neglected, the reaction of the bridge against the body, as already observed, will be represented by $Cy/(x - x^2)^2$, where C depends on the length and stiffness of the bridge. Since this expression becomes negative with y, the preceding solution will not be applicable beyond the value of x for which y first vanishes, unless we suppose the body held down to the bridge by some contrivance. If it be not so held, which in fact is the case, it will quit the bridge when y becomes nega-

---

* The dotted curve ought to have been drawn wholly outside the full curve. The two curves touch each other at the point where they are cut by the line ACB, as is represented in the figure.

tive. More properly speaking, the bridge will follow the body, in consequence of its inertia, for at least a certain distance above the horizontal line AB, and will exert a positive pressure against the body : but this pressure must be neglected for the sake of consistency, in consequence of the simplification adopted in Art. 1, and therefore the body may be considered to quit the bridge as soon as it gets above the line AB. The preceding solution shews that when $\beta > \frac{1}{4}$ the body will inevitably leap before it gets to the end of the bridge. The leap need not be high; and in fact it is evident that it must be very small when β is very large. In consequence of the change of conditions, it is only the first maximum value of T which corresponds to the problem, as has been already observed.

18. According to the preceding investigation, when $\beta < \frac{1}{4}$ the body does not leap, the tangent to its path at last becomes vertical, and T becomes infinite. The occurrence of this infinite value indicates the failure, in some respect, of the system of approximation adopted. Now the inertia of the bridge has been neglected throughout; and, consequently, in the system of the bridge and the moving body, that amount of labouring force which is requisite to produce the *vis viva* of the bridge has been neglected. If ξ, η be the co-ordinates of any point of the bridge on the same scale on which x, y represent those of the body, and ξ be less than x, it may be proved on the supposition that the bridge may be regarded at any instant as in equilibrium, that

$$\frac{2\eta}{y} = \left(\frac{2}{x} + \frac{1}{1-x}\right) \xi - \frac{\xi^3}{x^2(1-x)} \dots\dots\dots\dots(38).$$

When x becomes very nearly equal to 1, y varies ultimately as $(1-x)^{\frac{1}{4}-r}$, and therefore η contains terms involving $(1-x)^{-\frac{1}{4}-r}$, and $(d\eta/dx)^2$, and consequently $(d\eta/dt)^2$ contains terms involving $(1-x)^{-3-2r}$. Hence the expression for the *vis viva* neglected at last becomes infinite; and therefore however light the bridge may be, the mode of approximation adopted ceases to be legitimate before the body comes to the end of the bridge. The same result would have been arrived at if β had been supposed equal to or greater than $\frac{1}{4}$.

19. There is one practical result which seems to follow from the very imperfect solution of the problem which is obtained when

the inertia of the bridge is neglected. Since this inertia is the main cause which prevents the tendency to break from becoming enormously great, it would seem that of two bridges of equal length and equal strength, but unequal mass, the lighter would be the more liable to break under the action of a heavy body moving swiftly over it. The effect of the inertia may possibly be thought worthy of experimental investigation.

20.   The mass of a rail on a railroad must be so small compared with that of an engine, or rather with a quarter of the mass of an engine, if we suppose the engine to be a four-wheeled one, and the weight to be equally distributed between the four wheels, that the preceding investigation must be nearly applicable till the wheel is very near the end of the rail on which it was moving, except in so far as relates to regarding the wheel as a heavy point. Consider the motion of the fore wheels, and for simplicity suppose the hind wheels moving on a rigid horizontal plane. Then the fore wheels can only ascend or descend by the turning of the whole engine round the hind axle, or else the line of contact of the hind wheels with the rails, which comes to nearly the same thing.   Let M be the mass of the whole engine, l the horizontal distance between the fore and hind axles, h the horizontal distance of the centre of gravity from the latter axle, k the radius of gyration about the hind axle, x, y the coordinates of the centre of one of the fore wheels, and let the rest of the notation be as in Art. 1.   Then to determine the motion of this wheel we shall have

$$Mk^2 \frac{d^2}{dt^2}\left(\frac{y}{l}\right) = Mgh - \frac{2Cy}{(2cx - x^2)^2} \cdot l,$$

whereas to determine the motion of a single particle whose mass is $\frac{1}{4}M$ we should have had

$$\frac{M}{4}\frac{d^2y}{dt^2} = \frac{M}{4}g - \frac{Cy}{(2cx - x^2)^2}.$$

Now h must be nearly equal to $\frac{1}{2}l$, and k^2 must be a little greater than $\frac{1}{3}l^2$, say equal to $\frac{1}{2}l^2$, so that the two equations are very nearly the same.

Hence, β being the quantity defined by equation (5), where S denotes the central statical deflection due to a weight $\frac{1}{4}Mg$, it appears that the rail ought to be made so strong, or else so short, as to render β a good deal larger than $\frac{1}{4}$. In practice, however, a rail does not rest merely on the chairs, but is supported throughout its whole length by ballast rammed underneath.

21.   In the case of a long bridge, β would probably be large in practice. When β is so large that the coefficient $\beta^2 \pi / \rho \, (\epsilon^{\rho\pi} + \epsilon^{-\rho\pi})$, or $\pi\beta^{\frac{3}{2}}\epsilon^{-\pi\beta^{\frac{1}{2}}}$ nearly, in $\phi(x)$ may be neglected, the motion of the body is sensibly symmetrical with respect to the centre of the bridge, and consequently T, as well as y, is a maximum when $x = \frac{1}{2}$. For this value of x we have $4\,(x - x^2) = 1$, and therefore $z = T = y$. Putting C_i for the $(i + 1)^{\text{th}}$ term of the series (9), so that $C_i = A_i 2^{-i}$, we have for $x = \frac{1}{2}$

$$T = C_0 + C_1 + C_2 + \ldots \ldots \ldots \ldots \ldots \ldots (39)$$

where
$$C_0 = \frac{\beta}{2 + \beta}, \quad C_1 = \frac{6\,C_0}{6 + \beta},$$

and generally,

$$C_i = \frac{(i + 1)\,(i + 2)}{(i + 1)\,(i + 2) + \beta} \left\{ C_{i-1} - \tfrac{1}{4}\,C_{i-2} \right\},$$

whence T is easily calculated. Thus for $\beta = 5$ we have $\pi\beta^{\frac{3}{2}}\,\epsilon^{-\pi\beta^{\frac{1}{2}}} = \cdot 031$ nearly, which is not large, and we get from the series (39) $T = 1 \cdot 27$ nearly. For $\beta = 10$, the approximate value of the coefficient in $\phi(x)$ is $\cdot 0048$, which is very small, and we get $T = 1 \cdot 14$. In these calculations the inertia of the bridge has been neglected, but the effect of the inertia would probably be rather to diminish than to increase the greatest value of T.

22.   The inertia of a bridge such as one of those actually in use must be considerable: the bridge and a carriage moving over it form a dynamical system in which the inertia of all the parts ought to be taken into account. Let it be required *to construct the same dynamical system on a different scale.* For this purpose it will be necessary to attend to the dimensions of the different constants on which the unknown quantities of the problem depend, with respect to each of the independent units involved in the problem. Now if the thickness of the bridge be regarded as very small compared with its length, and the moving body be regarded as a heavy particle, the only constants which enter into the problem are M, the mass of the body, M', the mass of the bridge, $2c$, the length of the bridge, S, the central statical deflection, V, the

horizontal velocity of the body, and g, the force of gravity. The independent units employed in dynamics are three, the unit of length, the unit of time, and the unit of density, or, which is equivalent, and which will be somewhat more convenient in the present case, the unit of length, the unit of time, and the unit of mass. The dimensions of the several constants M, M', &c., with respect to each of these units are given in the following table.

	Unit of length.	Unit of time.	Unit of mass.
M and M'.	0	0	1
c and S.	1	0	0
V.	1	-1	0
g.	1	-2	0

Now any result whatsoever concerning the problem will consist of a relation between certain unknown quantities x', x'' ... and the six constants just written, a relation which may be expressed by

$$f(x', x'', ...M, M', c, S, V, g) = 0 \quad\text{...............} (40).$$

But by the principle of homogeneity and by the preceding table this equation must be of the form

$$F\left\{\frac{x'}{(x')}, \frac{x''}{(x'')} ..., \frac{M'}{M}, \frac{S}{c}, \frac{V^2}{cg}\right\} = 0 \quad\text{............}(41),$$

where (x'), (x'') ..., denote any quantities made up of the six constants in such a manner as to have with respect to each of the independent units the same dimensions as x', x'' ..., respectively. Thus, if (40) be the equation which gives the maximum value T, of T in terms of the six constants, we shall have but one unknown quantity x', where $x' = T_i$, and we may take for (x'), Mcg, or else $M'V^2$. If (40) be the equation to the trajectory of the body, we shall have two unknown constants, x', x'', where x' is the same as in Art. 1, and $x'' = y'$, and we may take $(x') = c$, $(x'') = c$. The equation (41) shews that in order to keep to the same dynami- cal system, only on a different scale, we must alter the quantities M, M', &c. in such a manner that

$$M' \propto M, \ S \propto c, \ V^2 \propto cg,$$

and consequently, since g is not a quantity which we can alter at pleasure in our experiments, V must vary as $\sqrt{c}$. A small system constructed with attention to the above variations forms an *exact dynamical model* of a larger system with respect to which it may

be desired to obtain certain results.   It is not even necessary for
the truth of this statement that the thickness of the large bridge
be small in comparison with its length, provided that the same
proportionate thickness be preserved in the model.

To take a numerical example, suppose that we wished, by
means of a model bridge five feet long and weighing 100 ounces,
to investigate the greatest central deflection produced by an
engine weighing 20 tons, which passes with the successive velocities of 30, 40, and 50 miles an hour over a bridge 50 feet long
weighing 100 tons, the central statical deflection produced by
the engine being one inch.   We must give to our model carriage
a weight of 20 ounces, and make the small bridge of such a stiffness that a weight of 20 ounces placed on the centre shall cause
a deflection of $\frac{1}{10}$th of an inch; and then we must give to the
carriage the successive velocities of $3\sqrt{10}$, $4\sqrt{10}$, $5\sqrt{10}$, or 9·49,
12·65, 15·81 miles per hour, or 13·91, 18·55, 23·19 feet per second.
If we suppose the observed central deflections in the model to be
·12, ·16, ·18 of an inch, we may conclude that the central deflections in the large bridge corresponding to the velocities of 30, 40,
and 50 miles per hour would be 1·2, 1·6, and 1·8 inch.

Addition to the preceding Paper.

Since the above was written, Professor Willis has informed me
that the values of β are much larger in practice than those which
are contained in Table I., on which account it would be interesting
to calculate the numerical values of the functions for a few larger
values of β.   I have accordingly performed the calculations for
the values 3, 5, 8, 12, and 20.   The results are contained in
Table II.   In calculating z from $x = 0$ to $x = \cdot5$, I employed the
formula (12), with the assistance occasionally of (15).   I worked
with four places of decimals, of which three only are retained.
The values of z for $x = \cdot5$, in which case the series are least convergent, have been verified by the formula (42) given below: the
results agreed within two or three units in the fourth place of
decimals.   The remaining values of z were calculated from the

TABLE II.

x	z					T					y'_s				
	$\beta=3$	$\beta=5$	$\beta=8$	$\beta=12$	$\beta=20$	$\beta=3$	$\beta=5$	$\beta=8$	$\beta=12$	$\beta=20$	$\beta=3$	$\beta=5$	$\beta=8$	$\beta=12$	$\beta=20$
·00	·600	·714	·800	·857	·909	·000	·000	·000	·000	·000	·000	·000	·000	·000	·000
·05	·640	·755	·835	·886	·931	·122	·143	·159	·168	·177	·023	·027	·030	·032	·034
·10	·689	·798	·872	·915	·950	·248	·287	·314	·330	·342	·089	·103	·113	·119	·123
·15	·751	·846	·910	·945	·970	·383	·431	·464	·482	·495	·195	·220	·237	·246	·252
·20	·799	·897	·950	·975	·989	·511	·574	·608	·624	·633	·327	·367	·389	·399	·405
·25	·863	·951	·991	1·004	1·016	·647	·714	·743	·753	·762	·486	·535	·558	·565	·572
·30	·936	1·010	1·023	1·032	1·023	·786	·849	·859	·867	·859	·661	·713	·722	·728	·721
·35	1·018	1·073	1·074	1·059	1·038	·926	·976	·977	·963	·944	·843	·883	·889	·877	·859
·40	1·110	1·133	1·114	1·081	1·049	1·066	1·092	1·069	1·038	1·007	1·023	1·049	1·026	·997	·966
·45	1·214	1·207	1·150	1·099	1·056	1·202	1·195	1·138	1·089	1·046	1·190	1·183	1·127	1·078	1·035
·50	1·331	1·274	1·180	1·111	1·060	1·331	1·274	1·180	1·111	1·060	1·331	1·274	1·180	1·111	1·060
·55	1·461	1·341	1·203	1·114	1·058	1·446	1·327	1·191	1·103	1·047	1·431	1·314	1·179	1·092	1·037
·60	1·602	1·390	1·202	1·105	1·051	1·538	1·334	1·154	1·060	1·009	1·486	1·281	1·108	1·018	·968
·65	1·748	1·417	1·179	1·081	1·038	1·590	1·289	1·072	·983	·945	1·446	1·173	·954	·895	·860
·70	1·891	1·393	1·107	1·039	1·021	1·588	1·170	·930	·873	·858	1·334	·983	·781	·733	·720
·75	1·974	1·273	1·003	·984	1·013	1·481	·955	·752	·738	·760	1·111	·716	·564	·554	·570
·80	1·885	·968	·832	·932	·939	1·206	·620	·532	·596	·633	·772	·396	·341	·382	·405
·85	1·286	·344	·660	·925	·976	·656	·176	·336	·472	·498	·335	·090	·172	·241	·254
·90	-·970	-·616	·802	1·013	·947	-·349	-·222	·289	·365	·341	-·126	-·080	·104	·131	·123
·95	-8·227	+1·248	1·884	·720	·943	-1·563	+·237	·358	·137	·179	-·297	+·045	·068	·026	·034

expression for $(x - x^2)^{-2}\phi(x)$. The values of T and y'/S were
deduced from those of z, by merely multiplying twice in succession by $4x(1-x)$. Professor Willis has laid down in curves the
numbers contained in the last five columns. In laying down
these curves several errors were detected in the latter half of the
Table, that is, from $x = \cdot55$ to $x \cdot95$. These errors were corrected
by re-examining the calculation; so that I feel pretty confident
that the table as it now stands contains no errors of importance.

The form of the trajectory will be sufficiently perceived by
comparing this table with the curves represented in the figure.
As β increases, the first point of intersection of the trajectory with
the equilibrium trajectory eee moves towards A. Since $z = 1$ at
this point, we get from the part of the table headed " z," for the
abscissa of the point of intersection, by taking proportional parts,
$\cdot34$, $\cdot29$, $\cdot26$, $\cdot24$, and $\cdot22$, corresponding to the respective values
3, 5, 8, 12, and 20 of β. Beyond this point of intersection the
trajectory passes below the equilibrium trajectory, and remains
below it during the greater part of the remaining course. As β
increases, the trajectory becomes more and more nearly symmetrical with respect to C: when $\beta = 20$ the deviation from symmetry may be considered insensible, except close to the extremities
A, B, where however the depression itself is insensible. The
greatest depression of the body, as appears from the column which
gives y', takes place a little beyond the centre; the point of
greatest depression approaches indefinitely to the centre as β
increases. This greatest depression of the body must be carefully
distinguished from the greatest depression of the bridge, which
is decidedly larger, and occurs in a different place, and at a different time. The numbers in the columns headed " T" shew that
T is a maximum for a value of x greater than that which renders
y' a maximum, as in fact immediately follows from a consideration
of the mode in which y' is derived from T. The first maximum
value of T, which according to what has been already remarked
is the only such value that we need attend to, is about $1\cdot59$ for
$\beta = 3$, $1\cdot33$ for $\beta = 5$, $1\cdot19$ for $\beta = 8$, $1\cdot11$ for $\beta = 12$, and $1\cdot06$ for
$\beta = 20$.

When β is equal to or greater than 8, the maximum value
of T occurs so nearly when $x = \cdot5$ that it will be sufficient to suppose $x = \cdot5$. The value of z, T, or y'/S for $x = \cdot5$ may be readily

calculated by the method explained in Art. 21. I have also obtained the following expression for this particular value

$$z = 2\beta - 4\beta^2 \left\{ \frac{1}{1 \cdot 2 + \beta} - \frac{1}{2 \cdot 3 + \beta} + \frac{1}{3 \cdot 4 + \beta} - \dots \right\} \dots \dots (42).$$

When β is small, or only moderately large, the series (42) appears more convenient for numerical calculation, at least with the assistance of a table of reciprocals, than the series (39), but when β is very large the latter is more convenient than the former. In using the series (42), it will be best to sum the series within brackets directly to a few terms, and then find the remainder from the formula

$$u_x - u_{x+1} + u_{x+2} - \dots = \tfrac{1}{2}u_x - \tfrac{1}{4}\Delta u_x + \tfrac{1}{8}\Delta^2 u_x - \dots$$

The formula (42) was obtained from equation (20) by a transformation of the definite integral.   In the transformation of Art. 8, the limits of s will be 1 and ∞, and the definite integral on which the result depends will be

$$\int_1^\infty \frac{s^{m-1} - s^{-m}}{1 + s} ds.$$

The formula (42) may be obtained by expanding the denominator, integrating, and expressing m in terms of β.

·   In practice the values of β are very large, and it will be con·venient to expand according to inverse powers of β.   This may be easily effected by successive substitutions.   Putting for shortness $x - x^2 = X$, equation (4) becomes by a slight transformation

$$y = X^2 - \beta^{-1} X^2 \frac{d^2y}{dx^2},$$

and we have for a first approximation $y = X^2$, for a second

$$y = X^2 - \beta^{-1} X^2 \cdot d^2 X^2 / dx^2,$$

and so on.   The result of the successive substitutions may be expressed as follows :

$$y = X^2 - \beta^{-1} X^2 \frac{d^2}{dx^2} X^2 + \beta^{-2} X^2 \frac{d^2}{dx^2} X^2 \frac{d^2}{dx^2} X^2 - \&c. \dots \dots (43),$$

where each term, taken positively, is derived from the preceding by differentiating twice, and then multiplying by $\beta^{-1} X^2$.

For such large values of β, we need attend to nothing but the value of z for $x = \tfrac{1}{2}$, and this may be obtained from (43) by putting

$x = \frac{1}{2}$, after differentiation, and multiplying by 16. It will however be more convenient to replace x by $\frac{1}{2}(1 + w)$, which gives
$d^2/dx^2 = 4 \cdot d^2/dw^2$; $X^2 = \frac{1}{16} W$, where $W = (1 - w^2)^2$. We thus get
from (43)

$$z = W - (4\beta)^{-1} W \frac{d^2}{dw^2} W + (4\beta)^{-2} W \frac{d^2}{dw^2} W \frac{d^2}{dw^2} W - \ldots,$$

where we must put $w = 0$ after differentiation, if we wish to get
the value of z for $x = \frac{1}{2}$. This equation gives, on performing the
differentiations and multiplications, and then putting $w = 0$,

$$z = 1 + \beta^{-1} + \tfrac{5}{2}\beta^{-2} + 13\beta^{-3} + \ldots\ldots\ldots\ldots (44).$$

In practical cases this series may be reduced to $1 + \beta^{-1}$. The
latter term is the same as would be got by taking into account the
centrifugal force, and substituting, in the small term involving that
force, the radius of curvature of the equilibrium trajectory for the
radius of curvature of the actual trajectory. The problem has
already been considered in this manner by others by whom it has
been attacked.

My attention has recently been directed by Professor Willis
to an article by Mr Cox *On the Dynamical Deflection and Strain
of Railway Girders*, which is printed in *The Civil Engineer and
Architect's Journal* for September, 1848. In this article the
subject is treated in a very original and striking manner. There
is, however, one conclusion at which Mr Cox has arrived which
is so directly opposed to the conclusions to which I have been led,
that I feel compelled to notice it. By reasoning founded on the
principle of *vis viva*, Mr Cox has arrived at the result that the
moving body cannot in any case produce a deflection greater than
double the central statical deflection, the elasticity of the bridge
being supposed perfect. But among the sources of labouring force
which can be employed in deflecting the bridge, Mr Cox has omitted
to consider the *vis viva* arising from the horizontal motion of the
body. It is possible to conceive beforehand that a portion of this
vis viva should be converted into labouring force, which is expended in deflecting the bridge. And this is, in fact, precisely
what takes place. During the first part of the motion, the horizontal component of the reaction of the bridge against the body
impels the body forwards, and therefore increases the *vis viva* due
to the horizontal motion; and the labouring force which produces
this increase being derived from the bridge, the bridge is less

deflected than it would have been had the horizontal velocity of the body been unchanged. But during the latter part of the motion the horizontal component of the reaction acts backwards, and a portion of the *vis viva* due to the horizontal motion of the body is continually converted into labouring force, which is stored up in the bridge. Now, on account of the asymmetry of the motion, the direction of the reaction is more inclined to the vertical when the body is moving over the second half of the bridge than when it is moving over the first half, and moreover the reaction itself is greater, and therefore, on both accounts, more *vis viva* depending upon the horizontal motion is destroyed in the latter portion of the body's course than is generated in the former portion; and therefore, on the whole, the bridge is more deflected than it would have been had the horizontal velocity of the body remained unchanged.

It is true that the change of horizontal velocity is small; but nevertheless, in this mode of treating the subject, it must be taken into account. For, in applying to the problem the principle of *vis viva*, we are concerned with the square of the vertical velocity, and we must not omit any quantities which are comparable with that square. Now the square of the absolute velocity of the body is equal to the sum of the squares of the horizontal and vertical velocities; and the change in the square of the horizontal velocity depends upon the product of the horizontal velocity and the change of horizontal velocity; but this product is not small in comparison with the square of the vertical velocity.

In Art. 22 I have investigated the changes which we are allowed by the general principle of homogeneous quantities to make in the parts of a system consisting of an elastic bridge and a travelling weight, without affecting the results, or altering anything but the scale of the system. These changes are the most general that we are at liberty to make by virtue merely of that general principle, and without examining the particular equations which relate to the particular problem here considered. But when we set down these equations, we shall see that there are some further changes which we may make without affecting our results, or at least without ceasing to be able to infer the results which would be obtained on one system from those actually obtained on another.

In an apparatus recently constructed by Professor Willis, which will be described in detail in the report of the commission, to which the reader has already been referred, the travelling weight moves over a single central trial bar, and is attached to a horizontal arm which is moveable, with as little friction as possible, about a fulcrum carried by the carriage.   In this form of the experiment, the carriage serves merely to direct the weight, and moves on rails quite independent of the trial bar.   For the sake of greater generality I shall suppose the travelling weight, instead of being free, to be attached in this manner to a carriage.

Let M be the mass of the weight, including the arm, k the radius of gyration of the whole about the fulcrum, h the horizontal distance of the centre of gravity from the fulcrum, l the horizontal distance of the point of contact of the weight with the bridge, x, y the co-ordinates of that point at the time t, ξ, η those of any element of the bridge, R the reaction of the bridge against the weight, M' the mass of the bridge, R', R'' the vertical pressures of the bridge at its two extremities, diminished by the statical pressures due to the weight of the bridge alone.   Suppose, as before, the deflection to be very small, and neglect its square.

By D'Alembert's principle the effective moving forces reversed will be in statical equilibrium with the impressed forces.   Since the weight of the bridge is in equilibrium with the statical pressures at the extremities, these forces may be left out in the equations of equilibrium, and the only impressed forces we shall have to consider will be the weight of the travelling body and the reactions due to the motion.   The mass of any element of the bridge will be $M'/2c \,.\, d\xi$ very nearly; the horizontal effective force of this element will be insensible, and the vertical effective force will be $M'/2c \,.\, d^2\eta/dt^2 \,.\, d\xi$, and this force, being reversed, must be supposed to act vertically upwards.

The curvature of the bridge being proportional to the moment of the bending forces, let the reciprocal of the radius of curvature be equal to K multiplied by that moment.   Let A, B be the extremities of the bridge, P the point of contact of the bridge with the moving weight, Q any point of the bridge between A and P.   Then by considering the portion AQ of the bridge we get, taking moments round Q,

$$- \frac{d^2\eta}{d\xi^2} = K \left\{ R'\xi + \frac{M'}{2c} \int_0^\xi \frac{d^2\eta'}{dt^2} (\xi - \xi') \, d\xi' \right\} \dots\dots (45),$$

η' being the same function of ξ' that η is of ξ. To determine K, let S be the central statical deflection produced by the weight Mg resting partly on the bridge and partly on the fulcrum, which is equivalent to a weight $h/l \cdot Mg$ resting on the centre of the bridge. In this case we should have

$$-\frac{d^2\eta}{d\xi^2} = K \frac{Mgh}{2l} \xi.$$

Integrating this equation twice, and observing that $d\eta/d\xi = 0$ when $\xi = c$, and $\eta = 0$ when $\xi = 0$, and that S is the value of η when $\xi = c$, we get

$$K = \frac{6lS}{Mghc^3} \quad \dots\dots\dots\dots\dots\dots\dots\dots (46).$$

Returning now to the bridge in its actual state, we get to determine R', by taking moments about B,

$$R' \cdot 2c - R(2c - x) + \frac{M'}{2c}\int_0^{2c} \frac{d^2\eta'}{dt^2}(2c - \xi')\,d\xi' = 0\dots\dots(47).$$

Eliminating R' between (45) and (47), putting for K its value given by (46), and eliminating t by the equation $dx/dt = V$, we get

$$-\frac{d^2\eta}{d\xi^2} = \frac{3lS}{Mghc^4}\left\{(2c - x)\,\xi R - \frac{M'V^2}{2c}\left[(2c - \xi)\int_0^\xi \frac{d^2\eta'}{dx^2}\,\xi'd\xi'\right.\right.$$

$$\left.\left. + \xi\int_\xi^{2c} \frac{d^2\eta'}{dx^2}(2c - \xi')\,d\xi'\right]\right\} \dots\dots(48),$$

This equation applies to any point of the bridge between A and P. To get the equation which applies to any point between P and B, we should merely have to write $2c - \xi$ for ξ, $2c - x$ for x.

If we suppose the fulcrum to be very nearly in the same horizontal plane with the point of contact, the angle through which the travelling weight turns will be y/l very nearly; and we shall have, to determine the motion of this weight,

$$Mk^2V^2 \frac{d^2y}{dx^2} = Mghl - Rl^2 \dots\dots\dots\dots\dots(49).$$

We have also the equations of condition,

$\eta = 0$ when $x = 0$, for any value of ξ from 0 to $2c$;

$\eta = y$ when $\xi = x$, for any value of x from 0 to $2c$; $\Big\}\dots(50).$

$\eta = 0$ when $\xi = 0$ or $= 2c$; $y = 0$ and $dy/dx = 0$ when $x = 0$

Now the general equations (48), (or the equation answering to it which applies to the portion PB of the bridge,) and (49), combined with the equations of condition (50), whether we can manage them or not, are sufficient for the complete determination of the motion, it being understood that η and $d\eta/d\xi$ vary continuously in passing from AP to PB, so that there is no occasion formally to set down the equations of condition which express this circumstance. Now the form of the equations shews that, being once satisfied, they will continue to be satisfied provided $\eta \propto y$, $\xi \propto x \propto c$, and

$$\frac{y}{c^2} \propto \frac{lSR}{Mghc^3} \propto \frac{lSM'V^2y}{Mghc^4}, \quad Mk^2V^2\frac{y}{c^3} \propto Mghl \propto Rl^2.$$

These variations give, on eliminating the variation of R,

$$y \propto S, \quad \frac{gc^2}{V^2S} \propto \frac{k^2}{hl}, \quad \frac{M}{M'} \propto \frac{l^2}{k^2} \quad \dots\dots\dots\dots\dots(51).$$

Although g is of course practically constant, it has been retained in the variations because it may be conceived to vary, and it is by no means essential to the success of the method that it should be constant. The variations (51) shew that if we have any two systems in which the ratio of Mk^2 to $M'l^2$ is the same, and we conceive the travelling weights to move over the two bridges respectively, with velocities ranging from 0 to ∞, the trajectories described in the one case, and the deflections of the bridge, correspond exactly to the trajectories and deflections in the other case, so that to pass from one to the other, it will be sufficient to alter all horizontal lines on the same scale as the length of the bridge, and all vertical lines on the same scale as the central statical deflection. The velocity in the one system which corresponds to a given velocity in the other is determined by the second of the variations (51).

We may pass at once to the case of a free weight by putting $h = k = l$, which gives

$$y \propto S, \quad V^2S \propto gc^2, \quad M \propto M' \dots\dots\dots\dots\dots(52).$$

The second of these variations shews that corresponding velocities in the two systems are those which give the same value to the constant β. When $S \propto c$ we get $V^2 \propto gc$, which agrees with Art. 22.

In consequence of some recent experiments of Professor Willis's, from which it appeared that the deflection produced by a given weight travelling over the trial bar with a given velocity was in some cases increased by connecting a balanced lever with the centre of the bar, so as to increase its inertia without increasing its weight, while in other cases the deflection was diminished, I have been induced to attempt an approximate solution of the problem, taking into account the inertia of the bridge. I find that when we replace each force acting on the bridge by a uniformly distributed force of such an amount as to produce the same mean deflection as would be produced by the actual force taken alone, which evidently cannot occasion any very material error, and when we moreover neglect the difference between the pressure exerted by the travelling mass on the bridge and its weight, the equation admits of integration in finite terms.

Let the notation be the same as in the investigation which immediately precedes; only, for simplicity's sake, take the length of the bridge for unity, and suppose the travelling weight a heavy particle. It will be easy in the end to restore the general unit of length if it should be desirable. It will be requisite in the first place to investigate the relation between a force acting at a given point of the bridge and the uniformly distributed force which would produce the same mean deflection.

Let a force F act vertically downwards at a point of the bridge whose abscissa is x, and let y be the deflection produced at that point. Then, ξ, η being the co-ordinates of any point of the bridge, we get from (38)

$$\int_0^x \eta \, d\xi = \frac{y}{2} \left\{ x + \frac{x^2}{4(1-x)} \right\}.$$

To obtain $\int_x^1 \eta \, d\xi$, we have only got to write $1-x$ in place of x. Adding together the results, and observing that, according to a formula referred to in Art. 1, $y = 16S \cdot F/Mg \cdot x^2 (1-x)^2$, we obtain

$$\int_0^1 \eta \, d\xi = \frac{2SF}{Mg} \left\{ x(1-x) + x^2(1-x)^2 \right\} \ldots\ldots\ldots\ldots(53);$$

and this integral expresses the mean deflection produced by the force F, since the length of the bridge is unity.

Now suppose the bridge subject to the action of a uniformly distributed force F'. In this case we should have

$$-\frac{d^2\eta}{d\xi^2} = K\{\tfrac{1}{2} F' \xi - \int_0^\xi (\xi - \xi') F' d\xi'\} = \tfrac{1}{2} KF' (\xi - \xi^2).$$

Integrating this equation twice, and observing that $d\eta/d\xi = 0$ when $\xi = \tfrac{1}{2}$, and $\eta = 0$ when $\xi = 0$, and that (46) gives, on putting $l = h$ and $c = \tfrac{1}{2}$, $K = 48S/Mg$, we obtain

$$\eta = \frac{2SF'}{Mg} (\xi - 2\xi^3 + \xi^4) \ldots\ldots\ldots\ldots\ldots(54).$$

This equation gives for the mean deflection

$$\int_0^1 \eta d\xi = \frac{2SF'}{5Mg} \ldots\ldots\ldots\ldots\ldots\ldots(55);$$

and equating the mean deflections produced by the force F acting at the point whose abscissa is x, and by the uniformly distributed force F', we get $F' = uF$, where

$$u = 5x(1 - x) + 5x^2(1 - x)^2 \ldots\ldots\ldots\ldots\ldots(56).$$

Putting μ for the mean deflection, expressing F' in terms of μ, and slightly modifying the form of the quantity within parentheses in (54), we get for the equation to the bridge when at rest under the action of any uniformly distributed force

$$\eta = 5\mu \{\xi(1 - \xi) + \xi^2(1 - \xi)^2\} \ldots\ldots\ldots\ldots(57).$$

If D be the central deflection, $\eta = D$ when $\xi = \tfrac{1}{2}$; so that $D : \mu :: 25 : 16$.

Now suppose the bridge in motion, with the mass M travelling over it, and let x, y be the co-ordinates of M. As before, the bridge would be in equilibrium under the action of the force $M(g - d^2y/dt^2)$ acting vertically downwards at the point whose abscissa is x, and the system of forces such as $M'd\xi \cdot d^2\eta/dt^2$ acting vertically upwards at the several elements of the bridge. According to the hypothesis adopted, the former force may be replaced by a uniformly distributed force the value of which will be obtained by multiplying by u, and each force of the latter system may be replaced by a uniformly distributed force obtained by multiplying by u', where u' is what u becomes when ξ is put for x. Hence if F_1 be the whole uniformly distributed force we have

$$F_1 = M\left(g - \frac{d^2y}{dt^2}\right) u - M' \int_0^1 \frac{d^2\eta}{dt^2} u' d\xi \ldots\ldots\ldots(58).$$

Now according to our hypothesis the bridge must always have the form which it would assume under the action of a uniformly distributed force; and therefore, if μ be the mean deflection at the time t, (57) will be the equation to the bridge at that instant. Moreover, since the point (x, y) is a point in the bridge, we must have $\eta = y$ when $\xi = x$, whence $y = \mu u$. We have also

$$\eta = \mu u', \quad \frac{d^2\eta}{dt^2} = \frac{d^2\mu}{dt^2} u', \quad \int_0^1 \frac{d^2\eta}{dt^2} u' d\xi = \frac{d^2\mu}{dt^2} \int_0^1 u'^2 d\xi = \frac{155}{126} \frac{d^2\mu}{dt^2}.$$

We get from (55), $F_1 = 5Mg\mu/2S$. Making these various substitutions in (58), and replacing d/dt by $V \cdot d/dx$, we get for the differential equation of motion

$$\frac{5Mg}{2S}\mu = Mgu - MV^2 u\frac{d^2\mu u}{dx^2} - \frac{155}{126} M'V^2\frac{d^2\mu}{dx^2}\ldots\ldots(59).$$

Since μ is comparable with S, the several terms of this equation are comparable with

$$Mg, \; Mg, \; MV^2S, \; M'V^2S,$$

respectively. If then V^2S be small compared with g, and likewise M small compared with M', we may neglect the third term, while we retain the others. This term, it is to be observed, expresses the difference between the pressure on the bridge and the weight of the travelling mass. Since $c = \frac{1}{2}$, we have $V^2S/g = 1/16\beta$, which will be small when β is large, or even moderately large. Hence the conditions under which we are at liberty to neglect the difference between the pressure on the bridge and the weight of the travelling mass are, *first*, that β be large, *secondly*, that the mass of the travelling body be small compared with the mass of the bridge. If β be large, but M be comparable with M', it is true that the third term in (59) will be small compared with the leading terms; but then it will be comparable with the fourth, and the approximation adopted in neglecting the third term alone would be faulty, in this way, that of two small terms comparable with each other, one would be retained while the other was neglected. Hence, although the absolute error of our results would be but small, it would be comparable with the difference between the results actually obtained and those which would be obtained on the supposition that the travelling mass moved with an infinitely small velocity.

Neglecting the third term in equation (59), and putting for u its value, we get

$$\frac{d^2\mu}{dx^2} + q^2\mu = 2q^2 S (x - 2x^3 + x^4)\ldots\ldots\ldots(60),$$

where

$$q^2 = \frac{63Mg}{31M'V^2S} = \frac{1008M\beta}{31M'}\ldots\ldots\ldots\ldots(61).$$

The linear equation (60) is easily integrated. Integrating, and determining the arbitrary constants by the conditions that $\mu = 0$, and $d\mu/dx = 0$, when $x = 0$, we get

$$\mu = 2S\left\{x^4 - 2x^3 - \frac{12x^2}{q^2} + \left(1 + \frac{12}{q^2}\right)\left(x - \frac{\sin qx}{q}\right)\right.$$
$$\left. + \frac{24}{q^4}(1 - \cos qx)\right\}\ldots\ldots\ldots(62);$$

and we have for the equation to the trajectory

$$y = 5\mu(x - 2x^3 + x^4) = 5\mu(X + X^2)\ldots\ldots\ldots(63),$$

where as before $X = x(1 - x)$.

When $V = 0$, $q = \infty$, and we get from (62), (63), for the approximate equation to the equilibrium trajectory,

$$y = 10S(X + X^2)^2\ldots\ldots\ldots\ldots\ldots(64);$$

whereas the true equation is

$$y = 16SX^2\ldots\ldots\ldots\ldots\ldots\ldots(65).$$

Since the forms of these equations are very different, it will be proper to verify the assertion that (64) is in fact an approximation to (65). Since the curves represented by these equations are both symmetrical with respect to the centre of the bridge, it will be sufficient to consider values of x from 0 to $\frac{1}{2}$, to which correspond values of X ranging from 0 to $\frac{1}{4}$. Denoting the error of the formula (64), that is the excess of the y in (64) over the y in (65), by $S\delta$, we have

$$\delta = -6X^2 + 20X^3 + 10X^4,$$

$$\frac{d\delta}{dx} = 4(-3 + 15X + 10X^2)X\frac{dX}{dx}.$$

Equating $d\delta/dx$ to zero, we get $X = 0$, $x = 0$, $\delta = 0$, a maximum; $X = \cdot1787$, $x = \cdot233$, $\delta = -\cdot067$, nearly, a minimum; and $x = \frac{1}{2}$, $\delta = -\cdot023$, nearly, a maximum. Hence the greatest error in the approximate value of the ordinate of the equilibrium trajectory is equal to about the one-fifteenth of S.

Putting $\mu = \mu_0 + \mu_1$, $y = y_0 + y_1$, where μ_0, y_0 are the values of μ, y for $q = \infty$, we have

$$\mu_1 = 2S \left\{ \frac{12}{q^2} x (1 - x) - \left(\frac{1}{q} + \frac{12}{q^3} \right) \sin qx + \frac{24}{q^4} (1 - \cos qx) \right\} \dots (66),$$

$$y_1 = 5x (1 - x) \{ 1 + x (1 - x) \} \mu_1 \dots\dots\dots\dots\dots\dots\dots\dots(67).$$

The values of μ_1 and y_1 may be calculated from these formulæ for different values of q, and they are then to be added to the values of μ_0, y_0, respectively, which have to be calculated once for all. If instead of the mean deflection μ we wish to employ the central deflection D, we have only got to multiply the second sides of equations (62), (66) by $\frac{25}{16}$, and those of (63), (67) by $\frac{16}{25}$, and to write D for μ. The following table contains the values of the ratios of D and y to S for ten different values of q, as well as for the limiting value $q = \infty$, which belongs to the equilibrium trajectory.

The numerical results contained in Table III. are represented graphically in figs. 2 and 3 of the woodcut on p. 216, where however some of the curves are left out, in order to prevent confusion in the figures. In these figures the numbers written against the several curves are the values of $2q/\pi$ to which the curves respectively belong, the symbol ∞ being written against the equilibrium curves. Fig. 2 represents the trajectory of the body for different values of q, and will be understood without further explanation. In the curves of fig. 3, the ordinate represents the deflection of the centre of the bridge when the moving body has travelled over a distance represented by the abscissa. Fig. 1, which represents the trajectories described when the mass of the bridge is neglected, is here given for the sake of comparison with fig. 2. The numbers in fig. 1 refer to the values of β. The equilibrium curve represented in this figure is the true equilibrium trajectory expressed by equation (65), whereas the equilibrium curve represented in fig. 2 is the approximate equilibrium trajectory expressed by equation (64). In fig. 1, the body is represented as flying off near the second extremity of the bridge, which is in fact the case. The numerous small oscillations which would take place if the body were held down to the bridge could not be

TABLE III.

x	\multicolumn{11}{c}{Values of $\frac{D}{S}$ when $\frac{2q}{\pi}$ is equal to}

x	1	2	3	4	5	6	8	10	12	16	∞
·00	·000	·000	·000	·000	·000	·000	·000	·000	·000	·000	·000
·05	·004	·004	·005	·006	·007	·008	·014	·019	·025	·041	·156
·10	·009	·013	·022	·027	·037	·053	·081	·117	·158	·239	·307
·15	·017	·028	·048	·075	·108	·146	·234	·327	·412	·530	·449
·20	·025	·052	·099	·159	·231	·309	·469	·607	·696	·707	·580
·25	·041	·093	·177	·285	·406	·531	·746	·871	·884	·707	·696
·30	·056	·144	·282	·451	·626	·787	1·003	1·031	·915	·689	·794
·35	·070	·214	·418	·650	·871	1·045	1·180	1·052	·845	·814	·873
·40	·100	·300	·578	·870	1·115	1·265	1·238	·967	·796	1·017	·930
·45	·134	·399	·757	1·097	1·332	1·412	1·178	·859	·856	1·097	·965
·50	·169	·516	·947	1·310	1·492	1·460	1·036	·812	1·004	·991	·977
·55	·213	·640	1·139	1·491	1·574	1·403	·870	·860	1·127	·862	·965
·60	·256	·776	1·321	1·619	1·562	1·250	·739	·969	1·115	·872	·930
·65	·306	·913	1·482	1·681	1·454	1·027	·682	1·054	·948	·959	·873
·70	·359	1·050	1·609	1·663	1·257	·769	·695	1·031	·718	·924	·794
·75	·419	1·181	1·691	1·560	·990	·517	·746	·869	·549	·707	·696
·80	·475	1·296	1·717	1·371	·677	·303	·777	·604	·499	·472	·580
·85	·533	1·399	1·681	1·106	·350	·149	·733	·325	·516	·384	·449
·90	·586	1·476	1·588	·776	·037	·064	·579	·117	·477	·385	·307
·95	·646	1·525	1·402	·400	−·234	·025	·321	·021	·296	·276	·156
1·00	·699	1·540	1·158	·000	−·446	·019	·000	·001	−·001	·000	·000

x	\multicolumn{11}{c}{Values of $\frac{y}{S}$ when $\frac{2q}{\pi}$ is equal to}

x	1	2	3	4	5	6	8	10	12	16	∞
·00	·000	·000	·000	·000	·000	·000	·000	·000	·000	·000	·000
·05	·001	·001	·001	·001	·001	·001	·002	·003	·004	·006	·025
·10	·003	·004	·007	·008	·012	·017	·025	·037	·050	·075	·096
·15	·008	·013	·022	·034	·050	·067	·108	·150	·190	·244	·207
·20	·015	·031	·059	·095	·137	·184	·279	·360	·414	·420	·344
·25	·029	·056	·126	·203	·290	·378	·532	·621	·630	·504	·496
·30	·045	·117	·230	·366	·509	·640	·814	·839	·744	·560	·646
·35	·063	·191	·374	·581	·778	·934	1·054	·940	·755	·727	·780
·40	·096	·285	·550	·828	1·062	1·205	1·178	·921	·759	·969	·886
·45	·133	·394	·748	1·085	1·316	1·395	1·164	·849	·846	1·084	·954
·50	·169	·516	·947	1·310	1·492	1·460	1·036	·812	1·004	·991	·977
·55	·210	·632	1·126	1·473	1·555	1·387	·860	·850	1·114	·852	·954
·60	·244	·739	1·258	1·542	1·487	1·191	·704	·923	1·062	·830	·886
·65	·274	·816	1·325	1·502	1·300	·917	·609	·942	·848	·857	·780
·70	·292	·854	1·308	1·352	1·022	·626	·565	·839	·584	·752	·646
·75	·298	·842	1·205	1·111	·705	·369	·532	·619	·391	·488	·496
·80	·282	·770	1·020	·814	·402	·180	·462	·359	·297	·280	·344
·85	·245	·644	·774	·509	·161	·069	·337	·149	·237	·178	·207
·90	·184	·463	·498	·244	·012	·020	·182	·037	·150	·121	·096
·95	·103	·243	·224	·064	−·037	·004	·051	·003	·047	·044	·025
1·00	·000	·000	·000	·000	·000	·000	·000	·000	·000	·000	·000

properly represented in the figure without using a much larger scale. The reader is however requested to bear in mind the existence of these oscillations, as indicated by the analysis, because,

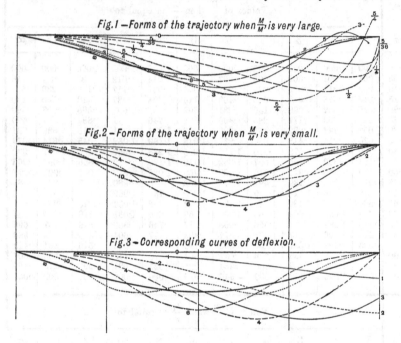

Fig.1 —Forms of the trajectory when $\frac{M}{M'}$ is very large.

Fig.2 — Forms of the trajectory when $\frac{M}{M'}$ is very small.

Fig.3 — Corresponding curves of deflexion.

if the ratio of M to M' altered continuously from ∞ to 0, they would probably pass continuously into the oscillations which are so conspicuous in the case of the larger values of q in fig. 2. Thus the consideration of these insignificant oscillations which, strictly speaking, belong to fig. 1, aids us in mentally filling up the gap which corresponds to the cases in which the ratio of M to M' is neither very small nor very large.

As everything depends on the value of q, in the approximate investigation in which the inertia of the bridge is taken into account, it will be proper to consider further the meaning of this constant. In the first place it is to be observed that although M appears in equation (61), q is really independent of the mass of the travelling body. For, when M alone varies, β varies inversely as S, and S varies directly as M, so that q remains constant. To get rid of the apparent dependence of q on M, let S_1 be the central statical deflection produced by a mass equal to that of the bridge, and at the same time restore the general unit of length. If x continue to denote the ratio of the abscissa of the body to the length of the bridge, q will be numerical, and therefore, to restore the general unit of length, it will be sufficient to take the general expression (5) for β. Let moreover τ be the time the body takes to travel over the bridge, so that $2c = V\tau$; then we get

$$q^2 = \frac{63g\tau^2}{31S_1} \dots\dots\dots\dots\dots\dots (68).$$

If we suppose τ expressed in seconds, and S_1 in inches, we must put $g = 32.2 \times 12 = 386$, nearly, and we get,

$$q = \frac{28\tau}{\sqrt{S_1}} \dots\dots\dots\dots\dots\dots (69).$$

Conceive the mass M removed; suppose the bridge depressed through a small space, and then left to itself. The equation of motion will be got from (59) by putting $M = 0$, where M is not divided by S, and replacing M/S by M'/S_1, and $V \cdot d/dx$ by d/dt. We thus get

$$\frac{d^2\mu}{dt^2} + \frac{63g}{31S_1}\mu = 0 ;$$

and therefore, if P be the period of the motion, or twice the time of oscillation from rest to rest,

$$P = 2\pi \sqrt{\frac{31S_1}{63g}} ; \quad q = 2\pi \frac{\tau}{P} \dots\dots\dots\dots (70).$$

Hence the numbers 1, 2, 3, &c., written at the head of Table III. and against the curves of figs. 2 and 3, represent the number of quarter periods of oscillation of the bridge which elapse during the passage of the body over it. This consideration will materially assist us in understanding the nature of the motion. It should be remarked too that q is increased by diminishing either the velocity of the body or the inertia of the bridge.

In the trajectory 1, fig. 2, the ordinates are small because the body passed over before there was time to produce much deflection in the bridge, at least except towards the end of the body's course, where even a large deflection of the bridge would produce only a small deflection of the body. The corresponding deflection curve, (curve 1, fig. 3,) shews that the bridge was depressed, and that its deflection was rapidly increasing, when the body left it. When the body is made to move with velocities successively one-half and one-third of the former velocity, more time is allowed for deflecting the bridge, and the trajectories marked 2, 3, are described, in which the ordinates are far larger than in that marked 1. The deflections too, as appears from fig. 3, are much larger than before, or at least much larger than any deflection which was produced in the first case while the body remained on the bridge. It appears from Table III., or from fig. 3, that the greatest deflection occurs in the case of the third curve, nearly, and that it exceeds the central statical deflection by about three-fourths of the whole. When the velocity is considerably diminished, the bridge has time to make several oscillations while the body is going over it. These oscillations may be easily observed in fig. 3, and their effect on the form of the trajectory, which may indeed be readily understood from fig. 3, will be seen on referring to fig. 2.

When q is large, as is the case in practice, it will be sufficient in equation (66) to retain only the term which is divided by the first power of q. With this simplification we get

$$\frac{D_1}{S} = \frac{25}{16}\frac{\mu_1}{S} = -\frac{25}{8q}\sin qx \dots\dots\dots\dots (71);$$

so that the central deflection is liable to be alternately increased and decreased by the fraction $25/8q$ of the central statical deflection. By means of the expressions (61), (69), we get

$$\frac{25}{8q} = \cdot 55 \sqrt{\frac{M'}{M\beta}} = \cdot 112 \frac{\sqrt{S_1}}{\tau} \dots\dots\dots\dots (72).$$

It is to be remembered that in the latter of these expressions the units of space and time are an inch and a second respectively. Since the difference between the pressure on the bridge and weight of the body is neglected in the investigation in which the inertia of the bridge is considered, it is evident that the result will be sensibly the same whether the bridge in its natural position be straight, or be slightly raised towards the centre, or, as it is technically termed, *cambered*. The increase of deflection in the case first investigated would be diminished by a camber.

In this paper the problem has been worked out, or worked out approximately, only in the two extreme cases in which the mass of the travelling body is infinitely great and infinitely small respectively, compared with the mass of the bridge. The causes of the increase of deflection in these two extreme cases are quite distinct. In the former case, the increase of deflection depends entirely on the difference between the pressure on the bridge and the weight of the body, and may be regarded as depending on the centrifugal force. In the latter, the effect depends on the manner in which the force, regarded as a function of the time, is applied to the bridge. In practical cases the masses of the body and of the bridge are generally comparable with each other, and the two effects are mixed up in the actual result. Nevertheless, if we find that each effect, taken separately, is insensible, or so small as to be of no practical importance, we may conclude without much fear of error that the actual effect is insignificant. Now we have seen that if we take only the most important terms, the increase of deflection is measured by the fractions $1/\beta$ and $25/8q$ of S. It is only when these fractions are both small that we are at liberty to neglect all but the most important terms, but in practical cases they are actually small. The magnitude of these fractions will enable us to judge of the amount of the actual effect.

To take a numerical example lying within practical limits, let the span of a given bridge be 44 feet, and suppose a weight equal to $\frac{4}{5}$ of the weight of the bridge to cause a deflection of $\frac{1}{5}$ inch. These are nearly the circumstances of the Ewell bridge, mentioned in the report of the commissioners. In this case, $S_1 = \frac{3}{4} \times {\cdot}2 = {\cdot}15$; and if the velocity be 44 feet in a second, or 30 miles an hour, we have $\tau = 1$, and therefore from the second of the formulæ (72),

$$\frac{25}{8q} = {\cdot}0434, \quad q = 72{\cdot}1 = 45{\cdot}9 \times \frac{\pi}{4}.$$

The travelling load being supposed to produce a deflection of ${\cdot}2$ inch, we have $\beta = 127$, $1/\beta = {\cdot}0079$. Hence in this case the deflection due to the inertia of the bridge is between 5 and 6 times as great as that obtained by considering the bridge as infinitely light, but in neither case is the deflection important. With a velocity of 60 miles an hour the increase of deflection ${\cdot}0434 S$ would be doubled.

In the case of one of the long tubes of the Britannia bridge β must be extremely large; but on account of the enormous mass of the tube it might be feared that the effect of the inertia of the tube itself would be of importance. To make a supposition every way disadvantageous, regard the tube as unconnected with the rest of the structure, and suppose the weight of the whole train collected at one point. The clear span of one of the great tubes is 460 feet, and the weight of the tube 1400 tons. When the platform on which the tube had been built was removed, the centre sank 10 inches, which was very nearly what had been calculated, so that the bottom became very nearly straight, since, in anticipation of the deflection which would be produced by the weight of the tube itself, it had been originally built curved upwards. Since a uniformly distributed weight produces the same deflection as $\frac{8}{5}$ ths of the same weight placed at the centre, we have in this case $S_1 = \frac{8}{5} \times 10 = 16$; and supposing the train to be going at the rate of 30 miles an hour, we have $\tau = 460 \div 44 = 10\cdot5$, nearly. Hence in this case $25/8q = \cdot043$, or $\frac{1}{23}$ nearly, so that the increase of deflection due to the inertia of the bridge is unimportant.

In conclusion, it will be proper to state that this "Addition" has been written on two or three different occasions, as the reader will probably have perceived. It was not until a few days after the reading of the paper itself that I perceived that the equation (16) was integrable in finite terms, and consequently that the variables were separable in (4). I was led to try whether this might not be the case in consequence of a remarkable numerical coincidence. This circumstance occasioned the complete remodelling of the paper after the first six articles. I had previously obtained for the calculation of z for values of x approaching 1, in which case the series (9) becomes inconvenient, series proceeding according to ascending powers of $1 - x$, and involving two arbitrary constants. The determination of these constants, which at first appeared to require the numerical calculation of five series, had been made to depend on that of three only, which were ultimately geometric series with a ratio equal to $\frac{1}{2}$.

The fact of the integrability of equation (4) in the form given in Art. 7, to which I had myself been led from the circumstance above mentioned, has since been communicated to me by Mr Cooper, Fellow of St John's College, through Mr Adams, and by Professors Malmsten and A. F. Svanberg of Upsala through Professor Thomson; and I take this opportunity of thanking these mathematicians for the communication.

[From the *Cambridge and Dublin Mathematical Journal*, Vol. IV. p. 219
(*November*, 1849)].

NOTES ON HYDRODYNAMICS.

IV.—*On Waves.*

THE theory of waves has formed the subject of two profound
memoirs by MM. Poisson and Cauchy, in which some of the
highest resources of analysis are employed, and the results deduced
from expressions of great complexity. This circumstance might
naturally lead to the notion that the subject of waves was unap-
proachable by one who was either unable or unwilling to grapple
with mathematical difficulties of a high order. The complexity,
however, of the memoirs alluded to arises from the nature of the
problem which the authors have thought fit to attack, which is the
determination of the motion of a mass of liquid of great depth
when a small portion of the surface has been slightly oisturbed in
a given arbitrary manner. But after all it is not such problems
that possess the greatest interest. It is seldom possible to realize
in experiment the conditions assumed in theory respecting the
initial disturbance. Waves are usually produced either by some
sudden disturbing cause, which acts at a particular part of the
fluid in a manner too complicated for calculation, or by the wind
exciting the surface in a manner which cannot be strictly investigated. What chiefly strikes our attention is the propagation of
waves already produced, no matter how : what we feel most desire
to investigate is the mechanism and the laws of such propagation.
But even here it is not every possible motion that may have been
excited that it is either easy or interesting to investigate ; there
are two classes of waves which appear to be especially worthy of
attention.

The first consists of those whose length is very great compared with the depth of the fluid in which they are propagated. To this class belongs the great tidal wave which, originally derived from the oceanic oscillations produced by the disturbing forces of the sun and moon, is propagated along our shores and up our channels. To this class belongs likewise that sort of wave propagated along a canal which Mr Russell has called a *solitary wave*. As an example of this kind of wave may be mentioned the wave which, when a canal boat is stopped, travels along the canal with a velocity depending, not on the previous velocity of the boat, but merely upon the form and depth of the canal.

The second class consists of those waves which Mr Russell has called *oscillatory*. To this class belong the waves produced by the action of wind on the surface of water, from the ripples on a pool to the long swell of the Atlantic. By the waves of the sea which are referred to this class must not be understood the surf which breaks on shore, but the waves produced in the open sea, and which, after the breeze that has produced them has subsided, travel along without breaking or undergoing any material change of form. The theory of oscillatory waves, or at least of what may be regarded as the type of oscillatory waves, is sufficiently simple, although not quite so simple as the theory of long waves.

Theory of Long Waves.

Conceive a long wave to travel along a uniform canal. For the sake of clear ideas, suppose the wave to consist entirely of an elevation. Let k be the greatest height of the surface above the plane of the surface of the fluid at a distance from the wave, where the fluid is consequently sensibly at rest; let λ be the length of the wave, measured suppose from the point where the wave first becomes sensible to where it ceases to be sensible on the opposite side of the ridge; let b be the breadth, and h the depth of the canal if it be rectangular, or quantities comparable with the breadth and depth respectively if the canal be not rectangular. Then the volume of fluid elevated will be comparable with $b\lambda k$. As the wave passes over a given particle, this volume (not however consisting of the same particles be it observed) will be transferred from the one side to the other of the particle in question. Consequently if we suppose the horizontal motions of the particles situated in the same vertical plane perpendicular to the length of the canal to be the same, a supposition which cannot possibly give the greatest horizontal motion too great, although previously to investigation it might be supposed to give it too small, the horizontal displacement of any particle will be comparable with $b\lambda k/bh$ or $\lambda k/h$. Hence if λ be very great compared with h, the horizontal displacements and horizontal velocities will be very great compared with the vertical displacements and vertical velocities. Hence we may neglect the vertical effective force, and therefore regard the fluid as in equilibrium, so far as vertical forces are concerned, so that the pressure at any depth δ below the actual surface will be $g\rho\delta$, g being the force of gravity, and ρ the density of the fluid, the atmospheric pressure being omitted. It is this circumstance that makes the theory of long waves so extremely simple. If the canal be not rectangular, there will be a slight horizontal motion in a direction perpendicular to the length of the canal; but the corresponding effective force may be neglected for the same reason as the vertical effective force, at least if the breadth of the canal be not very great compared with its depth, which is supposed to be the case; and therefore the fluid contained between any two infinitely close vertical planes drawn perpendicular to the length of the canal may be considered to be in equilibrium, except in so far as motion in the direction of the length of the canal is concerned. It need hardly be remarked that the investigation which applies to a rectangular canal will apply to an extended sheet of standing fluid, provided the motion be in two dimensions.

Let x be measured horizontally in the direction of the length of the canal; and at the time t draw two planes perpendicular to the axis of x, and passing through points whose abscissæ are x' and $x' + dx'$. Then if η be the elevation of the surface at any point of the horizontal line in which it is cut by the first plane, $\eta + d\eta/dx' \cdot dx'$ will be the elevation of the surface where it is cut by the second plane. Draw a right line parallel to the axis of x, and cutting the planes in the points P, P'. Then if δ be the depth of the line PP' below the surface of the fluid in equilibrium, the pressures at P, P' will be $g\rho\,(\delta + \eta)$ and $g\rho\,(\delta + \eta + d\eta/dx' \cdot dx')$ respectively; and therefore the difference of pressures will be $g\rho \, d\eta/dx' . \, dx'$. About the line PP' describe an infinitely thin cylindrical surface, with its generating lines perpendicular to the planes, and let κ be the area which it cuts from either plane; and consider the motion of fluid which is bounded by the cylindrical surface and the two planes. The difference of the pressures on the two ends is ultimately $g\rho\kappa \, d\eta/dx' . \, dx'$, and the mass being $\rho\kappa \, dx'$, the accelerating force is $g \, d\eta/dx'$. Hence the effective force is the same for all particles situated in the same vertical plane perpendicular to the axis of x; and since the particles are supposed to have no sensible motion before the wave reaches them, it follows that the particles once in a vertical plane perpendicular to the length of the canal remain in such a vertical plane throughout the motion.

Let x be the abscissa of any plane of particles in its position of equilibrium, $x + \xi$ the common abscissa of the same set of particles at the time t, so that ξ and η are functions of x and t. Then equating the effective to the impressed accelerating force, we get

$$\frac{d^2\xi}{dt^2} = -g \, \frac{d\eta}{dx'} \quad \dotfill (1);$$

and we have
$$x' = x + \xi \quad \dotfill (2).$$

Thus far the canal has been supposed to be not necessarily rectangular, nor even uniform, provided that its form and dimensions change very slowly, nor has the motion been supposed to be necessarily very small. If we adopt the latter supposition, and neglect the squares of small quantities, we shall get from (1) and (2)

$$\frac{d^2\xi}{dt^2} = -g \, \frac{d\eta}{dx} \dotfill (3).$$

It remains to form the equation of continuity. Suppose the canal to be uniform and rectangular, and let b be its breadth and h its depth. Consider the portion of fluid contained between two vertical planes whose abscissæ in the position of equilibrium are x and $x + dx$. The volume of this portion is expressed by $bh \, dx$. At the time t the abscissæ of the bounding planes of particles are $x + \xi$ and $x + \xi + (1 + d\xi/dx) \, dx$; the depth of the fluid contained between these planes is $h + \eta$; and therefore the expression for the volume is $b(h + \eta)(1 + d\xi/dx) \, dx$. Equating the two expressions for the volume, dividing by bdx, and neglecting the product of the two small quantities, we get

$$h \frac{d\xi}{dx} + \eta = 0 \ \dots\dots\dots\dots\dots\dots\dots (4) *.$$

Eliminating ξ between (3) and (4), we get

$$\frac{d^2\eta}{dt^2} = gh \frac{d^2\eta}{dx^2} \ \dots\dots\dots\dots\dots\dots\dots (5).$$

The complete integral of this equation is

$$\eta = f\{x - \sqrt{(gh)}\, t\} + F\{x + \sqrt{(gh)}\, t\} \ \dots\dots\dots\dots (6),$$

where f, F denote two arbitrary functions. This integral evidently represents two waves travelling, one in the positive, and the other in the negative direction, with a velocity equal to $\sqrt{(gh)}$, or to that acquired by a heavy body in falling through a space equal to half the depth of the fluid. It may be remarked that the velocity of propagation is independent of the density of the fluid.

It is needless to consider the determination of the arbitrary functions f, F by means of the initial values of η and $d\eta/dt$, supposed to be given, or the reflection of a wave when the canal is stopped by a vertical barrier, since these investigations are precisely the same as in the case of sound, or in that of a vibrating

---

* This equation is in fact a second integral of the ordinary equation of continuity, corrected so as to suit the particular case of motion which is under consideration. For motion in two dimensions the latter equation is

$$\frac{du}{dx} + \frac{dv}{dy} = 0 \ \dots\dots\dots\dots\dots\dots\dots\dots\dots (a) ;$$

and denoting by η' the vertical displacement of any particle, we have

$$u = \frac{d\xi}{dt}, \quad v = \frac{d\eta'}{dt} .$$

Substituting in (a), and integrating with respect to t, we get

$$\frac{d\xi}{dx} + \frac{d\eta'}{dy} = \psi(x, y) \dots\dots\dots\dots\dots\dots\dots (b),$$

$\psi(x, y)$ denoting an arbitrary function of x, y, that is, a quantity which may vary from one particle to another, but is independent of the time. To determine ψ we must observe that when any particle is not involved in the wave $\eta' = 0$, and ξ does not vary in passing from one particle to another, and therefore $\psi(x, y) = 0$. Integrating equation (b) with respect to y from $y = 0$ to $y = h + \eta$, observing that ξ is independent of y, and that the limits of η' are 0 and η, and neglecting $\eta \, d\xi/dx$, which is a small quantity of the second order, we get the equation in the text.

string. The only thing peculiar to the present problem consists in the determination of the motion of the individual particles.

It is evident that the particles move in vertical planes parallel to the length of the canal. Consider an elementary column of fluid contained between two such planes infinitely close to each other, and two vertical planes, also infinitely close to each other, perpendicular to the length of the canal. By what has been already shewn, this column of fluid will remain throughout the motion a vertical column on a rectangular base; and since there can be no vertical motion at the bottom of the canal, it is evident that the vertical displacements of the several particles in the column will be proportional to their heights above the base. Hence it will be sufficient to determine the motion of a particle at the surface; when the motion of a particle at a given depth will be found by diminishing in a given ratio the vertical displacement of the superficial particle immediately above it, without altering the horizontal displacement.

The motion of a particle at the surface is defined by the values of η and ξ. The former is given by (6), where the functions f, F are now supposed known, and the latter will be obtained from (4) by integration. Consider the case in which a single wave consisting of an elevation is travelling in the positive direction; let λ be the length of the wave, and suppose the origin taken at the posterior extremity of the wave in the position it occupies when $t = 0$: then we may suppress the second function in (6), and we shall have $f(x) = 0$ from $x = -\infty$ to $x = 0$, and from $x = \lambda$ to $x = +\infty$, and $f(x)$ will be positive from $x = 0$ to $x = \lambda$. Let

$$ c = \sqrt{(gh)} \dots\dots\dots\dots\dots\dots\dots\dots (7), $$

so that c is the velocity of propagation, and let the position of equilibrium of a particle be considered to be that which it occupies before the wave reaches it, so that ξ vanishes for $x = +\infty$. Then we have from (4) and (6)

$$ \xi = -\frac{1}{h}\int_{\infty}^{x} \eta \, dx = \frac{1}{h}\int_{x}^{\infty} f(x - ct) \, dx \dots\dots\dots\dots (8). $$

Consider a particle situated in front of the wave when $t = 0$, so that $x > \lambda$. Since $f(x) = 0$ when $x > \lambda$, we shall have $f(x - ct) = 0$, until $ct = x - \lambda$. Consequently from (6) and (8) there will be no motion until $t = x - \lambda / c$, when the motion will commence. Suppose now that a very small portion only of the wave, of length s, has passed over the particle considered. Then $x - ct = \lambda - s$; and we have from (6) and (8)

$$\eta = f(\lambda - s), \quad \xi = \frac{1}{h}\int_{-\infty}^{s} f(\lambda - s)\,ds = \frac{1}{h}\int_{0}^{s} f(\lambda - s)\,ds :$$

for since $f(x)$ vanishes when $x > \lambda$, we may replace the limits $-\infty$ and s by 0 and s. Since $\int_{0}^{s} f(\lambda - s)\,ds$ is equal to s multiplied by the mean value of $f(\lambda - s)$ from 0 to s, and this mean value is comparable with $f(\lambda - s)$, it follows that ξ is at first very small compared with η. Hence the particle begins to move vertically; and since η is positive the motion takes place upwards. As the wave advances, ξ becomes sensible, and goes on increasing positively. Hence the particle moves forwards as well as upwards. When the ridge of the waves reaches the particle, η is a maximum; the upward motion ceases, but it follows from (8) that ξ is then increasing most rapidly, so that the horizontal velocity is a maximum. As the wave still proceeds, η begins to decrease, and ξ to increase less rapidly. Hence the particle begins to descend, and at the same time its onward velocity is checked. As the wave leaves the particle, it may be shewn just as before that the final motion takes place vertically downwards. When the wave has passed, $\eta = 0$, so that the particle is at the same height from the bottom as at first; but ξ is a positive constant, equal to

$$\frac{1}{h}\int_{0}^{\lambda} f(x)\,dx, \quad \text{or to} \quad \frac{1}{bh}\int_{0}^{\lambda} bf(x)\,dx,$$

that is, to the volume elevated divided by the area of the section of the canal. Hence the particle is finally deposited in advance of its initial position by the space just named.

If the wave consists of a single depression, instead of a single elevation, everything is the same as before, except that the particle is depressed and then raised to its original height, in place of being first raised and then depressed, and that it is moved backwards, or in a direction contrary to that of propagation, instead of being moved forwards.

These results of theory with reference to the motions of the individual particles may be compared with Mr Russell's experiments described at page 342 of his second report on waves*.

* Report of the 14th meeting of the British Association. Mr Russell's first report is contained in the Report of the 7th meeting.

In the preceding investigation the canal has been supposed rectangular. A very trifling modification, however, of the preceding process will enable us to find the velocity of propagation in a uniform canal, the section of which is of any arbitrary contour. In fact, the dynamical equation (3) will remain the same as before; the equation of continuity alone will have to be altered. Let A be the area of a section of the canal, b the breadth at the surface of the fluid; and consider the mass of fluid contained between two vertical planes whose abscissæ in the position of equilibrium are x and $x + dx$, and which therefore has for its volume $A dx$. At the time t, the distance between the bounding planes of particles is $(1 + d\xi/dx)\,dx$, and the area of a section of the fluid is $A + b\eta$ nearly, so that the volume is

$$(A + b\eta)\left(1 + \frac{d\xi}{dx}\right) dx, \text{ or } \left(A + b\eta + A\frac{d\xi}{dx}\right) dx$$

nearly. Equating the two expressions for the volume, we get

$$A\frac{d\xi}{dx} + b\eta = 0.$$

Comparing this equation with (4), we see that it is only necessary to write A/b for h; so that if c be the velocity of propagation,

$$c = \sqrt{\left(\frac{gA}{b}\right)} \quad \text{......................} \quad (9).$$

The formula (9) of course includes (7) as a particular case. The latter was given long ago by Lagrange[*] : the more comprehensive formula (9) was first given by Prof. Kelland[†], though at the same time or rather earlier it was discovered independently

---

[*] *Berlin Memoirs*, 1786, p. 192. In this memoir Lagrange has obtained the velocity of propagation by very simple reasoning. Laplace had a little earlier (*Mém. de l'Académie* for 1776, p. 542) given the expression (see equation (29) of this note) for the velocity of propagation of oscillatory waves, which when h is very small compared with λ reduces itself to Lagrange's formula, but had made an unwarrantable extension of the application of the formula. In the *Mécanique Analytique* Lagrange has obtained analytically the expression (7) for the velocity of propagation when the depth is small, whether the motion take place in two or three dimensions, by assuming the result of an investigation relating to sound.

For a full account of the various theoretical investigations in the theory of waves, which had been made at the date of publication, as well as for a number of interesting experiments, the reader is referred to a work by the brothers Weber, entitled *Wellenlehre auf Experimente gegründet*, Leipzig, 1825.

[†] *Transactions of the Royal Society of Edinburgh*, Vol. xiv. pp. 524, 530.

by Green*, in the particular case of a triangular canal. These formulæ agree very well with experiment when the height of the waves is small, which has been supposed to be the case in the previous investigation, as may be seen from Mr Russell's reports. A table containing a comparison of theory and experiment in the case of a triangular canal is given in Green's paper. In this table the mean error is only about 1/60th of the whole velocity.

As the object of this note is merely to give the simplest cases of wave motion, the reader is referred to Mr Airy's treatise on tides and waves for the effect produced by a slow variation in the dimensions of the canal on the length and height of the wave†, as well as for the effect of the finite height of the wave on the velocity of propagation. With respect to the latter subject, however, it must be observed that in the case of a solitary wave artificially excited in a canal it does not appear to be sufficient to regard the wave as infinitely long when we are investigating the correction for the height; it appears to be necessary to take account of the finite length, as well as finite height of the wave.

Theory of Oscillatory Waves.

In the preceding investigation, the general equations of hydrodynamics have not been employed, but the results have been obtained by referring directly to first principles. It will now be convenient to employ the general equations. The problem which it is here proposed to consider is the following.

The surface of a mass of fluid of great depth is agitated by a series of waves, which are such that the motion takes place in two dimensions. The motion is supposed to be small, and the squares of small quantities are to be neglected. The motion of each particle being periodic, and expressed, so far as the time is concerned, by a circular function of given period, it is required to determine all the circumstance of the motion of the fluid. The case in which the depth is finite and uniform will be considered afterwards.

---

* *Transactions of the Cambridge Philosophical Society*, Vol. VII. p. 87.
† *Encyclopædia Metropolitana.* Art. 260 of the treatise.

It must be observed that the supposition of the periodicity of the motion is not, like the hypothesis of parallel sections, a mere arbitrary hypothesis introduced in addition to our general equations, which, whether we can manage them or not, are sufficient for the complete determination of the motion in any given case. On the contrary, it will be justified by the result, by enabling us to satisfy all the necessary equations; so that it is used merely to define, and select from the general class of possible motions, that particular kind of motion which we please to contemplate.

Let the vertical plane of motion be taken for the plane of xy. Let x be measured horizontally, and y vertically upwards from the mean surface of the fluid. If a, b be the co-ordinates of any particle in its mean position, the co-ordinates of the same particle at the time t will be $a + \int u dt$, $b + \int v dt$, respectively. Since the squares of small quantities are omitted, it is immaterial whether we conceive u and v to be expressed in terms of a, b, t, or in terms of x, y, t; and, on the latter supposition, we may consider x and y as constant in the integration with respect to t. Since the variable terms in the expressions for the co-ordinates are supposed to contain t under the form $\sin nt$ or $\cos nt$, the same must be the case with u and v. We may therefore assume

$$u = u_1 \sin nt + u_2 \cos nt, \quad v = v_1 \sin nt + v_2 \cos nt,$$

where u_1, u_2, v_1, v_2 are functions of x and y without t. Substituting these values of u and v in the general equations of motion, neglecting the squares of small quantities, and observing that the only impressed force acting on the fluid is that of gravity, we get

$$\left. \begin{aligned} \frac{1}{\rho} \frac{dp}{dx} &= -nu_1 \cos nt + nu_2 \sin nt, \\ \frac{1}{\rho} \frac{dp}{dy} &= -g - nv_1 \cos nt + nv_2 \sin nt \end{aligned} \right\} \dots\dots (10),$$

and the equation of continuity becomes

$$\left(\frac{du_1}{dx} + \frac{dv_1}{dy}\right) \sin nt + \left(\frac{du_2}{dx} + \frac{dv_2}{dy}\right) \cos nt = 0 \dots\dots (11).$$

Eliminating p by differentiation from the two equations (10), we get

$$\left(\frac{du_1}{dy} - \frac{dv_1}{dx}\right) \cos nt - \left(\frac{du_2}{dy} - \frac{dv_2}{dx}\right) \sin nt = 0 \dots\dots (12);$$

and in order that this equation may be satisfied, we must have separately

$$\frac{du_1}{dy} - \frac{dv_1}{dx} = 0, \quad \frac{du_2}{dy} - \frac{dv_2}{dx} = 0 \ldots\ldots\ldots\ldots(13).$$

The first of these equations requires that $u_1 dx + v_1 dy$ be an exact differential $d\phi_1$, and is satisfied merely by this supposition. Similarly the second requires that $u_2 dx + v_2 dy$ be an exact differential $d\phi_2$. The functions ϕ_1, ϕ_2 may be supposed not to contain t, provided that in integrating equations (10) we express explicitly an arbitrary function of t instead of an arbitrary constant. In order to satisfy (11) we must equate separately to zero the coefficients of $\sin nt$ and $\cos nt$. Expressing u_1, v_1, u_2, v_2 in terms of ϕ_1, ϕ_2 in the resulting equations, we get

$$\frac{d^2\phi_1}{dx^2} + \frac{d^2\phi_1}{dy^2} = 0 \ldots\ldots\ldots\ldots\ldots\ldots(14),$$

with a similar equation for ϕ_2. Integrating the value of dp given by (10), we get

$$\frac{p}{\rho} = -gy - n\phi_1 \cos nt + n\phi_2 \sin nt + \psi(t) \ldots\ldots (15).$$

It remains to form the equation of condition which has to be satisfied at the free surface. If we suppose the atmospheric pressure not to be included in p, we shall have $p = 0$ at the free surface; and we must have at the same time (Note II.)

$$\frac{dp}{dt} + u\frac{dp}{dx} + v\frac{dp}{dy} = 0 \ldots\ldots\ldots\ldots\ldots\ldots (16).$$

The second term in this equation is of the second order, and in the third we may put for dp/dy its approximate value $-g\rho$. Consequently at the free surface, which is defined by the equation

$$gy + n\phi_1 \cos nt - n\phi_2 \sin nt - \psi(t) = 0 \ldots\ldots\ldots (17),$$

we must have

$$n^2\phi_1 \sin nt + n^2\phi_2 \cos nt + \psi'(t) - g\left(\frac{d\phi_1}{dy}\sin nt + \frac{d\phi_2}{dy}\cos nt\right) = 0 \ (18):$$

and we have the further condition that the motion shall vanish at an infinite depth. Since the value of y given by (17) is a small quantity of the first order, it will be sufficient after differentiation to put $y = 0$ in (18).

Equations (18), (14), and the corresponding equation for ϕ_2 shew that the functions ϕ_1, ϕ_2 are independent of each other; and (15), (17) shew that the pressure at any point, and the ordinate of the free surface are composed of the sums of the parts due to these two functions respectively. Consequently we may temporarily suppress one of the functions ϕ_2, which may be easily restored in the end by writing $t + \pi/2n$ for t, and changing the arbitrary constants.

Equation (14) may be satisfied in the most general way by an infinite number of particular solutions of the form $A\epsilon^{m'x+my}$, where any one of the three constants A, m', m may be positive or negative, real or imaginary, and m', m are connected by the equation $m'^2 + m^2 = 0$.* Now m' cannot be wholly real, nor partly real and partly imaginary, since in that case the corresponding particular solution would become infinite either for $x = -\infty$ or for $x = +\infty$, whereas the fluid is supposed to extend indefinitely in the direction of x, and the expressions for the velocity, &c. must not become infinite for any point of space occupied by the fluid. Hence m' must be wholly imaginary, and therefore m wholly real. Moreover m must be positive, since otherwise the expression considered would become infinite for $y = -\infty$. The equation connecting m and m' gives $m' = \pm m\sqrt{(-1)}$. Uniting in one the two corresponding solutions with their different arbitrary constants, we have for the most general particular solution which we are at liberty to take $(A\epsilon^{m\sqrt{(-1)}} + B\epsilon^{-m\sqrt{(-1)}})\,\epsilon^{my}$, which becomes, on replacing the imaginary exponentials by circular functions, and changing the arbitrary constants,

$$(A \sin mx + B \cos mx)\,\epsilon^{my}.$$

Hence we must have

$$\phi_1 = \Sigma\,(A \sin mx + B \cos mx)\,\epsilon^{my} \quad \ldots\ldots\ldots\ldots (19),$$

the sign Σ denoting that we may take any number of positive values of m with the corresponding values of A and B.

Substituting now in (18), supposed to be deprived of the function ϕ_2, the value of ϕ_1 given by (19), and putting $y = 0$ after differentiation, we have

$$\sin nt\,\Sigma\,(n^2 - mg)\,(A \sin mx + B \cos mx) + \psi'(t) = 0.$$

---

* See Poisson, *Traité de Mécanique*, Tom. II. p. 347, or *Théorie de la Chaleur*, Chap. v.

Since no two terms such as $A \sin mx$ or $B \cos mx$ can destroy each other, or unite with the term $\psi'(t)$, we must have separately $\psi'(t) = 0$, and

$$n^2 - mg = 0 \quad \ldots\ldots\ldots\ldots\ldots\ldots\ldots (20).$$

The former of these equations gives $\psi(t) = k$, where k is a constant; but (17) shews that the mean value of the ordinate y of the free surface is k/g, inasmuch as ϕ_1 and ϕ_2 consist of circular functions so far as x is concerned, and therefore we must have $k = 0$, since we have supposed the origin of co-ordinates to be situated in the mean surface of the fluid. The latter equation restricts (19) to one particular value of m.

To obtain ϕ_2 it will be sufficient to take the expression for ϕ_1 with new arbitrary constants. If we put ϕ for

$$\phi_1 \sin nt + \phi_2 \cos nt, \text{ so that } \phi = \int (u\,dx + v\,dy),$$

we see that ϕ consists of four terms, each consisting of the product of an arbitrary constant, a sine or cosine of nt, a sine or cosine of mx and of the same function ϵ^{my} of y. By replacing the products of the circular functions by sines or cosines of sums or differences, and changing the arbitrary constants, we shall get four terms multiplied by arbitrary constants, and involving sines and cosines of $mx - nt$ and of $mx + nt$. The terms involving $mx - nt$ will represent a disturbance travelling in the positive direction, and those involving $mx + nt$ a disturbance travelling in the negative direction. If we wish to consider only the disturbance which travels in the positive direction, we must suppress the terms involving $mx + nt$, and we shall then have got only two terms left, involving respectively $\sin (mx - nt)$ and $\cos (mx - nt)$. One of these terms, whichever we please, may be got rid of by altering the origin of x; and we may therefore take

$$\phi = A \sin (mx - nt)\, \epsilon^{my} \quad \ldots\ldots\ldots\ldots (21);$$

and ϕ determines, by its partial differential coefficients with respect to x and y, the horizontal and vertical components of the velocity at any point. We have from (21), and the definitions of ϕ_1, ϕ_2,

$$\phi_1 = -A \cos mx \cdot \epsilon^{my}, \quad \phi_2 = A \sin mx \cdot \epsilon^{my}.$$

Substituting in (15) and (17), putting $\psi\,(t) = 0$, and replacing y by 0 in the second and third terms of (17), we get

$$\frac{p}{\rho} = g\,(-\,y) + nA \cos\,(mx - nt)\,\epsilon^{my} \,\ldots\ldots\ldots\ldots\, (22),$$

which gives the pressure at any point, and

$$y = \frac{nA}{g} \cos\,(mx - nt)\ldots\ldots\ldots\ldots\ldots\ldots(23)^{*},$$

which gives the equation to the free surface at any instant.

If λ be the length of a wave, T its period, c the velocity of propagation, we have $m = 2\pi/\lambda$, $n = 2\pi/T$, $n = cm$; and therefore from (20)

$$c = \sqrt{\left(\frac{g\lambda}{2\pi}\right)}\ldots\ldots\ldots\ldots\ldots\ldots (24).$$

Hence the velocity of propagation varies directly, and the period of the wave inversely, as the square root of the wave's length. Equation (23) shews that a section of the surface at any instant is the curve of sines.

It may be remarked that in consequence of the form of ϕ equation (18) is satisfied, not merely for $y = 0$, but for any value of y; and therefore (16) is satisfied, not merely at the free surface, but throughout the mass. Hence the pressure experienced by a given particle is constant throughout the motion. This is not true when the depth is finite, as may be seen from the value of ϕ adapted to that case, which will be given presently; but it may be shewn to be true when the depth is infinite, whether the motion take place in two, or three dimensions, and whether it be regular or irregular, provided it be small, and be such that $u\,dx + v\,dy + w\,dz$ is an exact differential.

It will be interesting to determine the motions of the individual particles. Let $x + \xi$, $y + \eta$ be the co-ordinates of the particle whose mean position has for co-ordinates x, y. Then we have

$$\frac{d\xi}{dt} = u = \frac{d\phi}{dx}, \quad \frac{d\eta}{dt} = v = \frac{d\phi}{dy};$$

and in the values of u, v we may take x, y to denote the actual

---

* Equations (22), (23) may be got at once from the equations

$$\frac{p}{\rho} = -\,gy - \frac{d\phi}{dt}, \quad 0 = gy + \frac{d\phi}{dt}.$$

co-ordinates of any particle or their mean values indifferently, on account of the smallness of the motion. Hence we get from (21) after differentiation and integration

$$\xi = -\frac{mA}{n}\sin(mx - nt)\,\epsilon^{my}, \quad \eta = \frac{mA}{n}\cos(mx - nt)\epsilon^{my} \dots (25).$$

Hence the particles describe circles about their mean places, with a uniform angular motion. Since η is a maximum at the same time with y in (23), and $d\xi/dt$ is then positive, any particle is in its highest position when the crest of the wave is passing over it, and is then moving horizontally forwards, that is, in the direction of propagation. Similarly any particle is in its lowest position when the middle of the trough is passing over it, and it is then moving horizontally backwards. The radius of the circle described is equal to $mA/n\,.\,\epsilon^{my}$, and it therefore decreases in geometric progression as the depth of the particle considered increases in arithmetic. The rate of decrease is such that at a depth equal to λ the displacement is to the displacement at the surface as $\epsilon^{-2\pi}$ to 1, or as 1 to 535 nearly.

If the depth of the fluid be finite, the preceding solution may of course be applied without sensible error, provided ϵ^{my} be insensible for a negative value of y equal to the depth of the fluid. This will be equally true whether the bottom be regular or irregular, provided that in the latter case we consider the depth to be represented by the least actual depth.

Let us now suppose the depth of the fluid finite and uniform. Let h be the mean depth of the fluid, that is, its depth as unaffected by the waves. It will be convenient to measure y from the bottom rather than from the mean surface. Consequently we must put $y = h$, instead of $y = 0$, in the values of ϕ_1, ϕ_2, and their differential coefficients, in (17) and (18). The only essential change in the equations of condition of the problem is, that the condition that the motion shall vanish at an infinite depth is replaced by the condition that the fluid shall not penetrate into, or separate from the bottom, a condition which is expressed by the equation

$$\frac{d\phi}{dy} = 0 \text{ when } y = 0 \quad \dots\dots\dots\dots (26).$$

Everything is the same as in the preceding investigation till we come to the selection of a particular integral of (14). As before,

y must appear in an exponential, and x under a circular function; but both exponentials must now be retained. Hence the only particular solution which we are at liberty to take is of the form

$$A\epsilon^{my} \cos mx + B\epsilon^{my} \sin mx + C\epsilon^{-my} \cos mx + D\epsilon^{-my} \sin mx,$$

or, which is the same thing, the coefficients only being altered,

$$(\epsilon^{my} + \epsilon^{-my}) (A \cos mx + B \sin mx)$$
$$+ (\epsilon^{my} - \epsilon^{-my}) (C \cos mx + D \sin mx).$$

Now (26) must be satisfied by ϕ_1 and ϕ_2 separately. Substituting then in this equation the value of ϕ_1 which is made up of an infinite number of particular values of the above form, we see that we must have for each value of m in particular $C = 0$, $D = 0$; so that

$$\phi_1 = \Sigma (\epsilon^{my} + \epsilon^{-my}) (A \cos mx + B \sin mx).$$

Substituting in equation (18), in which ϕ_2 is supposed to be suppressed, and y put equal to h after differentiation, we get

$$n^2 (\epsilon^{mh} + \epsilon^{-mh}) - mg (\epsilon^{mh} - \epsilon^{-mh}) = 0 \;...,...... \;(27),$$

and $\psi'(t) = 0$, which gives $\psi(t) = k$. The equation (17) shews that this constant k must be equal to h, which is the mean value of y at the surface. It is easy to prove that equation (27), in which m is regarded as the unknown quantity, has one and but one positive root. For, putting $mh = \mu$, and denoting by ν the function of μ defined by the equation

$$\nu (\epsilon^\mu + \epsilon^{-\mu}) = \mu (\epsilon^\mu - \epsilon^{-\mu}).................(28),$$

we get by taking logarithms and differentiating

$$\frac{1}{\nu} \frac{d\nu}{d\mu} = \frac{1}{\mu} + \frac{\epsilon^\mu + \epsilon^{-\mu}}{\epsilon^\mu - \epsilon^{-\mu}} - \frac{\epsilon^\mu - \epsilon^{-\mu}}{\epsilon^\mu + \epsilon^{-\mu}}.$$

Now the right-hand member of this equation is evidently positive when μ is positive; and since ν is also positive, as appears from (28), it follows that $d\nu/d\mu$ is positive; and therefore μ and ν increase together. Now (28) shews that ν passes from 0 to ∞ as μ passes from 0 to ∞, and therefore for one and but one positive value of μ, ν is equal to the given quantity n^2h/g, which proves the theorem enunciated. Hence as before the most general value of ϕ corresponds to two series of waves, of determinate length, which are propagated, one in the positive, and the other in the negative direction. If c be the velocity of propagation, we get from (27), since $n = cm = c \cdot 2\pi/\lambda$,

$$c = \left\{ \frac{g\lambda}{2\pi} \frac{1 - \epsilon^{-4\pi h/\lambda}}{1 + \epsilon^{-4\pi h/\lambda}} \right\} \dots\dots\dots\dots\dots (29).$$

If we consider only the series which is propagated in the positive direction, we may take for the same reason as before

$$\phi = A \left(\epsilon^{my} + \epsilon^{-my} \right) \sin (mx - nt) \dots\dots\dots (30);$$

which gives

$$\frac{p}{\rho} = g (h - y) + nA \left(\epsilon^{my} + \epsilon^{my} \right) \cos (mx - nt) \dots\dots (31),$$

and for the equation to the free surface

$$g (y - h) = nA \left(\epsilon^{mh} + \epsilon^{-mh} \right) \cos (mx - nt) \dots\dots (32).$$

Equations (21), (22), (23) may be got from (30), (31), (32) by writing $y + h$ for y, $A\epsilon^{-mh}$ for A, and then making h infinite. When λ is very small compared with h, the formula (29) reduces itself to (24): when on the contrary λ is very great it reduces itself to (7). It should be observed however that this mode of proving equation (7) for very long waves supposes a section of the surface of the fluid to be the curve of sines, whereas the equation has been already obtained independently of any such restriction.

The motion of the individual particles may be determined, just as before, from (30). We get

$$\xi = - \frac{mA}{n} \left(\epsilon^{my} + \epsilon^{my} \right) \sin (mx - nt),$$

$$\eta = \frac{mA}{n} \left(\epsilon^{my} - \epsilon^{my} \right) \cos (mx - nt) \dots\dots\dots (33).$$

Hence the particles describe elliptic orbits, the major axes of which are horizontal, and the motion in the ellipses is the same as in the case of a body describing an ellipse under the action of a force tending to the centre. The ratio of the minor to the major axis is that of $1 - \epsilon^{-2my}$ to $1 + \epsilon^{-2my}$, which diminishes from the surface downwards, and vanishes at the bottom, where the ellipses pass into right lines.

The ratio of the horizontal displacement at the depth $h - y$ to that at the surface is equal to the ratio of $\epsilon^{my} + \epsilon^{-my}$ to $\epsilon^{mh} + \epsilon^{-mh}$. The ratio of the vertical displacements is that of $\epsilon^{my} - \epsilon^{-my}$ to $\epsilon^{mh} - \epsilon^{-mh}$. The former of these ratios is greater, and the latter less than that of $\epsilon^{-m(h-y)}$ to 1. Hence, for a given length of wave, the horizontal displacements decrease less, and the vertical displacements more rapidly from the surface downwards when the depth of the fluid is finite, than when it is infinitely great.

In a paper " On the Theory of Oscillatory Waves* " I have considered these waves as mathematically defined by the character of uniform propagation in a mass of fluid otherwise at rest, so that the waves are such as could be propagated into a portion of fluid which had no previous motion, or excited in such a portion by means of forces applied to the surface. It follows from the latter character, by virtue of the theorem proved in Note IV, that $u\,dx + v\,dy$ is an exact differential. This definition is equally applicable whether the motion be or be not very small; but in the present note I have supposed the species of wave considered to be defined by the character of periodicity, which perhaps forms a somewhat simpler definition when the motion is small. In the paper just mentioned I have proceeded to a second approximation, and in the particular case of an infinite depth to a third approximation. The most interesting result, perhaps, of the second approximation is, that the ridges are steeper and narrower than the troughs, a character of these waves which must have struck everybody who has been in the habit of watching the waves of the sea, or even the ripples on a pool or canal. It appears also from the second approximation that in addition to their oscillatory motion the particles have a progressive motion in the direction of propagation, which decreases rapidly from the surface downwards. The factor expressing the rate of decrease in the case in which the fluid is very deep is ϵ^{-2my}, y being the depth of the particle considered below the surface. The velocity of propagation is the same as to a first approximation, as might have been seen a priori, because changing the sign of the coefficient denoted by A in equations (21) and (30) comes to the same thing as shifting the origin of x through a space equal to $\frac{1}{2}\lambda$, which does not alter the physical circumstances of the motion; so that the expression for the velocity of propagation cannot contain any odd powers of A. The third approximation in the case of an infinite depth gives an increase in the velocity of propagation depending upon the height of the waves. The velocity is found to be equal to

$c_0(1 + 2\pi^2 a^2/\lambda^2)$, c_0 being the velocity given by (24), and a the height of the waves above the mean surface, or rather the coefficient of the first term in the equation to the surface.

A comparison of theory and observation with regard to the velocity of propagation of waves of this last sort may be seen at pages 271 and 274 of Mr Russell's second report. The following table gives a comparison between theory and experiment in the case of some observations made by Capt. Stanley, R.N. The observations were communicated to the British Association at its late meeting at Swansea*.

In the following table

A is the length of a wave, in fathoms;

B is the velocity of propagation deduced from the observations, expressed in knots per hour;

C is the velocity given by the formula (24), the observations being no doubt made in deep water;

D is the difference between the numbers given in columns B and C.

In calculating the numbers in table C, I have taken $g = 32.2$ feet, and expressed the velocity in knots of 1000 fathoms or 6000 feet†.

A	B	C	D
55	27·0	24·7	2·3
43	24·5	21·8	2·7
50	24·0	23·5	0·5
35 to 40	22·1	20·4	1·7
33	22·1	19·1	3·0
57	26·2	25·1	1·1
35	22·0	19·7	2·3

The mean of the numbers in column D is 1·94, nearly, which is about the one-eleventh of the mean of those in column C. The quantity 1·94 appears to be less than the most probable error of any one observation, judging by the details of the experiments; but as all the errors lie in one direction, it is probable that the

---

* Report for 1848, Part II. p. 38.

† I have taken a knot to be 1000 fathoms rather than 2040 yards, because the former value appears to have been used in calculating the numbers in column B.

formula (24) gives a velocity a little too small to agree with observations under the circumstances of the experiments. The height of the waves from crest to trough is given in experiments No. 1, 2, 3, 6, 7, by numbers of feet ranging from 17 to 22. I have calculated the theoretical correction for the velocity of propagation depending upon the height of the waves, and found it to be ·5 or ·6 of a knot, by which the numbers in column C ought to be increased. But on the other hand, according to theory, the particles at the surface have a progressive motion of twice that amount; so that if the ship's velocity, as measured by the log-line, were the velocity relatively to the surface of the water, her velocity would be under-estimated to the amount of 1 or 1·2 knot, which would have to be added to the numbers in column B, or which is the same subtracted from those in column C, in order to compare theory and experiment; so that on the whole ·5 or ·6 would have to be subtracted from the numbers in column C. But on account of the depth to which the ship sinks in the sea, and the rapid decrease of the factor ϵ^{-2my} from the surface downwards, the correction 1 or 1·2 for the "heave of the sea*" would be too great; and therefore, on the whole, the numbers in column C may be allowed to stand. If the numbers given in Capt. Stanley's column, headed "Speed of Ship" already contain some such correction, the numbers in column C must be increased, and therefore those in column D diminished, by ·5 or ·6.

It has been supposed in the theoretical investigation that the surface of the fluid was subject to a uniform pressure. But in the experiments the wind was blowing strong enough to propel the ship at the rate of from 5 to 7·8 knots an hour. There is nothing improbable in the supposition that the wind might have slightly increased the velocity of propagation of the waves.

There is one other instance of wave motion which may be noticed before we conclude. Suppose that two series of oscillatory waves, of equal magnitude, are propagated in opposite directions. The value of ϕ which belongs to the compound motion will be

$$(\epsilon^{my} + \epsilon^{-my}) \, [A \cos (mx - nt) + A \cos (mx + nt + a)],$$

---

* I have been told by a naval friend that an allowance for the "heave of the sea" is sometimes actually made. As well as I recollect, this allowance might have been about 10 knots a day for waves of the magnitude of those here considered.

the squares of small quantities being neglected, as throughout this note. Since

$$\cos (mx - nt) + \cos (mx + nt + a) = 2 \cos (mx + \tfrac{1}{2}x) \cos (nt + \tfrac{1}{2}a),$$

we get by writing $\tfrac{1}{2}A$ for A, and altering the origins of x and t, so as to get rid of a,

$$\phi = A \left(\epsilon^{my} + \epsilon^{-my}\right) \cos mx \, . \, \cos nt \ldots\ldots (34).$$

This is in fact one of the elementary forms already considered, from which two series of progressive oscillatory waves were derived by merely replacing products of sines and cosines by sums and differences. Any one of these four elementary forms corresponds to the same kind of motion as any other, since any two may be derived from each other by merely altering the origins of x and t; and therefore it will be sufficient to consider that which has just been written. We get from (34)

$$\left. \begin{aligned} u &= - mA \left(\epsilon^{my} + \epsilon^{-my}\right) \sin mx \cos nt \\ v &= mA \left(\epsilon^{my} - \epsilon^{-my}\right) \cos mx \cos nt \end{aligned} \right\} \ldots\ldots (35).$$

We have also for the equation to the free surface

$$y - h = \frac{nA}{g} \left(\epsilon^{my} + \epsilon^{-my}\right) \cos mx \sin nt \ldots\ldots (36).$$

Equations (35) shew that for an infinite series of planes for which $mx = 0, = \pm \pi, = \pm 2\pi$, &c., i.e. $x = 0, = \pm \tfrac{1}{2}\lambda, = \pm \lambda$, &c., there is no horizontal motion, whatever be the value of t; and for planes midway between these the motion is entirely horizontal. When $t = 0$, (36) shews that the surface is horizontal; the particles are then moving with their greatest velocity. As t increases, the surface becomes elevated (A being supposed positive) from $x = 0$ to $x = \tfrac{1}{4}\lambda$, and depressed from $x = \tfrac{1}{4}\lambda$ to $x = \tfrac{1}{2}\lambda$, which sufficiently defines the form of the whole, since the planes whose equations are $x = 0$, $x = \tfrac{1}{2}\lambda$, are planes of symmetry. When $nt = \tfrac{1}{2}\pi$, the elevation or depression is the greatest; the whole fluid is then for an instant at rest, after which the direction of motion of each particle is reversed. When nt becomes equal to π, the surface again becomes horizontal; but the direction of each particle's motion is just the reverse of what it was at first, the magnitude of the velocity being the same. The previous motion of the fluid is now repeated in a reverse direction, those portions of the surface which were elevated becoming depressed, and vice versâ. When $nt = 2\pi$, everything is the same as at first.

Equations (35) shew that each particle moves backwards and forwards in a right line.

This sort of wave, or rather oscillation, may be seen formed more or less perfectly when a series of progressive oscillatory waves is incident perpendicularly on a vertical wall. By means of this kind of wave the reader may if he pleases make experiments for himself on the velocity of propagation of small oscillatory waves, without trouble or expense. It will be sufficient to pour some water into a rectangular box, and, first allowing the water to come to rest, to set it in motion by tilting the box, turning it round one edge. The oscillations may be conveniently counted by watching the bright spot on the wall or ceiling occasioned by the light of the sun reflected from the surface of the water, care being taken not to have the motion too great. The time of oscillation from rest to rest is half the period of a wave, and the length of the interior edge parallel to the plane of motion is half the length of a wave; and therefore the velocity of propagation will be got by dividing the length of the edge by the time of oscillation. This velocity is then to be compared with the formula (29).

[From the *Transactions of the Cambridge Philosophical Society*,
Vol. IX. p. 1.]

I. On the Dynamical Theory of Diffraction.

[Read *November* 26, 1849.]

WHEN light is incident on a small aperture in a screen, the
illumination at any point in front of the screen is determined, on
the undulatory theory, in the following manner. The incident
waves are conceived to be broken up on arriving at the aperture;
each element of the aperture is considered as the centre of an
elementary disturbance, which diverges spherically in all directions, with an intensity which does not vary rapidly from one
direction to another in the neighbourhood of the normal to the
primary wave; and the disturbance at any point is found by
taking the aggregate of the disturbances due to all the secondary
waves, the phase of vibration of each being retarded by a quantity
corresponding to the distance from its centre to the point where
the disturbance is sought. The square of the coefficient of vibration is then taken as a measure of the intensity of illumination.
Let us consider for a moment the hypotheses on which this process rests. In the first place, it is no hypothesis that we may
conceive the waves broken up on arriving at the aperture: it is
a necessary consequence of the dynamical principle of the superposition of small motions; and if this principle be inapplicable to
light, the undulatory theory is upset from its very foundations.
The mathematical resolution of a wave, or any portion of a wave,
into elementary disturbances must not be confounded with a physical breaking up of the wave, with which it has no more to do
than the division of a rod of variable density into differential

16—2

elements, for the purpose of finding its centre of gravity, has to do with breaking the rod in pieces. It *is* a hypothesis that we may find the disturbance in front of the aperture by merely taking the aggregate of the disturbances due to all the secondary waves, each secondary wave proceeding as if the screen were away; in other words, that the effect of the screen is *merely to stop* a certain portion of the incident light. This hypothesis, exceedingly probable *a priori*, when we are only concerned with points at no great distance from the normal to the primary wave, is confirmed by experiment, which shews that the same appearances are presented, with a given aperture, whatever be the nature of the screen in which the aperture is pierced, whether, for example, it consist of paper or of foil, whether a small aperture be divided by a hair or by a wire of equal thickness. It is a hypothesis, again, that the intensity in a secondary wave is nearly constant, at a given distance from the centre, in different directions very near the normal to the primary wave; but it seems to me almost impossible to conceive a mechanical theory which would not lead to this result. It is evident that the difference of phase of the various secondary waves which agitate a given point must be determined by the difference of their radii; and if it should afterwards be found necessary to add a constant to all the phases the results will not be at all affected. Lastly, good reasons may be assigned why the intensity should be measured by the square of the coefficient of vibration; but it is not necessary here to enter into them.

In this way we are able to calculate the relative intensities at different points of a diffraction pattern. It may be regarded as established, that the coefficient of vibration in a secondary wave varies, in a given direction, inversely as the radius, and consequently, we are able to calculate the relative intensities at different distances from the aperture. To complete this part of the subject, it is requisite to know the absolute intensity. Now it has been shewn that the absolute intensity will be obtained by taking the reciprocal of the wave length for the quantity by which to multiply the product of a differential element of the area of the aperture, the reciprocal of the radius, and the circular function expressing the phase. It appears at the same time that the phase of vibration of each secondary wave must be accelerated by a quarter of an undulation. In the investigations alluded to, it is supposed that the law of disturbance in a secondary wave is the same in all directions; but this will not affect the result, provided the solution be restricted to the neighbourhood of the normal to the primary wave, to which indeed alone the reasoning is applicable; and the solution so restricted is sufficient to meet all ordinary cases of diffraction.

Now the object of the first part of the following paper is, to determine, on purely dynamical principles, the law of disturbance in a secondary wave, and that, not merely in the neighbourhood of the normal to the primary wave, but in all directions. The occurrence of the reciprocal of the radius in the coefficient, the acceleration of a quarter of an undulation, and the absolute value of the coefficient in the neighbourhood of the normal to the primary wave, will thus appear as particular results of the general formula.

Before attacking the problem dynamically, it is of course necessary to make some supposition respecting the nature of that medium, or *ether*, the vibrations of which constitute light, according to the theory of undulations. Now, if we adopt the theory of transverse vibrations—and certainly, if the simplicity of a theory which conducts us through a multitude of curious and complicated phenomena, like a thread through a labyrinth, be considered to carry the stamp of truth, the claims of the theory of transverse vibrations seem but little short of those of the theory of universal gravitation—if, I say, we adopt this theory, we are obliged to suppose the existence of a tangential force in the ether, called into play by the continuous sliding of one layer, or film, of the medium over another. In consequence of the existence of this force, the ether must behave, so far as regards the luminous vibrations, like an elastic solid. We have no occasion to speculate as to the cause of this tangential force, nor to assume either that the ether does, or that it does not, consist of distinct particles; nor are we directly called on to consider in what manner the ether behaves with respect to the motion of solid bodies, such as the earth and planets.

Accordingly, I have assumed, as applicable to the luminiferous ether in vacuum, the known equations of motion of an elastic medium, such as an elastic solid. These equations contain two arbitrary constants, depending upon the nature of the medium. The argument which Green has employed to shew that the luminiferous ether must be regarded as sensibly incompressible, in treating of the motions which constitute light*, appears to me of great force. The supposition of incompressibility reduces the two arbitrary constants to one; but as the equations are not thus rendered more manageable, I have retained them in their more general shape.

The first problem relating to an elastic medium of which the object that I had in view required the solution was, to determine the disturbance at any time, and at any point of an elastic medium, produced by a given initial disturbance which was confined to a finite portion of the medium. This problem was solved long ago by Poisson, in a memoir contained in the tenth volume of the Memoirs of the Academy of Sciences. Poisson indeed employed equations of motion with but one arbitrary constant, which are what the general equations of motion become when a certain numerical relation is assumed to exist between the two constants which they involve. This relation was the consequence of a particular physical supposition which he adopted, but which has since been shewn to be untenable, inasmuch as it leads to results which are contradicted by experiment. Nevertheless nothing in Poisson's method depends for its success on the particular numerical relation assumed; and in fact, to save the constant writing of a radical, Poisson introduced a second constant, which made his equations identical with the general equations, so long as the particular relation supposed to exist between the two constants was not employed. I might accordingly have at once assumed Poisson's results. I have however begun at the beginning, and given a totally different solution of the problem, which will I hope be found somewhat simpler and more direct than Poisson's. The solution of this problem and the discussion of the result occupy the first two sections of the paper.

Having had occasion to solve the problem in all its generality, I have in one or two instances entered into details which have no immediate relation to light. I have also occasionally considered some points relating to the theory of light which have no immediate bearing on diffraction. It would occupy too much room to enumerate these points here, which will be found in their proper place. I will merely mention one very general theorem at which I have arrived by considering the physical interpretation of a

---

* *Camb. Phil. Trans.* Vol. VII. p. 2.

certain step of analysis, though, properly speaking, this theorem is a digression from the main object of the paper. The theorem may be enunciated as follows.

If any material system in which the forces acting depend only on the positions of the particles be slightly disturbed from a position of equilibrium, and then left to itself, the part of the subsequent motion which depends on the initial displacements may be obtained from the part which depends on the initial velocities by replacing the arbitrary functions, or arbitrary constants, which express the initial velocities by those which express the corresponding initial displacements, and differentiating with respect to the time.

Particular cases of this general theorem occur so frequently in researches of this kind, that I think it not improbable that the theorem may be somewhere given in all its generality. I have not however met with a statement of it except in particular cases, and even then the subject was mentioned merely as a casual result of analysis.

In the third section of this paper, the problem solved in the second section is applied to the determination of the law of disturbance in a secondary wave of light. This determination forms the whole of the dynamical part of the theory of diffraction, at least when we confine ourselves to diffraction in vacuum, or, more generally, within a homogeneous singly refracting medium : the rest is a mere matter of integration ; and whatever difficulties the solution of the problem may present for particular forms of aperture, they are purely mathematical.

In the investigation, the incident light is supposed to be plane-polarized, and the following results are arrived at. Each diffracted ray is plane-polarized, and the plane of polarization is determined by this law ; *The plane of vibration of the diffracted ray is parallel to the direction of vibration of the incident ray.* The expression *plane of vibration* is here used to denote the plane passing through the ray and the direction of vibration. The direction of vibration in any diffracted ray being determined by the law above mentioned, the phase and coefficient of vibration at that part of a secondary wave are given by the formulæ of Art. 33.

The law just enunciated seems to lead to a crucial experiment for deciding between the two rival theories respecting the directions of vibration in plane-polarized light. Suppose the plane of polarization, and consequently the plane of vibration, of the incident light to be turned round through equal angles of say 5° at a time. Then, according to theory, the planes of vibration of the diffracted ray will not be distributed uniformly, but will be crowded towards the plane perpendicular to the plane of diffraction, or that which contains the incident and diffracted rays. The law and amount of the crowding will in fact be just the same as if the planes of vibration of the incident ray were represented in section on a plane perpendicular to that ray, and then projected on a plane perpendicular to the diffracted ray. Now experiment will enable us to decide whether the *planes of polarization* of the diffracted ray are crowded towards the plane of diffraction or towards the plane perpendicular to the plane of diffraction, and we shall accordingly be led to conclude, either that the direction of vibration is perpendicular, or that it is parallel to the plane of polarization.

In ordinary cases of diffraction, the light is insensible at such a small distance from the direction of the incident ray produced that the crowding indicated by theory is too small to be detected by experiment. It is only by means of a fine grating that we can obtain light of considerable intensity which has been diffracted at a large angle.

On mentioning to my friend, Professor Miller, the result at which I had arrived, and making some inquiries about the fineness, &c. of gratings, he urged me to perform the experiment myself, and kindly lent me for the purpose a fine glass grating, which he has in his possession. For the use of two graduated instruments employed in determining the positions of the planes of polarization of the incident and diffracted rays I am indebted to the kindness of my friend Professor O'Brien.

The description of the experiments, and the discussion of the results, occupies Part II. of this Paper. Since in a glass grating the diffraction takes place at the common surface of two different media, namely, air and glass, the theory of Part. I. does not quite meet the case. Nevertheless it does not fail to point out whereabouts the plane of polarization of the diffracted ray ought to lie, according as we adopt one or other of the hypotheses respecting the direction of vibration. For theory assigns exact results on the two extreme suppositions, *first*, that the diffraction takes place before the light reaches the grooves; *secondly*, that it takes place after the light has passed between them; and these results are very different, according as we suppose the vibrations to be perpendicular or parallel to the plane of polarization. Most of the experiments were made on light which was diffracted in passing through the grating. The results appeared to be decisive in favour of Fresnel's hypothesis. In fact, theory shews that diffraction at a large angle is a powerful cause of crowding of the planes of vibration of the diffracted ray towards the perpendicular to the plane of diffraction, and experiment pointed out the existence of a powerful cause of crowding of the planes of polarization towards the plane of diffraction; for not only was the crowding in the contrary direction due to refraction overcome, but a considerable crowding was actually produced towards the plane of diffraction, especially when the grooved face of the glass plate was turned towards the incident light.

The experiments were no doubt rough, and are capable of being repeated with a good deal more accuracy by making some small changes in the apparatus and method of observing. Nevertheless the quantity with respect to which the two theories are at issue is so large that the experiments, such as they were, seem amply sufficient to shew which hypothesis is discarded by the phenomena.

The conclusive character of the experimental result with regard to the question at issue depends, I think, in a great measure on the simplicity of the law which forms the only result of theory that it is necessary to assume. This law in fact merely asserts that, whereas the direction of vibration in the diffracted ray cannot be parallel to the direction of vibration in the incident ray, being obliged to be perpendicular to the diffracted ray, it makes with it as small an angle as is consistent with the above restriction. This law seems only just to lie beyond the limits of the geometrical part of the theory of undulations. At the same time I may be permitted to add that, for my own part, I feel very great confidence in the equations of motion of the luminiferous ether in vacuum, and in that view of the nature of the ether which would lead to these equations, namely, that in the propagation of light, the ether, from whatever reason, behaves like an elastic solid. But when we consider the mutual action of the luminiferous ether and ponderable matter, a wide field, as it seems to me, is thrown open to conjecture. Thus, to take the most elementary of all the phenomena which relate to the action of transparent media on light, namely, the diminution of the velocity of propagation, this diminution seems capable of being accounted for on several different hypotheses. And if this elementary phenomenon leaves so much room for conjecture, much more may we form various hypotheses as to the state of things at the confines of two media, such as air and glass. Accordingly, conclusions in favour of either hypothesis which are derived from the comparison of theoretical and experimental results relating to the effects of reflection and refraction on the polarization of light, appear to me much more subject to doubt than those to which we are led by the experiments here described.

In commencing the theoretical investigation of diffraction, I naturally began with the simpler case of sound. As, however, the results which I have obtained for sound are of far less interest than those which relate to light, I have here omitted them, more especially as the paper has already swelled to a considerable size. I may, perhaps, on some future occasion bring them before the notice of this Society.

PART I.

THEORETICAL INVESTIGATION.

SECTION I.   *Preliminary Analysis.*

1. In what follows there will frequently be occasion to express a triple integration which has to be performed with respect to all space, or at least to all points of space for which the quantity to be integrated has a value different from zero. The conception of such an integration, regarded as a limiting summation, presents itself clearly and readily to the mind, without the consideration of co-ordinates of any kind. A system of co-ordinates forms merely the machinery by which the integration is to be effected in particular cases; and when the function to be integrated is arbitrary, and the nature of the problem does not point to one system rather than another, the employment of some particular system, and the

analytical expression thereby of the function to be integrated, serves only to distract the attention by the introduction of a foreign element, and to burden the pages with a crowd of unnecessary symbols. Accordingly, in the case mentioned above, I shall merely take dV to represent an element of volume, and write over it the sign $\iiint$, to indicate that the integration to be performed is in fact triple. Integral signs will be used in this manner without limits expressed when the integration is to extend to all points of space for which the function to be integrated differs from zero.

There will frequently be occasion too to represent a double integration which has to be performed with reference to the surface of a sphere, of radius r, described round the point which is regarded as origin, or else a double integration which has to be performed with reference to all angular space. In this case the sign $\iint$ will be used, and dS will be taken to represent an element of the surface of the sphere, and $d\sigma$ to represent an elementary solid angle, measured by the corresponding element of the surface of a sphere described about its vertex with radius unity. Hence, if dV, dS, $d\sigma$ denote corresponding elements, $dS = r^2 d\sigma$, $dV = dr dS = r^2 dr d\sigma$. When the signs $\iiint$ and $\iint$, referring to differentials which are denoted by a single symbol, come together, or along with other integral signs, they will be separated by a dot, as for example $\iiint . \iint U dV d\sigma$.

2. As the operation denoted by $\frac{d^2}{dx^2} + \frac{d^2}{dy^2} + \frac{d^2}{dz^2}$ will be perpetually recurring in this paper, I shall denote it for shortness by ∇. This operation admits of having assigned to it a geometrical meaning which is independent of co-ordinates. For if P be the point (x, y, z), T a small space containing P, which will finally be supposed to vanish, dn an element of a normal drawn outwards at the surface of T, U the function which is the subject of the operation, and if ∇ be defined as the equivalent of $\frac{d^2}{dx^2} + \frac{d^2}{dy^2} + \frac{d^2}{dz^2}$, it is easy to prove that

$$\nabla U = \text{limit of} \ \frac{\iint \frac{dU}{dn} dS}{T} \ \dots\dots\dots\dots\dots\dots(1),$$

the integration extending throughout the surface of T, of which

dS is an element. In fact, if l, m, n be the direction-cosines of the normal, we shall have

$$\iint \frac{dU}{dn}\, dS = \iint \left(l\,\frac{dU}{dx} + m\,\frac{dU}{dy} + n\,\frac{dU}{dz} \right) dS$$

$$= \iint \frac{dU}{dx}\, dy\, dz + \iint \frac{dU}{dy}\, dz\, dx + \iint \frac{dU}{dz}\, dx\, dy\ldots\ldots\ldots(2).$$

We have also, supposing the origin of co-ordinates to be at the point P, as we may without loss of generality,

$$\frac{dU}{dx} = \left(\frac{dU}{dx}\right) + \left(\frac{d^2U}{dx^2}\right) x + \left(\frac{d^2U}{dxdy}\right) y + \left(\frac{d^2U}{dxdz}\right) z$$

$$+ \text{ terms of the 2nd order, \&c.}\ldots\ldots\ldots\ldots\ldots\ldots(3),$$

where the parentheses denote that the differential coefficients which are enclosed in them have the values which belong to the point P. In the integral $\iint \dfrac{dU}{dz}\, dy\, dz$, each element must be taken positively or negatively, according as the normal which relates to it makes an acute or an obtuse angle with the positive direction of the axis of x. If we combine in pairs the elements of the integral which relate to opposite elements of the surface of T, we must write $\iint \left(\dfrac{dU_{\prime\prime}}{dx_{\prime\prime}} - \dfrac{dU_{\prime}}{dx_{\prime}} \right) dy\, dz$, where the single and double accents subscribed refer respectively to the first and second points in which the surface of T is cut by an indefinite straight line drawn parallel to the axis of x, and in the positive direction, through the point $(0, y, z)$. We thus get by means of (3), omitting the terms of a higher order than the first, which vanish in the limit,

$$\iint \left(\frac{dU_{\prime\prime}}{dx_{\prime\prime}} - \frac{dU_{\prime}}{dx_{\prime}} \right) dy\, dz = \left(\frac{d^2U}{dx^2}\right) \iint (x_{\prime\prime} - x_{\prime})\, dy\, dz.$$

But $\iint (x_{\prime\prime} - x_{\prime})\, dy\, dz$ is simply the volume T. Treating in the same manner the two other integrals which appear on the right-hand side of equation (2), we get

$$\iint \frac{dU}{dn}\, dS = T \left\{ \left(\frac{d^2U}{dx^2}\right) + \left(\frac{d^2U}{dy^2}\right) + \left(\frac{d^2U}{dz^2}\right) \right\}, \text{ ultimately.}$$

Dividing by T and passing to the limit, and omitting the parentheses, which are now no longer necessary, we obtain the theorem enunciated.

If in equation (1) we take for T the elementary volume $r^2 \sin \theta \, dr \, d\theta \, d\phi$, or $r \, dr \, d\theta \, dz$, according as we wish to employ polar co-ordinates, or one of three rectangular co-ordinates combined with polar co-ordinates in the plane of the two others, we may at once form the expression for ∇U, and thus pass from rectangular co-ordinates to either of these systems without the trouble of the transformation of co-ordinates in the ordinary way.

3. Let f be a quantity which may be regarded as a function of the rectangular co-ordinates of a point of space, or simply, without the aid of co-ordinates, as having a given value for each point of space. It will be supposed that f vanishes outside a certain portion T of infinite space, and that within T it does not become infinite. It is required to determine a function U by the conditions that it shall satisfy the partial differential equation

$$\nabla U = f. \quad \quad \quad (4)$$

at all points of infinite space, that it shall nowhere become infinite, and that it shall vanish at an infinite distance.

These conditions are precisely those which have to be satisfied by the potential of a finite mass whose density is $-f/4\pi$; and we shall have accordingly, if O be the point for which the value of U is required, and r be the radius vector of any element drawn from O,

$$U = -\frac{1}{4\pi} \iiint \frac{f}{r} \, dV. \quad \quad \quad (5).$$

In fact, it may be proved, just as in the theory of potentials, that the expression for U given by (5) does really satisfy (4) and the given conditions; and consequently, if $U + U'$ be the most general solution, U' must satisfy the equation $\nabla U' = 0$ at all points, must nowhere become infinite, and must vanish at an infinite distance. But this being the case it is easy to prove that U' cannot be different from zero.

The solution will still hold good in certain cases when f is infinite at some points, or when it is not confined to a finite space T, but only vanishes at an infinite distance. But such instances may be regarded as limiting cases of the problem restricted as above, and therefore need not be supposed to be excluded by those restrictions.

4. Let U be a quantity depending upon the time t, as well as upon the position of the point of space to which it relates, and satisfying the partial differential equation

$$\frac{d^2 U}{dt^2} = a^2 \nabla U \dots\dots\dots\dots\dots\dots\dots(6).$$

It is required to determine U by the above equation and the conditions that when $t = 0$, U and dU/dt shall have finite values given arbitrarily within a finite space T, and shall vanish outside T.

Let O be the point for which the value of U is sought, r the radius vector of any element drawn from O; $f(r), F(r)$ the initial values of $U, dU/dt$. By this notation it is not meant that these values are functions of r alone, for they will depend likewise upon the two angles which determine the direction of r; but there will be no occasion to express analytically their dependence on those angles. The solution of the problem is

$$U = \frac{t}{4\pi} \iint F(at)\, d\sigma + \frac{1}{4\pi} \frac{d}{dt} t \iint f(at)\, d\sigma \dots\dots\dots\dots\dots(7).$$

See a memoir by Poisson *Mém. de l'Académie*, Tom. III. p. 130, or *Gregory's Examples*, p. 499.

5. Let δ be a function which has given finite values within a finite portion of space, and vanishes elsewhere; and let it be required to determine three functions ξ, η, ζ by the conditions

$$\frac{d\zeta}{dy} - \frac{d\eta}{dz} = \frac{d\xi}{dz} - \frac{d\zeta}{dx} = \frac{d\eta}{dx} - \frac{d\xi}{dy} = 0 \dots\dots\dots\dots(8),$$

$$\frac{d\xi}{dx} + \frac{d\eta}{dy} + \frac{d\zeta}{dz} = \delta \dots\dots\dots\dots\dots(9).$$

The functions ξ, η, ζ are further supposed not to become infinite, and to vanish at an infinite distance. To save repetition, it will here be remarked, once for all, that the same supposition will be made in similar cases.

By virtue of equations (8), $\xi dx + \eta dy + \zeta dz$ is an exact differential $d\psi$, and (9) gives $\nabla\psi = \delta$. Hence we have by the formula (5)

$$\psi = -\frac{1}{4\pi} \iiint \frac{\delta}{r}\, dV \dots\dots\dots\dots(10),$$

and ψ being known, ξ, η, ζ will be obtained by mere differentiation. To differentiate ψ with respect to x, it will be sufficient to differentiate δ under the integral sign. For draw OO' parallel to the axis of x, and equal to Δx, let P, P' be two points similarly situated with respect to O, O', respectively, and consider the part of ψ and that of $\psi + \Delta\psi$ due to equal elements of volume dV situated at P, P' respectively. For these two elements r has the same value, since $OP = O'P'$, and in passing from the first to the second δ is changed into $\delta + \Delta\delta$, and therefore the increment of ψ is simply $-\Delta\delta/4\pi r \cdot dV$. To get the complete increment of ψ we have only to perform the triple integration, an integration which is always real, even though r vanishes in the denominator, as may be readily seen on passing momentarily to polar co-ordinates. Dividing now by Δx and passing to the limit, we get

$$\xi = \frac{d\psi}{dx} = -\frac{1}{4\pi}\iiint \frac{d\delta}{dx}\frac{dV}{r} \quad\ldots\ldots\ldots\ldots(11).$$

By employing temporarily rectangular co-ordinates in the triple integration, integrating by parts with respect to x, and observing that the quantity free from the integral sign vanishes at the limits, we get

$$\xi = -\frac{1}{4\pi}\iiint \frac{\delta}{r^2}\cos(rx)\,dV\ldots\ldots\ldots\ldots(12),$$

as might have been readily proved from (10), by referring O to a fixed origin, and then differentiating with respect to x. The expressions for η and ζ may be written down from symmetry.

6. Let ϖ', ϖ'', ϖ''' be three functions which have given finite values throughout a finite space and vanish elsewhere; it is required to determine three other functions, ξ, η, ζ, by the conditions

$$\frac{d\zeta}{dy} - \frac{d\eta}{dz} = 2\varpi', \quad \frac{d\xi}{dz} - \frac{d\zeta}{dx} = 2\varpi'', \quad \frac{d\eta}{dx} - \frac{d\xi}{dy} = 2\varpi''' \ldots(13),$$

$$\frac{d\xi}{dx} + \frac{d\eta}{dy} + \frac{d\zeta}{dz} = 0 \quad\ldots\ldots\ldots\ldots\ldots(14).$$

It is to be observed that ϖ', ϖ'', ϖ''' are not independent. For differentiating equations (13) with respect to x, y, z, and adding, we get

$$\frac{d\varpi'}{dx} + \frac{d\varpi''}{dy} + \frac{d\varpi'''}{dz} = 0 \quad\ldots\ldots\ldots\ldots(15).$$

Hence ϖ', ϖ'', ϖ''' must be supposed given arbitrarily only in so far as is consistent with the above equation.

Eliminating ζ from (14), and the second of equations (13), we get

$$\frac{d}{dx}\left(\frac{d\xi}{dx}+\frac{d\eta}{dy}\right)+\frac{d^2\xi}{dz^2}=2\frac{d\varpi''}{dz},$$

or

$$\nabla\,\xi+\frac{d}{dy}\left(\frac{d\eta}{dx}-\frac{d\xi}{dy}\right)=2\frac{d\varpi''}{dz},$$

which becomes by the last of equations (13)

$$\nabla\,\xi=2\left(\frac{d\varpi''}{dz}-\frac{d\varpi'''}{dy}\right).$$

Consequently, by equation (5),

$$\xi=\frac{1}{2\pi}\iiint\left(\frac{d\varpi'''}{dy}-\frac{d\varpi''}{dz}\right)\frac{dV}{r}.$$

Transforming this equation in the same manner as (11), supposing x, y, z measured from O, and writing down the two equations found by symmetry, we have finally,

$$\left.\begin{aligned}\xi&=\frac{1}{2\pi}\iiint(y\varpi'''-z\varpi'')\frac{dV}{r^3}\\[4pt]\eta&=\frac{1}{2\pi}\iiint(z\varpi'\ -x\varpi''')\frac{dV}{r^3}\\[4pt]\zeta&=\frac{1}{2\pi}\iiint(x\varpi''-y\varpi')\frac{dV}{r^3}\end{aligned}\right\}\ \ldots\ldots\ldots\ldots(16).$$

7. Let δ, ϖ', ϖ'', ϖ''' be as before; and let it be required to determine three functions ξ, η, ζ from the equations (9) and (13).

From the linearity of the equations it is evident that we have merely to add together the expressions obtained in the last two articles.

8. Let ξ_0, η_0, ζ_0 be three functions given arbitrarily within a finite space outside of which they are equal to zero: it is required to decompose these functions into two parts ξ_1, η_1, ζ_1 and ξ_2, η_2, ζ_2 such that $\xi_1 dx+\eta_1 dy+\zeta_1 dz$ may be an exact differential $d\psi_1$, and ξ_2, η_2, ζ_2 may satisfy (14).

Observing that $\xi_2 = \xi_0 - \xi_1$, $\eta_2 = \eta_0 - \eta_1$, $\zeta_2 = \zeta_0 - \zeta_1$, expressing ξ_1, η_1, ζ_1 in terms of ψ_1, and substituting in (14), we get

$$\nabla \psi_1 = \delta_0,$$

where δ_0 is what δ becomes when ξ_0, η_0, ζ_0 are written for ξ, η, ζ. The above equation gives

$$\psi_1 = -\frac{1}{4\pi} \iiint \frac{\delta_0}{r} \, dV,$$

whence ξ_1, η_1, ζ_1, and consequently ξ_2, η_2, ζ_2, are known.

SECTION II.

Propagation of an Arbitrary Disturbance in an Elastic Medium.

9. THE equations of motion of a homogeneous uncrystallized elastic medium, such as an elastic solid, in which the disturbance is supposed to be very small, are well known. They contain two distinct arbitrary constants, which cannot be united in one without adopting some particular physical hypothesis. These equations may be obtained by supposing the medium to consist of ultimate molecules, but they by no means require the adoption of such a hypothesis, for the same equations are arrived at by regarding the medium as continuous.

Let x, y, z be the co-ordinates of any particle of the medium in its natural state; ξ, η, ζ the displacements of the same particle at the end of the time t, measured in the directions of the three axes respectively. Then the first of the equations may be put under the form

$$\frac{d^2\xi}{dt^2} = b^2 \left(\frac{d^2\xi}{dx^2} + \frac{d^2\xi}{dy^2} + \frac{d^2\xi}{dz^2} \right) + (a^2 - b^2) \frac{d}{dx} \left(\frac{d\xi}{dx} + \frac{d\eta}{dy} + \frac{d\zeta}{dz} \right),$$

where a^2, b^2, denote the two arbitrary constants. Put for shortness

$$\frac{d\xi}{dx} + \frac{d\eta}{dy} + \frac{d\zeta}{dz} = \delta \dots\dots\dots\dots\dots(17),$$

and as before represent by $\nabla\xi$ the quantity multiplied by b^2. According to this notation, the three equations of motion are

$$\left.\begin{array}{l} \dfrac{d^2\xi}{dt^2} = b^2\nabla\xi + (a^2 - b^2)\dfrac{d\delta}{dx} \\[2mm] \dfrac{d^2\eta}{dt^2} = b^2\nabla\eta + (a^2 - b^2)\dfrac{d\delta}{dy} \\[2mm] \dfrac{d^2\zeta}{dt^2} = b^2\nabla\zeta + (a^2 - b^2)\dfrac{d\delta}{dz} \end{array}\right\} \dots\dots\dots\dots(18).$$

It is to be observed that δ denotes the dilatation of volume of the element situated at the point (x, y, z). In the limiting case in which the medium is regarded as absolutely incompressible δ vanishes; but in order that equations (18) may preserve their generality, we must suppose a at the same time to become infinite, and replace $a^2\delta$ by a new function of the co-ordinates. If we take $-p$ to denote this function, we must replace the last terms in these equations by $-\dfrac{dp}{dx}$, $-\dfrac{dp}{dy}$, $-\dfrac{dp}{dz}$, respectively, and we shall thus have a fourth unknown function, as well as a fourth equation, namely that obtained by replacing the second member of (17) by zero. But the retention of equations (18) in their present more general form does not exclude the supposition of incompressibility, since we may suppose a to become infinite in the end just as well as at first.

10.   Suppose the medium to extend infinitely in all directions, and conceive a portion of it occupying the finite space T to receive any arbitrary small disturbance, and then to be left to itself, the whole of the medium outside the space T being initially at rest; and let it be required to determine the subsequent motion.

Differentiating equations (18) with respect to x, y, z, respectively, and adding, we get by virtue of (17)

$$\frac{d^2\delta}{dt^2} = a^2\nabla\delta \dots\dots\dots\dots\dots\dots(19).$$

Again, differentiating the third of equations (18) with respect to y, and the second with respect to z, and subtracting the latter of the two resulting equations from the former, and treating in a similar manner the first and third, and then the second and first of equations (18), we get

$$\frac{d^2\varpi'}{dt^2} = b^2 \nabla \varpi', \qquad \frac{d^2\varpi''}{dt^2} = b^2 \nabla \varpi'', \qquad \frac{d^2\varpi'''}{dt^2} = b^2 \nabla \varpi'''\ldots(20),$$

where ϖ', ϖ'', ϖ''' are the quantities defined by equations (13). These quantities express the rotations of the element of the medium situated at the point (x, y, z) about axes parallel to the three co-ordinate axes respectively.

Now the formula (7) enables us to express δ, ϖ', ϖ'', and ϖ''' in terms of their initial values and those of their differential coefficients with respect to t, which are supposed known ; and these functions being known, we shall determine ξ, η, and ζ as in Art. 7. Our equations being thus completely integrated, nothing will remain but to simplify and discuss the formulæ obtained.

11.   Let O be the point of space at which it is required to determine the disturbance, r the radius vector of any element drawn from O ; and let the initial values of δ, $d\delta/dt$ be represented by $f(r)$, $F(r)$, respectively, with the same understanding as in Art. 4.   By the formula (7), we have

$$\delta = \frac{t}{4\pi} \iint F(at)\, d\sigma + \frac{1}{4\pi} \frac{d}{dt}\, t \iint f(at)\, d\sigma \ldots\ldots\ldots(21).$$

The double integrals in this expression vanish except when a spherical surface described round O as centre, with a radius equal to at, cuts a portion of the space T.   Hence, if O be situated outside the space T, and if r_1, r_2 be respectively the least and greatest values of the radius vector of any element of that space, there will be no dilatation at O until $at = r_1$.   The dilatation will then commence, will last during an interval of time equal to $a^{-1}(r_2 - r_1)$, and will then cease for ever.   The dilatation here spoken of is understood to be either positive or negative, a negative dilatation being the same thing as a condensation.

Hence a *wave of dilatation* will be propagated in all directions from the originally disturbed space T, with a velocity a.   To find the portion of space occupied by the wave, we have evidently only got to conceive a spherical surface, of radius at, described about each point of the space T as centre.   The space occupied by the assemblage of these surfaces is that in which the wave of dilatation is comprised. To find the limits of the wave, we need evidently only attend to those spheres which have their centres situated in the surface of the space T. When t is small, this system of spheres will have an exterior envelope of two sheets, the outer of these sheets being exterior and the inner interior to the shell formed by the assemblage of the spheres. The outer sheet forms the outer limit to the portion of the medium in which the dilatation is different from zero. As t increases, the inner sheet contracts, and at last its opposite sides cross, and it changes its character from being exterior, with reference to the spheres, to interior. It then expands, and forms the inner boundary of the shell in which the wave of condensation is comprised. It is easy to shew geometrically that each envelope is propagated with a velocity a in a normal direction.

12. It appears in a similar manner from equations (20) that there is a similar wave, propagated with a velocity b, to which are confined the rotations ϖ', ϖ'', ϖ'''. This wave may be called for the sake of distinction, the *wave of distortion*, because in it the medium is not dilated nor condensed, but only distorted in a manner consistent with the preservation of a constant density. The condition of the stability of the medium requires that the ratio of b to a be not greater than that of $\sqrt{3}$ to 2^*.

13. If the initial disturbance be such that there is neither dilatation nor velocity of dilatation initially, there will be no wave of dilatation, but only a wave of distortion. If it be such that the expressions $\xi dx + \eta dy + \zeta dz$ and $d\xi/dt \cdot dx + d\eta/dt \cdot dy + d\zeta/dt \cdot dz$ are initially exact differentials, there will be no wave of distortion, but only a wave of dilatation. By making $b = 0$ we pass to the case of an elastic fluid, such as air. By supposing $a = \infty$ we pass to the case of an incompressible elastic solid. In this case we must have initially $\delta = 0$ and $d\delta/dt = 0$; but in order that the results obtained by at once putting $a = \infty$ may have the same degree of generality as those which would be obtained by retaining a as a finite quantity, which in the end is supposed to increase indefinitely, we must not suppose the initial disturbance confined

---

* See a memoir by Green *On the reflection and refraction of Light. Camb. Phil. Trans.* Vol. VII. p. 2. See also *Camb. Phil. Trans.* Vol. VIII. p. 319. [*Ante*, Vol. I. p. 128.]

to the space T, but only the initial rotations and the initial
angular velocities. Consequently, outside T the expression

$$\xi dx + \eta dy + \zeta dz$$

must be initially an exact differential $d\psi$, where ψ satisfies the
equation $\nabla\psi = 0$ derived from (14), and the expression

$$\frac{d\xi}{dt} dx + \frac{d\eta}{dt} dy + \frac{d\zeta}{dt} dz$$

must be initially an exact differential $d\psi_1$, where ψ_1 satisfies the
equation $\nabla\psi_1 = 0$. So long as a is finite, it comes to the same
thing whether we regard the medium as animated initially by
certain velocities given arbitrarily throughout the space T, or as
acted on by impulsive accelerating forces capable of producing
those velocities; and the latter mode of conception is equally
applicable to the case of an incompressible medium, for which a
is infinite, although we cannot in that case conceive the initial
velocities as given arbitrarily, but only arbitrarily in so far as is
compatible with their satisfying the condition of incompressibility.
It is not so easy to see what interpretation is to be given, in the
case of an incompressible medium, to the initial displacements
which are considered in the general case, in so far as these displacements involve dilatation or condensation. As no simplicity
worth mentioning is gained by making a at once infinite, this
constant will be retained in its present shape, more especially as
the results arrived at will thus have greater generality.

14. The expressions for the disturbance of the medium at the
end of the time t are linear functions of the initial displacements
and initial velocities; and it appears from (21), and the corresponding equations which determine ϖ', ϖ'', and ϖ''', that the part
of the disturbance which is due to the initial displacements may
be obtained from the part which is due to the initial velocities by
differentiating with respect to t, and replacing the arbitrary functions which represent the initial velocities by those which represent
the initial displacements. The same result constantly presents
itself in investigations of this nature: on considering its physical
interpretation it will be found to be of extreme generality.

Let any material system whatsoever, in which the forces acting
depend only on the positions of the particles, be slightly disturbed

from a position of equilibrium, and then left to itself. In order to represent the most general initial disturbance, we must suppose small initial displacements and small initial velocities, the most general possible consistent with the connexion of the parts of the system, communicated to it. By the principle of the superposition of small motions, the subsequent disturbance will be compounded of the disturbance due to the initial velocities and that due to the initial displacements. It is immaterial for the truth of this statement whether the equilibrium be stable or unstable; only, in the latter case, it is to be observed that the time t which has elapsed since the disturbance must be sufficiently small to allow of our neglecting the square of the disturbance which exists at the end of that time. Still, as regards the purely mathematical question, for any previously assigned interval t, however great, it will be possible to find initial displacements and velocities so small that the disturbance at the end of the time t shall be as small as we please; and in this sense the principle of superposition, and the results which flow from it, will be equally true whether the equilibrium be stable or unstable.

Suppose now that no initial displacements were communicated to the system we are considering, but only initial velocities, and that the disturbance has been going on during the time t. Let $f(t)$ be the type of the disturbance at the end of the time t, where $f(t)$ may represent indifferently a displacement or a velocity, linear or angular, or in fact any quantity whereby the disturbance may be defined. In the case of a rigid body, or a finite number of rigid bodies, there will be a finite number of functions $f(t)$ by which the motion of the system will be defined: in the cases of a flexible string, a fluid, an elastic solid, &c., there will be an infinite number of such functions, or, in other words, the motion will have to be defined by functions which involve one or more independent variables besides the time. Let v_0 be in a similar manner the type of the initial velocities, and let τ be an increment of t, which in the end will be supposed to vanish. The disturbance at the end of the time $t + \tau$ will be represented by $f(t + \tau)$; but since by hypothesis the forces acting on the system do not depend explicitly on the time, this disturbance is the same as would exist at the end of the time t in consequence of the system of velocities v_0 communicated to the material system at the commencement of the time $- \tau$, the system being at that instant in its position of equilibrium. Suppose then the system of velocities v_0 communicated in this manner, and in addition suppose the system of velocities $-v_0$ communicated at the time 0. On account of the smallness of the motion, the disturbance produced by the system of velocities v_0 will be expressed by linear functions of these velocities; and consequently, if $f(t)$ represent the disturbance due to the system of velocities v_0, $-f(t)$ will represent the disturbance due to the system $-v_0$. Hence the disturbance at the end of the time t will be represented by $f(t+\tau) - f(t)$. Now we may evidently regard the state of the material system immediately after the communication of the system of velocities $-v_0$ as its initial state, and then seek the disturbance which would be produced by the initial disturbance. The velocities v_0 going on during the time τ will have produced by the end of that time a system of displacements represented by τv_0. By hypothesis, the system was in a position of equilibrium at the commencement of the time $-\tau$; and since the forces are supposed not to depend on the velocities, but only on the positions of the particles, the effective forces during the time τ vary from zero to small quantities of the order τ, and therefore the velocities generated by the end of the time $-\tau$ are small quantities of the order τ^2. Hence the velocities $-v_0$ communicated at the time 0 destroy the previously existing velocities, except so far as regards small quantities of the order τ^2, which vanish in the limit, and therefore we have nothing to consider but the system of displacements τv_0. Hence the disturbance produced by a system of initial displacements τv_0 is represented by $f(t+\tau) - f(t)$, ultimately; and therefore the disturbance produced by a system of initial displacements v_0 is represented by the limit of $\{f(t+\tau) - f(t)\}/\tau$, or by $f'(t)$. Hence, to get the disturbance due to the initial displacements from that due to the initial velocities, we have only to differentiate with respect to t, and to replace the arbitrary constants or arbitrary functions which express the initial velocities by those which express the corresponding initial displacements. Conversely, to get the disturbance due to the initial velocities from that due to the initial displacements, we have only to change the arbitrary constants or functions, and to integrate with respect to t, making the integral vanish with t if the disturbance is expressed by displacements, or correcting it so as to give the initial velocities when $t = 0$ if the disturbance is expressed by velocities.

The reader may easily, if he pleases, verify this theorem on some dynamical problem relating to small oscillations.

15. Let us proceed now to determine the general values of ξ, η, ζ in terms of their initial values, and those of their differential coefficients with respect to t. By the formulæ of Section I., ξ, η, ζ are linear functions of δ, ϖ', ϖ'', and ϖ''', and we may therefore first form the part which depends upon δ, and afterwards the part which depends upon ϖ', ϖ'', ϖ''', and then add the results together. Moreover, it will be unnecessary to retain the part of the expressions which depends upon initial displacements, since this can be supplied in the end by the theorem of the preceding article.

Omitting then for the present ϖ', ϖ'', ϖ''', as well as the second term in equations (21), we get from equations (10) and (21),

$$\psi = -\frac{t}{16\pi^2} \iiint \cdot \iint \frac{F(at)}{r} \, dV \, d\sigma \dots\dots\dots\dots\dots(22).$$

To understand the nature of the integration indicated in this equation, let O be the point of space for which the value of ψ is sought; from O draw in an arbitrary direction OP equal to r, and from P draw, also in an arbitrary direction, PQ equal to at. Then $F(at)$ denotes the value of the function F, or the initial rate of dilatation, at the point Q of space, and we have first to perform a double integration referring to all such points as Q, P being fixed, and then a triple integration referring to all such points as P. To facilitate the transformation of the integral (22), conceive PQ produced to Q', let $PQ' = s$, let dV' be an element of volume, and replace the double integral $\iint F' . d\sigma$ by the triple integral $h^{-1} \iiint F . s^{-2} \, dV'$, taken between the limits defined by the imparities $at < s < at + h$, which may be done, provided h be finally made to vanish. We shall thus have two triple integrations to perform, each of which we may conceive to extend to all space, provided we regard the quantity to be integrated as equal to zero when PQ', (or as it may now be denoted PQ, Q being a point taken generally,) lies beyond the limits at and $at + h$, as well as when the point Q falls outside the space T, to which the disturbance was originally confined. Now perform the first of the two triple integrations on the supposition that Q remains fixed while P is variable, instead of supposing P to remain fixed while Q is variable. We shall thus have F constant and r variable, instead of having F variable and r

constant. This first triple integration must evidently extend throughout the spherical shell which has Q for centre and $at, at + h$ for radii of the interior and exterior surfaces. We get, on making h vanish,

$$\text{limit of } \frac{1}{h} \iiint \frac{dV}{r} = \iint \frac{dS}{r},$$

dS being an element of the surface of a sphere described with Q for centre and at for radius. Now if $OQ = r'$, the integral $\iint r^{-1} dS$, which expresses the potential of a spherical shell, of radius at and density unity, at a point situated at a distance r' from the centre, is equal to $4\pi at$ or $4\pi a^2 t^2/r'$, according as $r' < > at$. Substituting in (22), and omitting the accents, which are now no longer necessary, we get

$$\psi = -\frac{1}{4\pi a} \iiint F . dV \, (r < at) - \frac{t}{4\pi} \iiint \frac{F}{r} . dV \, (r > at) \ldots (23),$$

where the limits of integration are defined by the imparities written after the integrals, as will be done in similar cases.

16. Let u_0, v_0, w_0, be the initial velocities; then

$$F = \frac{du_0}{dx} + \frac{dv_0}{dy} + \frac{dw_0}{dz}.$$

Substituting in the first term of the right-hand member of equation (23), and integrating by parts, exactly as in Art. 5, we get

$$-\frac{1}{4\pi a} \iiint F . dV \, (r < at) = -\frac{1}{4\pi a} \Sigma \iint \{(u_0)_{,,} - (u_0)_{,}\} \, dy \, dz,$$

where the Σ denotes that we must take the sum of the expression written down and the two formed from it by passing from x to y and from y to z, and the single and double accents refer respectively to the first and second point in which the surface of a sphere having O for centre, and at for radius, is cut by an indefinite line drawn parallel to the axis of x, and in the positive direction, through the point $(0, y, z)$. Treating the last term in equation (23) in the same way, and observing that the quantities once integrated vanish at an infinite distance, or, to speak more properly, at the limits of the space T, we get

$$-\frac{t}{4\pi} \iiint \frac{F}{r} \, dV \, (r > at) = \frac{1}{4\pi a} \Sigma \iint \{(u_0)_{,,} - (u_0)_{,}\} \, dy \, dz$$

$$-\frac{t}{4\pi} \iiint (u_0 x + v_0 y + w_0 z) \frac{dV}{r^3} \, (r > at).$$

The double integrals arising from the transformation of the second member of equation (23) destroy one another, and we get finally

$$\psi = - \frac{t}{4\pi} \iiint (u_0 x + v_0 y + w_0 z) \frac{dV}{r^3} \ (r > at) \ \ldots\ldots\ldots(24).$$

17. To obtain the part of the displacement ξ due to the initial velocity of dilatation, we have only to differentiate ψ with respect to x, and this will be effected by differentiating u_0, v_0, w_0 under the integral signs, as was shewn in Art. 5. Treating the resulting expression by integration by parts, as before, and putting l, m, n for the direction-cosines of the radius vector drawn to the point to which the accents refer, and ξ_1 for the part of ξ due to F, we get

$$\xi_1 = \frac{1}{4\pi a^2 t} \iint \{(lu_0 + mv_0 + nw_0)_{,,} - (lu_0 + mv_0 + nw_0)_{,}\} \, dy\, dz$$
$$+ \frac{t}{4\pi} \iiint \left(u_0 \frac{d}{dx}\frac{x}{r^3} + v_0 \frac{d}{dx}\frac{y}{r^3} + w_0 \frac{d}{dx}\frac{z}{r^3} \right) dV \ (r > at).$$

Let q_0 be the initial velocity resolved along the radius vector, so that $q_0 = lu_0 + mv_0 + nw_0$, and $(q_0)_{at}$ be the value of q_0 at a distance at from O; then

$$\iint \{(lu_0 + mv_0 + nw_0)_{,,} - (lu_0 + mv_0 + nw_0)_{,}\} \, dy\, dz$$
$$= \iint l\,(q_0)_{at}\, dS = a^2 t^2 \iint l\,(q_0)_{at}\, d\sigma,$$

and
$$u_0 \frac{d}{dx}\frac{x}{r^3} + v_0 \frac{d}{dx}\frac{y}{r^3} + w_0 \frac{d}{dx}\frac{z}{r^3} = \frac{u_0 - 3lq_0}{r^3}.$$

Substituting in the expression for ξ_1, we get finally

$$\xi_1 = \frac{t}{4\pi} \iint l\,(q_0)_{at}\, d\sigma + \frac{t}{4\pi} \iiint (u_0 - 3lq_0) \frac{dV}{r^3} \ (r > at) \ \ldots..(25).$$

18. Let us now form the part of ξ which depends on the initial rotations and angular velocities, and which may be denoted by ξ_2. The theorem of Art. 14 allows us to omit for the present the part due to the initial rotations, which may be supplied in the end. Let $\omega_0', \omega_0'', \omega_0'''$ be the initial angular velocities. Then ξ_2 is given in terms of ϖ'' and ϖ''' by the first of equations (16), and ϖ'', ϖ''' are given in terms of ω_0'', ω_0''' by the formula (7), in which however b must be put for a. We thus get

$$\xi_2 = \frac{t}{8\pi^2} \iiint . \iint (y\omega_0''' - z\omega_0'') \frac{dV d\sigma}{r^3}.$$

The integrations in this expression are to be understood as in Art. 15, and ω_0'', ω_0''' are supposed to have the values which belong to the point Q, but PQ is now equal to bt instead of at. The quintuple integral may be transformed into a triple integral just as before. We get in the first place

$$\xi_2 = \frac{t}{8\pi^2}\left(\iiint \omega_0''' dV' \iint \frac{y\,d\sigma}{r^3} - \iiint \omega_0'' dV' \iint \frac{z\,d\sigma}{r^3}\right) \dots \dots (26).$$

The double integration in this expression refers to all angular space, considered as extending round Q; x, y, z are the co-ordinates, measured from O, of a point P situated at a distance bt from Q, and $r = OP$. If $dS = (bt)^2 d\sigma$, the expressions for the integrals

$$\iint xr^{-3}\,dS,\quad \iint yr^{-3}\,dS,\quad \iint zr^{-3}\,dS$$

may be at once written down by observing that these integrals express the components of the attraction of a spherical shell, of radius bt and density 1, having Q for centre, on a particle situated at O. Hence if x', y', z' be the co-ordinates of Q, measured from O, and $r' = OQ$, the integrals vanish when $r' < bt$, and are equal to

$$4\pi (bt)^2 x' r'^{-3},\quad 4\pi (bt)^2 y' r'^{-3},\quad 4\pi (bt)^2 z' r'^{-3},$$

respectively, when $r' > bt$. Hence we get from (26), omitting the accents, which are now no longer necessary, since we have done with the point P,

$$\xi_2 = \frac{t}{2\pi}\iiint (\omega_0''' y - \omega_0'' z)\frac{dV}{r^3}\ (r > bt) \dots \dots \dots (27).$$

Now

$$2\omega_0' = \frac{dw_0}{dy} - \frac{dv_0}{dz};\quad 2\omega_0'' = \frac{du_0}{dz} - \frac{dw_0}{dx};\quad 2\omega_0''' = \frac{dv_0}{dx} - \frac{du_0}{dy}.$$

Substituting in (27), and adding and subtracting $x\,.\,du_0/dx$ under the integral signs, we get

$$\xi_2 = \frac{t}{4\pi}\iiint\left\{ -\left(x\frac{d}{dx} + y\frac{d}{dy} + z\frac{d}{dz}\right)u_0 \right.$$
$$\left. + \left(x\frac{du_0}{dx} + y\frac{dv_0}{dx} + z\frac{dw_0}{dx}\right)\right\}\frac{dV}{r^3}\ (r > bt).$$

But $x\,.\,d/dx + y\,.\,d/dy + z\,.\,d/dz$ is the same thing as $r\,.\,d/dr$, and we get accordingly

$$\iiint\left(x\frac{d}{dx} + y\frac{d}{dy} + z\frac{d}{dz}\right)u_0\frac{dV}{r^3}\ (r > bt)$$

$$= \int . \iint \frac{du_0}{dr}\,dr\,d\sigma\ (r > bt) = -\iint (u_0)_{bt}\,d\sigma.$$

The second part of ξ_2 is precisely the expression transformed in the preceding article, except that the sign is changed, and b put for a.   Hence we have

$$\xi_2 = \frac{t}{4\pi} \iint (u_0 - lq_0)_{bt} d\sigma - \frac{t}{4\pi} \iiint (u_0 - 3lq_0) \frac{dV}{r^3} \ (r > bt) \ ...(28).$$

19.   Adding together the expressions for ξ_1 and ξ_2, we get for the disturbance due to the initial velocities

$$\xi = \frac{t}{4\pi} \iint l\,(q_0)_{at} d\sigma + \frac{t}{4\pi} \iint (u_0 - lq_0)_{bt} d\sigma$$

$$+ \frac{t}{4\pi} \iiint (3lq_0 - u_0) \frac{dV}{r^3} \ (bt < r < at) \ ....(29).$$

The part of the disturbance due to the initial displacements may be obtained immediately by the theorem of Art. 14.   Let ξ_0, η_0, ζ_0 be the initial displacements, ρ_0 the initial displacement resolved along a radius vector drawn from O.   The last term in equation (29), it will be observed, involves t in two ways, for t enters as a coefficient, and likewise the limits depend upon t.   To find the part of the differential coefficient which relates to the variation of the limits, we have only to replace dV by $r^2 dr\, d\sigma$, and treat the integral in the usual way.   We get for the part of the disturbance due to the initial displacements

$$\xi = \frac{1}{4\pi} \iint \left\{ l\left(4\rho_0 + at\,\frac{d\rho_0}{dr}\right) - \xi_0 \right\}_{at} d\sigma$$

$$+ \frac{1}{4\pi} \iint \left\{ 2\xi_0 + bt\,\frac{d\xi_0}{dr} - l\left(4\rho_0 + bt\,\frac{d\rho_0}{dr}\right) \right\}_{bt} d\sigma$$

$$+ \frac{1}{4\pi} \iiint (3l\rho_0 - \xi_0) \frac{dV}{r^2} \ (bt < r < at)..................(30).$$

It is to be recollected that in this and the preceding equation l denotes the cosine of the angle between the axis of x and an arbitrary radius vector drawn from O, whose direction varies from one element $d\sigma$ of angular space to another, and that the at or bt subscribed denotes that r is supposed to be equal to at or bt after differentiation.   To obtain the whole displacement parallel to x which exists at the end of the time t at the point O, we have only to add together the second members of equations (29) and (30). The expressions for η and ζ may be written down from symmetry, or rather the axis of x may be supposed to be measured in the direction in which we wish to estimate the displacement.

20. The first of the double integrals in equations (29), (30) vanishes outside the limits of the wave of dilatation, the second vanishes outside the limits of the wave of distortion. The triple integrals vanish outside the outer limit of the wave of dilatation, and inside the inner limit of the wave of distortion, but have finite values within the two waves and between them. Hence a particle of the medium situated outside the space T does not begin to move till the wave of dilatation reaches it. Its motion then commences, and does not wholly cease till the wave of distortion has passed, after which the particle remains absolutely at rest.

21. If the initial disturbance be such that there is no wave of distortion, the quantities ϖ', ϖ'', ϖ''', ω', ω'', ω''' must be separately equal to zero, and the expression for ξ will be reduced to ξ_1, given by (25), and the expression thence derived which relates to the initial displacements. The triple integral in the expression for ξ_1 vanishes when the wave of dilatation has passed, and the same is the case with the corresponding integral which depends upon the initial displacements. Hence the medium returns to rest as soon as the wave of dilatation has passed; and since even in the general case each particle remains at rest until the wave of dilatation reaches it, it follows that when the initial disturbance is such that no wave of distortion is formed the disturbance at any time is confined to the wave of dilatation. The same conclusion might have been arrived at by transforming the triple integral.

22. When the initial motion is such that there is no wave of dilatation, as will be the case when there is initially neither dilatation nor velocity of dilatation, ξ will be reduced to ξ_2, given by (28), and the corresponding expression involving the initial displacements. By referring to the expression in Art. 17, from which the triple integral in equation (28) was derived, we get

$$\iiint (u_0 - 3lq_0) \frac{dV}{r^3} = \iiint \left(u_0 \frac{d}{dx}\frac{x}{r^3} + v_0 \frac{d}{dx}\frac{y}{r^3} + w_0 \frac{d}{dx}\frac{z}{r^3} \right) dV \ldots (31).$$

Now

$$\iiint u_0 \frac{d}{dx}\frac{x}{r^3} dV = \iiint u_0 \frac{d}{dx}\frac{x}{r^3} dx\, dy\, dz$$

$$= \iint \left(\frac{u_0 x}{r^3} \right) dy\, dz - \iiint \frac{du_0}{dx}\frac{x}{r^3} dx\, dy\, dz,$$

the parentheses denoting that the quantity enclosed in them is to be taken between limits. By the condition of the absence of initial velocity of dilatation we have

$$- \iiint \frac{du_0}{dx} \frac{x}{r^3} \, dx \, dy \, dz = \iiint \left(\frac{dv_0}{dy} + \frac{dw_0}{dz} \right) \frac{x}{r^3} \, dx \, dy \, dz.$$

Substituting in the second member of equation (31), and writing down for the present only the terms involving v_0, we obtain

$$\iiint \left(\frac{dv_0}{dy} \frac{x}{r^3} + v_0 \frac{d}{dx} \frac{y}{r^3} \right) dx \, dy \, dz,$$

which, since $d/dx \, . \, y/r^3 = d/dy \, . \, x/r^3$, becomes

$$\iiint \frac{d}{dy} \frac{v_0 x}{r^3} \, dx \, dy \, dz \text{ or } \iint \left(\frac{v_0 x}{r^3} \right) dx \, dz.$$

Treating the terms involving w_0 in the same manner, and substituting in (31), we get

$$\iiint (u_0 - 3lq_0) \frac{dV}{r^3} = \iint \left(\frac{u_0 x}{r^3} \right) dy \, dz + \iint \left(\frac{v_0 x}{r^3} \right) dz \, dx + \iint \left(\frac{w_0 x}{r^3} \right) dx \, dy.$$

Now the integration is to extend from $r = bt$ to $r = \infty$. The quantities once integrated vanish at the second limit, and the first limit relates to the surface of a sphere described round O as centre with a radius equal to bt. Putting dS or $b^2 t^2 \, d\sigma$ for an element of the surface of this sphere, we obtain for the value of the second member of the last equation

$$- (bt)^{-2} \iint (lu_0 + mv_0 + nw_0)_{bt} l dS, \quad \text{or} \; - \iint l \, (q_0)_{bt} d\sigma \; ;$$

and therefore the triple integral in equation (28) destroys the second part of the double integral in the same equation. Hence, writing down also the terms depending upon the initial displacements, we obtain for ξ the very simple expression

$$\xi = \frac{t}{4\pi} \iint (u_0)_{bt} d\sigma + \frac{1}{4\pi} \frac{d}{dt} t \iint (\xi_0)_{bt} \, d\sigma.$$

This expression might have been obtained at once by applying the formula (7) to the first of equations (18), which in this case take the form (6), since $\delta = 0$.

23. Let us return now to the general case, and consider especially the terms which alone are important at a great distance from the space to which the disturbance was originally confined;

and, first, let us take the part of ξ which is due to the initial velocities, which is given by equation (29).

Let the three parts of the second member of this equation be denoted by ξ_a, ξ_b, ξ_c, respectively, and replace $d\sigma$ by $(at)^{-2} dS$ or $(bt)^{-2} dS$, as the case may be; then

$$\xi_a = \frac{1}{4\pi a^2 t} \iint l \, (q_0)_{at} dS \dots\dots\dots\dots\dots\dots(32).$$

Let O_1 be a fixed point, taken within the space T, and regarded as the point of reference for all such points as O. Then when O is at such a distance from O_1 that the radius vector, drawn from O, of any element of T makes but a very small angle with OO_1, we may regard l as constant in the integration, and equal to the cosine of the angle between OO_1 and the direction in which we wish to estimate the displacement at O. Moreover the portion of the surface of a sphere having O for centre which lies within T will be ultimately a plane perpendicular to OO_1, and q_0 will be ultimately the initial velocity resolved in the direction OO_1. Hence we have ultimately

$$\xi_a = \frac{l}{4\pi a^2 t} \iint (q_0)_{at} dS,$$

where, for a given direction of O_1O, the integral receives the same series of values, as at increases through the value OO_1, whatever be the distance of O from O_1. Since the direction of the axis of x is arbitrary, and the component of the displacement in that direction is found by multiplying by l a quantity independent of the direction of the axes, it follows that the displacement itself is in the direction OO_1, or in the direction of a normal to the wave. For a given direction of O_1O, the law of disturbance is the same at one distance as at another, and the magnitude of the displacements varies inversely as at, the distance which the wave has travelled in the time t.

We get in a similar manner

$$\xi_b = \frac{1}{4\pi b^2 t} \iint (u_0 - l q_0)_{bt} dS \dots\dots\dots\dots(33),$$

where l, and the direction of the resolved part, q_0, of the initial velocity are ultimately constant, and the surface of which dS is an element is ultimately plane. To find the resolved part of the displacement in the direction OO_1, we must suppose x measured in that direction, and therefore put $l = 1$, $q_0 = u_0$, which gives $\xi_b = 0$. Hence the displacement now considered takes place in a direction perpendicular to OO_1, or is *transversal*.

For a given direction of O_1O, the law of disturbance is constant, but the magnitude of the displacements varies inversely as bt, the distance to which the wave has been propagated. To find the displacement in any direction, OE, perpendicular to OO_1, we have only to take OE for the direction of the axis of x, and therefore put $l = 0$, and suppose u_0 to refer to this direction.

Consider, lastly, the displacement, ξ_c, expressed by the last term in equation (29). The form of the expression shews that ξ_c will be a small quantity of the order t/r^3 or $1/r^2$, since t is of the same order as r; for otherwise the space T would lie outside the limits of integration, and the triple integral would vanish. But ξ_a and ξ_b are of the order $1/r$, and therefore ξ_c may be neglected, except in the immediate neighbourhood of T.

To see more clearly the relative magnitudes of these quantities, let v be a velocity which may be used as a standard of comparison of the initial velocities, R the radius of a sphere whose volume is equal to that of the space T, and compare the displacements ξ_a, ξ_b, ξ_c which exist, though at different times, at the same point O, where $O_1O = r$. These displacements are comparable with

$$\frac{vR^2}{ar}, \quad \frac{vR^2}{br}, \quad \frac{vR^3 t}{r^3},$$

which are proportional to

$$\frac{1}{a}, \quad \frac{1}{b}, \quad \frac{R}{r} \cdot \frac{t}{r}.$$

But, in order that the triple integral in (29) may not wholly vanish, t/r must lie between the limits $1/a$ and $1/b$, or at most lie a very little outside these limits, which it may do in consequence of the finite thickness of the two waves. Hence the quantity neglected in neglecting ξ_c is of the order R/r compared with the quantities retained.

The important terms in the disturbance due to the initial displacements might be got from equation (30), but they may be deduced immediately from the corresponding terms in the disturbance due to the initial velocities by the theorem of Art. 14.

24. If we confine our attention to the terms which vary ultimately inversely as the distance, and which alone are sensible at a great distance from T, we shall be able, by means of the formulæ of the preceding article, to obtain a clear conception of the motion which takes place, and of its connexion with the initial disturbance.

From the fixed point O_1, draw in any direction the right line O_1O equal to r, r being so large that the angle subtended at O by any two elements of T is very small; and let it be required to consider the disturbance at O. Draw a plane P perpendicular to OO_1, and cutting OO_1 produced at a distance p from O_1. Let $-p_1$, $+p_2$ be the two extreme values of p for which the plane P cuts the space T. Conceive the displacements and velocities resolved in three rectangular directions, the first of these, to which ξ and u relate, being the direction OO_1. Let $f_u(p), f_v(p), f_w(p)$ be three functions of p defined by the equations

$$f_u(p) = \iint u_0 dS, \quad f_v(p) = \iint v_0 dS, \quad f_w(p) = \iint w_0 dS, \ldots\ldots(34),$$

and $f_\xi(p), f_\eta(p), f_\zeta(p)$ three other functions depending on the initial displacements as the first three do on the initial velocities, so that

$$f_\xi(p) = \iint \xi_0 dS, \quad f_\eta(p) = \iint \eta_0 dS, \quad f_\zeta(p) = \iint \zeta_0 dS \ldots\ldots(35).$$

These functions, it will be observed, vanish when the variable lies outside of the limits $-p_1$ and $+p_2$. They depend upon the direction O_1O, so that in passing to another direction their values change, as well as the limits of the variable between which they differ from zero. It may be remarked however that in passing from any one direction to its opposite the functions receive the same values, as the variable decreases from $+p_1$ to $-p_2$, that they before received as the variable increased from $-p_1$ to $+p_2$, provided the directions in which the displacements are resolved, as well as the sides towards which the resolved parts are reckoned positive, are the same in the two cases.

The medium about O remains at rest until the end of the time $(r-p_1)/a$, when the wave of dilatation reaches O. During the passage of this wave, the displacements and velocities are given by the equations

$$\left. \begin{aligned} \xi &= \frac{1}{4\pi a r} f_u(at-r) + \frac{1}{4\pi r} f_\xi'(at-r), \quad \eta = 0, \quad \zeta = 0 \\[2mm] u &= \frac{1}{4\pi r} f_u'(at-r) + \frac{a}{4\pi r} f_\xi''(at-r), \quad v = 0, \quad w = 0 \end{aligned} \right\} \ldots(36).$$

The first term in the right-hand member of the first of these equations is got from (32) by putting $l = 1$, introducing the function f_n, and replacing at in the denominator by r, which may be done, since at differs from r only by a small quantity depending upon the finite dimensions of the space T. The second term is derived from the first by the theorem of Art. 14, and u is of course got from ξ by differentiating with respect to t. Had t been retained in the denominator, the differentiation would have introduced terms of the order t^{-2}, and therefore of the order r^{-2}, but such terms are supposed to be neglected.

The wave of dilatation will have just passed over O at the end of the time $(r + p_2)/a$. The medium about O will then remain sensibly at rest in its position of equilibrium till the wave of distortion reaches it, that is, till the end of the time $(r - p_1)/b$. During the passage of this wave, the displacements and velocities will be given by the equations

$$\left. \begin{aligned} \xi = 0, \quad \eta &= \frac{1}{4\pi br} f_v \, (bt - r) + \frac{1}{4\pi r} f_\eta{}' \, (bt - r) \\[2mm] \zeta &= \frac{1}{4\pi br} f_w \, (bt - r) + \frac{1}{4\pi r} f_\zeta' \, (bt - r) \\[2mm] u = 0, \quad v &= \frac{1}{4\pi r} f_v{}' \, (bt - r) + \frac{b}{4\pi r} f_\eta{}'' \, (bt - r) \\[2mm] w &= \frac{1}{4\pi r} f_w{}' \, (bt - r) + \frac{b}{4\pi r} f_\zeta{}'' \, (bt - r) \end{aligned} \right\} \quad \ldots\ldots (37).$$

After the passage of the wave of distortion, which occupies an interval of time equal to $(p_1 + p_2)/b$, the medium will return absolutely to rest in its position of equilibrium.

25.   A caution is here necessary with reference to the employment of equation (30). If we confine our attention to the important terms, we get

$$\xi = \frac{1}{4\pi at} \iint l \left(\frac{d\rho_0}{dr} \right)_{at} dS + \frac{1}{4\pi bt} \iint \left\{ \frac{d\xi_0}{dr} - l \frac{d\rho_0}{dr} \right\}_{bt} dS \ \ldots (38).$$

Now the initial displacements and velocities are supposed to have finite, but otherwise arbitrary, values within the space T, and to vanish outside. Consequently we cannot, without unwarrantably limiting the generality of the problem, exclude from consideration the cases in which the initial displacements and velocities alter abruptly in passing across the surface of T. In particular, if we wish to determine the disturbance at the end of the time t due to the initial disturbance in a part only of the space throughout which the medium was originally disturbed, we are obliged to consider such abrupt variations; and this is precisely what occurs in treating the problem of diffraction. In applying equation (38) to such a case, we must consider the abrupt variation as a limiting case of a continuous, but rapid, variation, and we shall have to add to the double integrals found by taking for $d\rho_0/dr$ and $d\xi_0/dr$ the finite values which refer to the space T, certain single integrals referring to the perimeter of that portion of the plane P which lies within T. The easiest way of treating the integrals is, to reserve the differentiation with respect to t from which the differential coefficients just written have arisen until after the double integration, and we shall thus be led to the formulæ of the preceding article, where the correct values of the terms in question were obtained at once by the theorem of Art. 14.

26. It appears from Arts. 11 and 12, that in the wave of distortion the density of the medium is strictly the same as in equilibrium; but the result obtained in Art. 23, that the displacements in this wave are transversal, that is, perpendicular to the radius of the wave, is only approximate, the approximation depending upon the largeness of the radius, r, of the wave compared with the dimensions of the space T, or, which comes to the same, compared with the thickness of the wave. In fact, if it were strictly true that the displacement at O due to the original disturbance in each element of the space T was transversal, it is evident that the crossing at O of the various waves corresponding to the various elements of T under finite, though small angles, would prevent the whole displacement from being strictly perpendicular to the radius vector drawn to O from an arbitrarily chosen point, O_1, within T. But it is not mathematically true that the disturbance proceeding from even a single point O_1, when a disturbing force is supposed to act, or rather that part of the disturbance which is propagated with the velocity

b, is perpendicular to OO_1, as will be seen more clearly in the next article. It is only so nearly perpendicular that it may be re-garded as strictly so without sensible error. As the wave grows larger, the inclination of the direction of displacement to the wave's front decreases with great rapidity.

Thus the motion of a layer of the medium in the front of a wave may be compared with the tidal motion of the sea, or rather with what it would be if the earth were wholly covered by water. In both cases the density of the medium is unchanged, and there is a slight increase or decrease of thickness in the layer, which allows the motion along the surface to take place without change of density : in both cases the motion in a direction perpendicular to the surface is very small compared with the motion along the surface.

27. From the integral already obtained of the equations of motion, it will be easy to deduce the disturbance due to a given variable force acting in a given direction at a given point of the medium.

Let O_1 be the given point, T a space comprising O_1. Let the time t be divided into equal intervals τ; and at the beginning of the nth interval let the velocity $\tau F(n\tau)$ be communicated, in the given direction, to that portion of the medium which occupies the space T. Conceive velocities communicated in this manner at the beginning of each interval, so that the disturbances produced by these several velocities are superposed. Let D be the den-sity of the medium in equilibrium; and let $F(n\tau) = (DT)^{-1} f(n\tau)$, so that $\tau f(n\tau)$ is the momentum communicated at the beginning of the nth interval. Now suppose the number of intervals τ indefinitely increased, and the volume T indefinitely dimin-ished, and we shall pass in the limit to the case of a moving force which acts continuously.

The disturbance produced by given initial velocities is ex-pressed, without approximation, by equation (29), that is, without any approximation depending on the largeness of the distance OO_1; for the square of the disturbance has been neglected all along. Let $OO_1 = r$; refer the displacement at O to the rect-angular axes of x, y, z; let l, m, n be the direction-cosines of OO_1; l', m', n' those of the given force, and put for shortness k for the cosine of the angle between the direction of the force and the line OO_1 produced, so that

$$k = ll' + mm' + nn'.$$

Consider at present the first term of the right-hand side of (29). Since the radius vector drawn from O to any element of T ultimately coincides with OO_1, we may put l outside the integral signs, and replace $d\sigma$ by $r^{-2}dS$. Moreover, since this term vanishes except when at lies between the greatest and least values of the radius vector drawn from O to any element of T, we may replace t outside the integral signs by r/a. Conceive a series of spheres, with radii $a\tau$, $2a\tau...na\tau,...$ described round O, and let the nth of these be the first which cuts T. Let S_1, S_2... be the areas of the surfaces of the spheres, beginning with the nth, which lie within T; then

$$\iint (q_0)_{at} dS = k\tau F(t - n\tau) S_1 + k\tau F\{t - (n+1)\tau\} S_2 + ...$$

But $F(t - n\tau)$, $F\{t - (n+1)\tau\} ...$ are ultimately equal to each other, and to

$$F\left(t - \frac{r}{a}\right), \text{ or } (DT)^{-1} f\left(t - \frac{r}{a}\right);$$

and $a\tau S_1 + a\tau S_2 + ...$ is ultimately equal to T. Hence we get, for the part of ξ which arises from the first of the double integrals,

$$\frac{lk}{4\pi D a^2 r} f\left(t - \frac{r}{a}\right).$$

The second of the double integrals is to be treated in exactly the same way.

To find what the triple integral becomes, let us consider first only the impulse which was communicated at the beginning of the time $t - n\tau$, where $n\tau$ lies between the limits r/a and r/b, and is not so nearly equal to one of these limits that any portion of the space T lies beyond the limits of integration. Then we must write $n\tau$ for t in the coefficient, and $3lq_0 - u_0$ becomes ultimately $(3lk - l')\,\tau F(t - n\tau)$, and, as well as r, is ultimately constant in the triple integration. Hence the triple integral ultimately becomes

$$\frac{(3lk - l')\,T}{4\pi r^3} n\tau \cdot \tau F(t - n\tau),$$

and we have now to perform a summation with reference to

different values of n, which in the limit becomes an integration. Putting $n\tau = t'$, we have ultimately

$$\tau = dt', \quad \Sigma n\tau \cdot \tau F(t - n\tau) = \int_{\frac{r}{a}}^{\frac{r}{b}} t'\, F(t - t')\, dt'.$$

It is easily seen that the terms arising from the triple integral when it has to be extended over a part only of the space T vanish in the limit. Hence we have, collecting all the terms, and expressing $F(t)$ in terms of $f(t)$,

$$\xi = \frac{lk}{4\pi Da^2 r} f\left(t - \frac{r}{a}\right) + \frac{l' - lk}{4\pi Db^2 r} f\left(t - \frac{r}{b}\right)$$

$$+ \frac{3lk - l'}{4\pi Dr^3} \int_{\frac{r}{a}}^{\frac{r}{b}} t' f(t - t')\, dt' \ldots\ldots\ldots\ldots(39).$$

To get η and ζ, we have only to pass from l, l' to m, m' and then to n, n'. If we take OO_1 for the axis of x, and the plane passing through OO_1 and the direction of the force for the plane xz, and put α for the inclination of the direction of the force to OO_1 produced, we shall have

$$l = 1, \quad m = 0, \quad n = 0, \quad l' = k = \cos \alpha, \quad m' = 0, \quad n' = \sin \alpha;$$

whence

$$\left. \begin{aligned} \xi &= \frac{\cos \alpha}{4\pi Da^2 r} f\left(t - \frac{r}{a}\right) + \frac{\cos \alpha}{2\pi Dr^3} \int_{\frac{r}{a}}^{\frac{r}{b}} t' f(t - t')\, dt' \\ \eta &= 0 \\ \zeta &= \frac{\sin \alpha}{4\pi Db^2 r} f\left(t - \frac{r}{b}\right) - \frac{\sin \alpha}{4\pi Dr^3} \int_{\frac{r}{a}}^{\frac{r}{b}} t' f(t - t')\, dt' \end{aligned} \right\} \ldots\ldots (40).$$

In the investigation, it has been supposed that the force began to act at the time 0, before which the fluid was at rest, so that $f(t) = 0$ when t is negative. But it is evident that exactly the same reasoning would have applied had the force begun to act at any past epoch, as remote as we please, so that we are not obliged to suppose $f(t)$ equal to zero when t is negative, and we may even suppose $f(t)$ periodic, so as to have finite values from $t = -\infty$ to $t = +\infty$.

By means of the formula (39), it would be very easy to write down the expressions for the disturbance due to a system of forces acting throughout any finite portion of the medium, the disturbing force varying in any given manner, both as to magnitude and direction, from one point of the medium to another, as well as from one instant of time to another.

The first term in ξ represents a disturbance which is propagated from O_1 with a velocity a. Since there is no corresponding term in η or ζ, the displacement, as far as relates to this disturbance, is strictly normal to the front of the wave. The first term in ζ represents a disturbance which is propagated from O_1 with a velocity b, and as far as relates to this disturbance the displacement takes place strictly in the front of the wave. The remaining terms in ξ and ζ represent a disturbance of the same kind as that which takes place in an incompressible fluid in consequence of the motion of solid bodies in it. If $f(t)$ represent a force which acts for a short time, and then ceases, $f(t - t')$ will differ from zero only between certain narrow limits of t, and the integral contained in the last terms of ξ and ζ will be of the order r, and therefore the terms themselves will be of the order r^{-2}, whereas the leading terms are of the order r^{-1}. Hence in this case the former terms will not be sensible beyond the immediate neighbourhood of O_1. The same will be true if $f(t)$ represent a periodic force, the mean value of which is zero. But if $f(t)$ represent a force always acting one way, as for example a constant force, the last terms in ξ and ζ will be of the same order, when r is large, as the first terms.

28. It has been remarked in the introduction that there is strong reason for believing that in the case of the luminiferous ether the ratio of a to b is extremely large, if not infinite. Consequently the first term in ξ, which relates to normal vibrations, will be insensible, if not absolutely evanescent. In fact, if the ratio of a to b were no greater than 100, the denominator in this term would be 10000 times as great as the denominator of the first term in ζ. Now the molecules of a solid or gas in the act of combustion are probably thrown into a state of violent vibration, and may be regarded, at least very approximately, as centres of disturbing forces. We may thus see why transversal vibrations should alone be produced, unaccompanied by normal vibrations, or at least by any which are of sufficient magnitude to be sensible. If we could be sure that the ether was strictly incompressible, we should of course be justified in asserting that normal vibrations are impossible.

29. If we suppose $a = \infty$, and $f(t) = c \sin 2\pi bt/\lambda$, we shall get from (40)

$$
\left.
\begin{aligned}
\xi &= \frac{c\lambda \cos\alpha}{4\pi^2 D b^2 r^2} \cos\frac{2\pi}{\lambda}(bt - r) - \frac{c\lambda^2 \cos\alpha}{4\pi^3 D b^2 r^3} \sin\frac{\pi r}{\lambda}\cos\frac{2\pi}{\lambda}\left(bt - \frac{r}{2}\right) \\
\eta &= 0 \\
\zeta &= \frac{c \sin\alpha}{4\pi D b^2 r} \sin\frac{2\pi}{\lambda}(bt - r) - \frac{c\lambda \sin\alpha}{8\pi^2 D b^2 r^2}\cos\frac{2\pi}{\lambda}(bt - r) \\
&\quad + \frac{c\lambda^2 \sin\alpha}{8\pi^3 D b^2 r^3}\sin\frac{\pi r}{\lambda}\cos\frac{2\pi}{\lambda}\left(bt - \frac{r}{2}\right)
\end{aligned}
\right\}
\quad (41);
$$

and we see that the most important term in ξ is of the order $\lambda/\pi r$ compared with the leading term in ζ, which represents the transversal vibrations properly so called.   Hence ξ, and the second and third terms in ζ, will be insensible, except at a distance from O_1 comparable with λ, and may be neglected; but the existence of terms of this nature, in the case of a spherical wave whose radius is not regarded as infinite, must be borne in mind, in order to understand in what manner transversal vibrations are compatible with the absence of dilatation or condensation.

30.   The integration of equations (18) might have been effected somewhat differently by first decomposing the given functions ξ_0, η_0, ζ_0, and u_0, v_0, w_0 into two parts, as in Art. 8, and then treating each part separately.   We should thus be led to consider separately that part of the initial disturbance which relates to a wave of dilatation and that part which relates to a wave of distortion.   Either of these parts, taken separately, represents a disturbance which is not confined to the space T, but extends indefinitely around it.   Outside T, the two disturbances are equal in magnitude and opposite in sign.

SECTION III.

Determination of the Law of the Disturbance in a Secondary Wave of Light.

31.   Conceive a series of plane waves of plane-polarized light propagated in vacuum in a direction perpendicular to a fixed mathematical plane P.   According to the undulatory theory of light, as commonly received, that is, including the doctrine of

transverse vibrations, the light in the case above supposed consists in the vibrations of an elastic medium or *ether*, the vibrations being such that the ether moves in sheets, in a direction perpendicular to that of propagation, and the vibration of each particle being symmetrical with respect to the plane of polarization, and therefore rectilinear, and either parallel or perpendicular to that plane. In order to account for the propagation of such vibrations, it is necessary to suppose the existence of a tangential force, or tangential pressure, called into play by the continuous sliding of the sheets one over another, and proportional to the amount of the displacement of sliding. There is no occasion to enter into any speculation as to the cause of this tangential force, nor to entertain the question whether the luminiferous ether consists of distinct molecules or is mathematically continuous, just as there is no occasion to speculate as to the cause of gravity in calculating the motions of the planets. But we are absolutely obliged to suppose the existence of such a force, unless we are prepared to throw overboard the theory of transversal vibrations, as usually received, notwithstanding the multitude of curious, and otherwise apparently inexplicable phenomena which that theory explains with the utmost simplicity. Consequently we are led to treat the ether as an elastic solid *so far as the motions which constitute light are concerned.* It does not at all follow that the ether is to be regarded as an elastic solid when large displacements are considered, such as we may conceive produced by the earth and planets, and solid bodies in general, moving through it. The mathematical theories of fluids and of elastic solids are founded, or at least may be founded, on the consideration of internal pressures. In the case of a fluid, these pressures are supposed normal to the common surface of the two portions whose mutual action is considered : this supposition forms in fact the mathematical definition of a fluid. In the case of an elastic solid, the pressures are in general oblique, and may even in certain directions be wholly tangential. The treatment of the question by means of pressures presupposes the absence of any sensible *direct* mutual action of two portions of the medium which are separated by a small but sensible interval. The state of constraint or of motion of any element affects the pressures in the surrounding medium, and in this way one element exerts an *indirect* action on another from which it is separated by a sensible interval.

Now the absence of prismatic colours in the stars, depending upon aberration, the absence of colour in the disappearance and reappearance of Jupiter's Satellites in the case of eclipses, and, still more, the absence of change of colour in the case of certain periodic stars, especially the star Algol, shew that the velocity of light of different colours is, if not mathematically, at least sensibly the same. According to the theory of undulations, this is equivalent to saying that in vacuum the velocity of propagation is independent of the length of the waves. Consequently the direct action of two elements of ether separated by a sensible interval must be sensibly if not mathematically equal to zero, or at least must be independent of the disturbance; for, were this not the case, the expression for the velocity of propagation would involve the length of a wave. An interval is here considered sensible which is comparable with the length of a wave. We are thus led to apply to the luminiferous ether in vacuum the ordinary equations of motion of an elastic solid, provided we are only considering those disturbances which constitute light.

Let us return now to the case supposed at the beginning of this section. According to the preceding explanation, we must regard the ether as an elastic solid, in which a series of rectilinear transversal vibrations is propagated in a direction perpendicular to the plane P. The disturbance at any distance in front of this plane is really produced by the disturbance continually transmitted across it; and, according to the general principle of the superposition of small motions, we have a perfect right to regard the disturbance in front as the aggregate of the elementary disturbances due to the disturbance continually transmitted across the several elements into which we may conceive the plane P divided. Let it then be required to determine the disturbance corresponding to an elementary portion only of this plane.

In practical cases of diffraction at an aperture, the breadth of the aperture is frequently sensible, though small, compared with the radius of the incident waves. But in determining the law of disturbance in a secondary wave we have nothing to do with an aperture; and in order that we should be at liberty to regard the incident waves as plane all that is necessary is, that the radius of the incident wave should be very large compared with the wave's length, a condition always fulfilled in experiment.

32. Let O_1 be any point in the plane P; and refer the medium to rectangular axes passing through O_1, x being measured in the direction of propagation of the incident light, and z in the direction of vibration. Let $f(bt-x)$ denote the displacement of the medium at any point behind the plane P, x of course being negative. Let the time t be divided into small intervals, each equal to τ, and consider separately the effect of the disturbance which is transmitted across the plane P during each separate interval. The disturbance transmitted during the interval τ which begins at the end of the time t' occupies a film of the medium, of thickness $b\tau$, and consists of a displacement $f(bt')$ and a velocity $bf'(bt')$. By the formulæ of Section II. we may find the effect, over the whole medium, of the disturbance which exists in so much only of the film as corresponds to an element dS of P adjacent to O_1. By doing the same for each interval τ, and then making the number of such intervals increase and the magnitude of each decrease indefinitely, we shall ultimately obtain the effect of the disturbance which is continually propagated across the element dS.

Let O be the point of the medium at which the disturbance is required; l, m, n the direction-cosines of O_1O measured from O_1, and therefore $-l$, $-m$, $-n$ those of OO_1 measured from O; and let $OO_1 = r$. Consider first the disturbance due to the velocity of the film. The displacements which express this disturbance are given without approximation by (29) and the two other equations which may be written down from symmetry. The first terms in these equations relate to normal vibrations, and on that account alone might be omitted in considering the diffraction of light. But, besides this, it is to be observed that t in the coefficient of these terms is to be replaced by r/a. Now there seems little doubt, as has been already remarked in the introduction, that in the case of the luminiferous ether a is incomparably greater than b, if not absolutely infinite*; so that the terms in question are insensible, if not absolutely evanescent. The third terms are insensible, except at a distance from O_1 comparable with λ, as has been already observed, and they may therefore be omitted if we suppose r very large compared with the length of a wave. Hence it will be sufficient to consider the second terms only. In the

* I have explained at full my views on this subject in a paper *On the constitution of the luminiferous ether*, printed in the 32nd volume of the *Philosophical Magazine*, p. 349. [*Ante*, p. 12.]

coefficient of these terms we must replace t by r/b; we must put $u_0 = 0$, $v_0 = 0$, $w_0 = bf'\,(bt-r)$, write $-l$, $-m$, $-n$ for l, m, n, and put $q_0 = -nw_0 = -nbf'\,(bt-r)$. The integral signs are to be omitted, since we want to get the disturbance which corresponds to an elementary portion only of the plane P.

It is to be observed that $d\sigma$ represents the elementary solid angle subtended at O by an element of the riband formed by that portion of the surface of a sphere described round O, with radius r, which lies between the plane yz and the parallel plane whose abscissa is $b\tau$. To find the aggregate disturbance at O corresponding to a small portion, S, of the plane P lying about O_1, we must describe spheres with radii $\ldots\, r-2b\tau,\; r-b\tau,\; r,\; r+b\tau,\; r+2b\tau\,\ldots,$ describing as many as cut S. These spheres cut S into ribands, which are ultimately equal to the corresponding ribands which lie on the spheres. For, conceive a plane drawn through OO_1 perpendicular to the plane yz. The intersections of this plane by two consecutive spheres and the two parallel planes form a quadrilateral, which is ultimately a rhombus; so that the breadths of corresponding ribands on a sphere and on the plane are equal, and their lengths are also equal, and therefore their areas are equal. Hence we must replace $d\sigma$ by $r^{-2}dS$, and we get accordingly

$$\xi = -\frac{lndS}{4\pi r}f'\,(bt-r), \qquad \eta = -\frac{mndS}{4\pi r}f'\,(bt-r),$$

$$\zeta = \frac{(1-n^2)\,dS}{4\pi r}f'\,(bt-r) \;\ldots\ldots(42).$$

Since $l\xi + m\eta + n\zeta = 0$, the displacement takes place in a plane through O perpendicular to O_1O. Again, since $\xi : \eta :: l : m$, it takes place in a plane through O_1O and the axis of z. Hence it takes place along a line drawn in the plane last mentioned perpendicular to OO_1. The direction of displacement being known, it remains only to determine the magnitude. Let ζ_1 be the displacement, and ϕ the angle between O_1O and the axis of z, so that $n = \cos\phi$. Then $\zeta_1 \sin\phi$ will be the displacement in the direction of z, and equating this to ζ in (42) we get

$$\zeta_1 = \frac{dS}{4\pi r}\sin\phi f'\,(bt-r) \;\ldots\ldots\ldots\ldots(43).$$

The part of the disturbance due to the successive displacements of the films may be got in the same way from (30) and the two other equations of the same system. The only terms which it will be necessary to retain in these equations are those which involve the differential coefficients of ξ_0, η_0, ζ_0, and ρ_0 in the second of the double integrals. We must put as before r for bt, and write $r^{-2}dS$ for $d\sigma$. Moreover we have for the incident vibrations

$$\xi = 0, \quad \eta = 0, \quad \zeta = f(bt' - x), \quad \rho = -nf(bt' - x).$$

To find the values of the differential coefficients which have to be used in (30) and the two other equations of that system, we must differentiate on the supposition that ξ, η, ζ, ρ are functions of r in consequence of being functions of x, and after differentiation we must put $x = 0$, $t' = t - r/b$. Since $d/dr = -l \cdot d/dx$, we get

$$\left(\frac{d\zeta_0}{dr}\right)_{bt} = lf'(bt - r), \quad \left(\frac{d\rho_0}{dr}\right)_{bt} = -lnf'(bt - r),$$

whence we get, remembering that the signs of l, m, n in (30) have to be changed,

$$\xi = -\frac{l^2 n\, dS}{4\pi r} f'(bt - r), \quad \eta = -\frac{lmn\, dS}{4\pi r} f'(bt - r),$$

$$\zeta = \frac{l(1 - n^2)\, dS}{4\pi r} f'(bt - r).$$

The displacement represented by these equations takes place along the same line as before; and if we put ζ_2 for the displacement, and write $\cos\theta$ for l, we get

$$\zeta_2 = \frac{dS}{4\pi r} \cos\theta \sin\phi f'(bt - r) \dots\dots\dots\dots\dots(44).$$

33. By combining the partial results obtained in the preceding article, we arrive at the following theorem.

Let $\xi = 0$, $\eta = 0$, $\zeta = f(bt - x)$ be the displacements corresponding to the incident light; let O_1 be any point in the plane P, dS an element of that plane adjacent to O_1; and consider the disturbance due to that portion only of the incident disturbance which passes continually across dS. Let O be any point in the medium situated at a distance from the point O_1 which is large in comparison with the length of a wave; let $O_1O = r$, and let this line make angles θ with the direction of propagation of the incident light, or the axis of x, and ϕ with the direction of vibration, or the axis of z. Then the displacement at O will take place in a direction perpendicular to O_1O, and lying in the plane zO_1O; and if ζ' be the displacement at O, reckoned positive in the direction nearest to that in which the incident vibrations are reckoned positive,

$$\zeta' = \frac{dS}{4\pi r}\, (1 + \cos\theta)\sin\phi f'\, (bt - r)*\ldots\ldots\ldots(45).$$

In particular, if

$$f(bt - x) = c\sin\frac{2\pi}{\lambda}\, (bt - x),$$

we shall have

$$\zeta' = \frac{cdS}{2\lambda r}(1 + \cos\theta)\sin\phi\cos\frac{2\pi}{\lambda}(bt - r)\ldots\ldots(46).$$

34. On finding by means of this formula the aggregate disturbance at O due to all the elements of the plane P, O being supposed to be situated at a great distance from P, we ought to arrive at the same result as if the waves had not been broken up.

To verify this, let fall from O the perpendicular OO' on the plane P, and let $OO' = p$, or $= -p$, according as O is situated in front of the plane P or behind it. Through O' draw $O'x'$, $O'y'$, parallel to O_1x, O_1y, and let $O'O_1 = r'$, $O_1O'y' = \omega$. Then

$$dS = r'dr'd\omega = rdrd\omega,$$

since $r^2 = p^2 + r'^2$, and p is constant. Let $\zeta' = s\sin\phi$. The displacement ζ' takes place in the plane zO_1O, and perpendicular to O_1O; and resolving it along and perpendicular to O_1z, we get for resolved parts $s\sin^2\phi$, $s\sin\phi\cos\phi$, of which the latter is estimated in the direction OM, where M is the projection of O_1 on $O'y'$. Let $MOO' = \chi$, χ being reckoned positive when M falls on that side of O' on which y' is reckoned positive; then, resolving the displacement along OM parallel to $O'x'$, $O'y'$, we get for resolved parts $-s\sin\phi\cos\phi\cos\chi$, $s\sin\phi\cos\phi\sin\chi$. Hence we get for the displacements ξ, η, ζ at O

$$\xi = -s\sin\phi\cos\phi\cos\chi, \quad \eta = s\sin\phi\cos\phi\sin\chi, \quad \zeta = s\sin^2\phi.$$

Now produce $O'O_1$ to O_2, and refer O_1x, O_1y, O_1z, O_1O_2, O_1O to a sphere described round O_1 with radius unity. Then zO_2O forms a spherical triangle, right-angled at O_2, and

$$zO_2 = \tfrac{1}{2}\pi - \omega, \quad O_2O = \tfrac{1}{2}\pi + \theta, \quad Oz = \phi, \quad OzO_2 = \tfrac{1}{2}\pi + \chi,$$

---

* The corresponding expression which I have obtained for sound differs from this only in having $\cos\theta$ in place of $\sin\phi$, provided we suppose b to be the velocity of propagation of sound, and ζ' to represent a displacement in the direction O_1O.

whence we get from spherical trigonometry,

$$\cos \phi = - \sin \theta \sin \omega, \quad \sin \phi \cos \chi = \cos \theta,$$

$$\sin \phi \sin \chi = \cos \theta \tan \chi = \sin \theta \cos \omega.$$

We have therefore

$$\xi = s \sin \theta \cos \theta \sin \omega, \quad \eta = - s \sin^2 \theta \sin \omega \cos \omega,$$

$$\zeta = s (1 - \sin^2 \theta \sin^2 \omega).$$

To find the aggregate disturbance at O, we must put for s its value, and perform the double integrations, the limits of ω being 0 and 2π, and those of r being $\sqrt{p^2}$ and ∞. The positive and negative parts of the integrals which give ξ and η will evidently destroy each other, and we need therefore only consider ζ. Putting for s its value, and expressing θ in terms of r, we get

$$\zeta = \frac{c}{2\lambda} \iint (r + p) (r^2 \cos^2 \omega + p^2 \sin^2 \omega) \cos \frac{2\pi}{\lambda} (bt - r) \frac{dr d\omega}{r^3} \dots (47).$$

Let us first conceive the integration performed over a large area A surrounding O', which we may afterwards suppose to increase indefinitely. Perform the integration with respect to r first, put for shortness $F(r)$ for the coefficient of the cosine under the integral signs, and let R, a function of ω, be the superior limit of r. We get by integration by parts

$$\int F(r) \cos \frac{2\pi}{\lambda} (bt - r) \, dr$$

$$= - \frac{\lambda}{2\pi} F(r) \sin \frac{2\pi}{\lambda} (bt - r) + \left(\frac{\lambda}{2\pi} \right)^2 F'(r) \cos \frac{2\pi}{\lambda} (bt - r) + \dots$$

Now the terms after the first must be neglected for consistency's sake, because the formula (46) is not exact, but only approximate, the approximation depending on the neglect of terms which are of the order λ compared with those retained. The first term, taken between limits, gives

$$\frac{\lambda}{2\pi} F(\pm p) \sin \frac{2\pi}{\lambda} (bt \mp p) - \frac{\lambda}{2\pi} F(R) \sin \frac{2\pi}{\lambda} (bt - R),$$

where the upper or lower sign has to be taken according as O lies in front of the plane P or behind it. We thus get from (47)

$$\zeta = \frac{c}{2} (1 \pm 1) \sin \frac{2\pi}{\lambda} (bt \mp p) - \frac{c}{4\pi} \int_0^{2\pi} F(R) \sin \frac{2\pi}{\lambda} (bt - R) \, d\omega.$$

When R becomes infinite, $F(R)$ reduces itself to $\cos^2 \omega$, and the last term in ζ becomes

$$- \frac{c}{4\pi} \int_0^{2\pi} \cos^2 \omega \sin \frac{2\pi}{\lambda} (bt - R)\, d\omega.$$

Suppose that no finite portion of the perimeter of A is a circular arc with O' for centre, and let this perimeter be conceived to expand indefinitely, remaining similar to itself. Then, for any finite interval, however small, in the integration with respect to ω, the function $\sin 2\pi \lambda^{-1}(bt - R)$ will change sign an infinite number of times, having a mean value which is ultimately zero, and the limit of the above expression will be rigorously zero. Hence we get in the limit

$$\zeta = c \sin \frac{2\pi}{\lambda} (bt - p), \text{ or } = 0,$$

according as p is positive or negative. Hence the disturbance continually transmitted across the plane P produces the same disturbance in front of that plane as if the wave had not been broken up, and does not produce any back wave, which is what it was required to verify.

It may be objected that the supposition that the perimeter of A is free from circular arcs having O' for centre is an arbitrary restriction. The reply to this objection is, that we have no right to assume that the disturbance at O which corresponds to an area A approaches in all cases to a limit as A expands, remaining similar to itself. All we have a right to assert *a priori* is, that *if* it approaches a limit that limit must be the disturbance which would exist if the wave had not been broken up.

It is hardly necessary to observe that the more general formula (45) might have been treated in precisely the same way as (46).

35. In the third Volume of the *Cambridge Mathematical Journal*, p. 46, will be found a short paper by Mr Archibald Smith, of which the object is to determine the intensity in a secondary wave of light. In this paper the author supposes the intensity at a given distance the same in all directions, and assumes the coefficient of vibration to vary, in a given direction, inversely as the radius of the secondary wave. The intensity is determined on the principle that when an infinite plane wave is conceived to be broken up, the aggregate effect of the secondary waves must

be the same as that of the primary wave. In the investigation, the difference of direction of the vibrations corresponding to the various secondary waves which agitate a given point is not taken into account, and moreover a term which appears under the form cos ∞ is assumed to vanish. The correctness of the result arrived at by the latter assumption may be shewn by considerations similar to those which have just been developed. If we suppose the distance from the primary wave of the point which is agitated by the secondary waves to be large in comparison with λ, it is only those secondary waves which reach the point in question in a direction nearly coinciding with the normal to the primary wave that produce a sensible effect, since the others neutralize each other at that point by interference. Hence the result will be true for a direction nearly coinciding with the normal to the primary wave, independently of the truth of the assumption that the disturbance in a secondary wave is equal in all directions, and notwithstanding the neglect of the mutual inclination of the directions of the disturbances corresponding to the various secondary waves. Accordingly, when the direction considered is nearly that of the normal to the primary wave, cos θ and sin ϕ in (46) are each nearly equal to 1, so that the coefficient of the circular function becomes $cdS(\lambda r)^{-1}$, nearly, and in passing from the primary to the secondary waves it is necessary to accelerate the phase by a quarter of an undulation. This agrees with Mr Smith's results.

The same subject has been treated by Professor Kelland in a memoir *On the Theoretical Investigation of the Absolute Intensity of Interfering Light*, printed in the fifteenth Volume of the *Transactions of the Royal Society of Edinburgh*, p. 315. In this memoir the author investigates the case of a series of plane waves which passes through a parallelogram in front of a lens, and is received on a scieen at the focus of the lens, as well as several other particular cases. By equating the total illumination on the screen to the area of the aperture multiplied by the illumination of the incident light, the author arrives in all cases at the conclusion that in the coefficient of vibration of a secondary wave the elementary area dS must be divided by λr. In consequence of the employment of intensities, not displacements, the necessity for the acceleration of the phase by a quarter of an undulation does not appear from this investigation.

In the investigations of Mr Smith and Professor Kelland, as well as in the verification of the formula (46) given in the last article, we are only concerned with that part of a secondary wave which lies near the normal to the primary. The correctness of this formula for all directions must rest on the dynamical theory.

36. In any given case of diffraction, the intensity of the illumination at a given point will depend mainly on the mode of interference of the secondary waves. If however the incident light be polarized, and the plane of polarization be altered, every thing else remaining the same, the mode of interference will not be changed, and the coefficient of vibration will vary as $\sin \phi$, so that the intensity will vary between limits which are as 1 to $\cos^2 \theta$. If common light of the same intensity be used, the intensity of the diffracted light at the given point will be proportional to $\frac{1}{2}(1 + \cos^2 \theta)$.

PART II.

EXPERIMENTS ON THE ROTATION OF THE PLANE OF POLARIZATION OF DIFFRACTED LIGHT.

SECTION I.

Description of the Experiments.

IF a plane passing through a ray of plane-polarized light, and containing the direction of vibration, be called the *plane of vibration*, the law obtained in the preceding section for the nature of the polarization of diffracted light, when the incident light is plane-polarized, may be expressed by saying, that any diffracted ray is plane-polarized, and the plane of vibration of the diffracted ray is parallel to the direction of vibration of the incident ray. Let the angle between the incident ray produced and the diffracted ray be called the *angle of diffraction*, and the plane containing these two rays the *plane of diffraction*; let α_i, α_d be the angles which the planes of vibration of the incident and diffracted rays respectively make with planes drawn through those rays perpendicular to the plane of diffraction, and θ the angle of diffraction. Then we easily get by a spherical triangle

$$\tan \alpha_d = \cos \theta \tan \alpha_i.$$

If then the plane of vibration of the incident ray be made to turn round with a uniform velocity, the plane of vibration of the diffracted ray will turn round with a variable velocity, the law connecting these velocities being the same as that which connects the sun's motions in right ascension and longitude, or the motions of the two axes of a Hook's joint. The angle of diffraction answers to the obliquity of the ecliptic in the one case, or the supplement of the angle between the axes in the other. If we suppose a series of equidifferent values given to α_i, such as $0°$, $5°$, $10°,...355°$, the planes of vibration of the diffracted ray will not be distributed uniformly, but will be crowded towards the plane perpendicular to the plane of diffraction, according to the law expressed by the above equation.

Now the angles which the *planes of polarization* of the incident and diffracted rays, (if the diffracted ray prove to be really plane-polarized,) make with planes perpendicular to the plane of diffraction can be measured by means of a pair of graduated instruments furnished with Nicol's prisms. Suppose the plane of polarization of the incident light to be inclined at the angles $0°$, $5°$, $10°...$, successively to the perpendicular to the plane of diffraction; then the readings of the instrument which is used as the analyzer will shew whether the planes of polarization of the diffracted ray are crowded towards the plane of diffraction or towards the plane perpendicular to the plane of diffraction. If ϖ, α be the azimuths of the planes of polarization of the incident and diffracted rays, both measured from planes perpendicular to the plane of diffraction, we should expect to find these angles connected by the equation $\tan \alpha = \sec \theta \tan \varpi$ in the former event, and $\tan \alpha = \cos \theta \tan \varpi$ in the latter. If the law and amount of the crowding agree with theory as well as could reasonably be expected, some allowance being made for the influence of modifying causes, (such as the direct action of the edge of the diffracting body,) whose exact effect cannot be calculated, then we shall be led to conclude that the vibrations in plane-polarized light are perpendicular or parallel to the plane of polarization, according as the crowding takes place towards or from the plane of diffraction.

In all ordinary cases of diffraction, the light becomes insensible at such a small angle from the direction of the incident ray produced that the crowding indicated by theory is too small to be sensible in experiment, except perhaps in the mean of a very great number of observations. It is only by means of a fine grating that we can obtain strong light which has been diffracted at a large angle. I doubt whether a grating properly so called, that is, one consisting of actual wires, or threads of silk, has ever been made which would be fine enough for the purpose. The experiments about to be described have accordingly been performed with the glass grating already mentioned, which consisted of a glass plate on which parallel and equidistant lines had been ruled with a diamond at the rate of about 1300 to an inch.

Although the law enunciated at the beginning of this section has been obtained for diffraction in vacuum, there is little doubt that the same law would apply to diffraction within a homogeneous uncrystallized medium, at least to the degree of accuracy that we employ when we speak of the refractive index of a substance, neglecting the dispersion. This is rendered probable by the simplicity of the law itself, which merely asserts that the vibrations in the diffracted light are rectilinear, and agree in direction with the vibrations in the incident light as nearly as is consistent with the necessary condition of being perpendicular to the diffracted ray. Moreover, when dispersion is neglected, the same equations of motion of the luminiferous ether are obtained, on mechanical theories, for singly refracting media as for vacuum; and if these equations be assumed to be correct, the law under consideration, which is deduced from the equations of motion, will continue to hold good. In the case of a glass grating however the diffraction takes place neither in air nor in glass, but at the confines of the two media, and thus theory fails to assign exact values to α. Nevertheless it does not fail to assign limits within which, or at least not far beyond which, α must reasonably be supposed to lie; and as the values comprised within these limits are very different according as one or other of the two rival theories respecting the direction of vibration is adopted, experiments with a glass grating may be nearly as satisfactory, so far as regards pointing to one or other of the two theories, as experiments would be which were made with a true grating.

The glass grating was mounted for me by Prof. Miller in a

small frame fixed on a board which rested on three screws, by means of which the plane of the plate and the direction of the grooves could be rendered perpendicular to the plane of a table on which the whole rested.

The graduated instruments lent to me by Prof. O'Brien consisted of small graduated brass circles, mounted on brass stands, so that when they stood on a horizontal table the planes of the circles were vertical, and the zeros of graduation vertically over the centres. The circles were pierced at the centre to admit doubly refracting prisms, which were fixed in brass collars which could be turned round within the circles, the axes of motion being perpendicular to the planes of the circles, and passing through their centres. In one of the instruments, which I used for the polarizer, the circle was graduated to degrees from 0° to 360°, and the collar carried simply a pointer. To stop the second pencil, I attached a wooden collar to the brass collar, and inserted in it a Nicol's prism, which was turned till the more refracted pencil was extinguished. In a few of the latest experiments the Nicol's prism was dispensed with, and the more refracted pencil stopped by a screen with a hole which allowed the less refracted pencil to pass. In the other instrument, which I used for the analyzer, the brass collar carried a vernier reading to 5′. In this instrument the doubly refracting prism admitted of being removed, and I accordingly removed it, and substituted a Nicol's prism, which was attached by a wooden collar. The Nicol's prism was usually inserted into the collar at random, and the index error was afterwards determined from the observations themselves.

The light employed in all the experiments was the sun light reflected from a mirror placed at the distance of a few feet from the polarizer. On account of the rotation of the earth, the mirror required re-adjustment every three or four minutes. The continual change in the direction of the incident light was one of the chief sources of difficulty in the experiments and inaccuracy in the results; but lamplight would, I fear, be too weak to be of much avail in these experiments.

The polarizer, the grating, and the analyzer stood on the same table, the grating a few inches from the polarizer, and the analyzer about a foot from the grating. The plane of diffraction was assumed to be parallel to the table, which was nearly the case; but the change in the direction of the incident light produced continual small changes in the position of this plane. In most experiments the grating was placed perpendicular to the incident light, by making the light reflected from the surface go back into the hole of the polarizer. The angle of diffraction was measured at the conclusion of each experiment by means of a protractor, lent to me for the purpose by Prof. Miller. The grating was removed, and the protractor placed with its centre as nearly as might be under the former position of the bright spot formed on the grating by the incident light. The protractor had a pair of opposite verniers moveable by a rack; and the directions of the incident and diffracted light were measured by means of sights attached to the verniers. The angle of diffraction in the different experiments ranged from about 20° to 60°.

The deviation of the less refracted pencil in the doubly refracting prism of the polarizer, though small, was very sensible, and was a great source both of difficulty and of error. To understand this, let AB be a ray incident at B on a slip of the surface of the plate contained between two consecutive grooves, BC a diffracted ray. On account of the interference of the light coming from the different parts of the slip, if a small pencil whose axis is AB be incident on the slip, the diffracted light will not be sensible except in a direction BC, determined by the condition that $AB + BC$ shall be a minimum, A and C being supposed fixed. Hence AB, BC must make equal angles with the slip, regarded as a line, the acute angles lying towards opposite ends of the slip, and therefore C must lie in the surface of a cone formed by the revolution of the produced part of AB about the slip. If AB represent the pencil coming through the polarizer, it will describe a cone of small angle as the pointer moves round, and therefore both the position of the vertex and the magnitude of the vertical angle of the cone which is the locus of C will change. Hence the sheet of the cone may sometimes fall above or below the eye-hole of the analyzer. In such a case it is necessary either to be content to miss one or more observations, corresponding to certain readings of the polarizer, or else to alter a little the direction of the incident light, or, by means of the screws, to turn the grating through a small angle round a horizontal axis. The deviation of the light which passed through the polarizer, and the small changes in the direction of the incident light, I regard as the chief causes of error in my experiments. In repeating the experiments so as to get accurate results, these causes of error would have to be avoided.

At first 1 took for granted that the instrument-maker had inserted the doubly refracting prism in the polarizer in such a manner that the plane of polarization of the less refracted pencil was either vertical or horizontal, (the instrument being supposed to stand on a horizontal table,) when the pointer stood at $0°$, having reason to know that it was not inserted at random; and having determined which, by an exceedingly rough trial, I concluded it was vertical. Meeting afterwards with some results which were irreconcileable with this supposition, I was led to make an actual measurement, and found that the plane of polarization was vertical when the pointer stood at $25°$. Consequently $25°$ is to be regarded as the index error of the polarizer, to be subtracted from the reading of the pointer. The circumstance just mentioned accounts for the apparently odd selection of values of ϖ in the earlier experiments, the results of which are given in the tables at the end of this section.

On viewing a luminous point or line through the grating, the central colourless image was seen accompanied by side spectra, namely, the spectra which Fraunhofer called *Spectra of the second class.* After a little, these spectra overlapped in such a manner that the individual spectra could no longer be distinguished, and nothing was to be seen but two tails of light, which extended, one on each side, nearly $90°$ from the central image. On viewing the flame of a spirit lamp through the grating, the individual spectra of the second class could be seen, where, with sun-light, nothing could be perceived but a tail of light. The tails themselves were not white, but exhibited very broad impure spectra; about two such could be made out on each side. These spectra are what were called *spectra of the first class* by Fraunhofer, who shewed that their breadth depended on the smaller of the two quantities, the breadth of a groove, and the breadth of the polished interval between two consecutive grooves. In the grating, the breadth of the grooves was much smaller than the breadth of the intervals between*.

* On viewing the grating under a microscope, the grooves were easily seen to be much narrower than the intervals between; their breadth was too small to be measured. On looking at the flame of a spirit lamp through the grating, I counted sixteen images on one side, then several images were too faint to be seen, and

In the experiments, the diffracted light observed belonged to a bright, though not always the brightest, part of a spectrum of the first class. The compound nature of the light was easily put in evidence by placing a screen with a vertical slit between the grating and the eye, and then viewing the slit through a prism with its edge vertical*. A spectrum was then seen which consisted of bright bands separated by dark intervals, strongly resembling the appearance presented when a pure spectrum is viewed through a pinhole, or narrow slit, which is half covered by a plate of mica, placed on the side at which the blue is seen. At a considerable angle of diffraction as many as 15 or 20 bands might be counted.

In the first experiment the grating was placed with its plane perpendicular to the light which passed through the polarizer, the grooved face being turned from the polarizer. The light observed was that which was diffracted at emergence from the glass. It is only when the eye is placed close to the grating, or when, if the eye be placed a few inches off, the whole of the grating is illuminated, that a large portion of a tail of light can be seen at once. When only a small portion of the grating is illuminated, and the eye is placed at the distance of several inches, as was the case in the experiments, it is only a small portion of a tail which can enter the pupil. The appearance presented is that of a bright spot on the grooved face of the glass. The angle of diffraction in the first experiment was large, 57° 5′ by measurement. Besides the principal image, or bright spot, a row of images were seen to the left: the regularly transmitted light lay to the right, right and left being estimated with reference to the position of the observer. These images were due to internal diffraction and reflection, as will be better understood further on.

---

further still the images again appeared, though they were fainter than before. I estimated the direction of zero illumination to be situated about the eighteenth image. If we take this estimation as correct, it follows from the theory of these gratings that the breadth of a groove was the eighteenth part of the interval between any point of one groove and the corresponding point of its consecutive, an interval which in the case of the present grating was equal to the 1-1300th part of an inch. Hence the breadth of a groove was equal to the 1-23400th part of an inch.

  * To separate the different spectra, Fraunhofer used a small prism with an angle of about 20°, fixed with its edge horizontal in front of the eye-piece of the telescope through which, in his experiments, the spectra were viewed.

They were separated by small angles, depending on the thickness of the glass, but sufficient to allow of one image being observed by itself. The observations were confined to the principal or right-hand image.

In the portion of a spectrum of the first class which was observed there was a predominance of red light. In most positions of the pointer of the polarizer the diffracted light did not wholly vanish on turning round the analyzer, but only passed through a minimum. In passing through the minimum the light rapidly changed colour, being blue at the minimum. This shews that the different colours were polarized in different planes, or perhaps not strictly plane-polarized. Nevertheless, as the intensity of the light at the minimum was evidently very small compared with its intensity at the maximum, and the change of colour was rapid, it is allowable to speak in an approximate way of the plane of polarization of the diffracted light, just as it is allowable to speak of the refractive index of a substance, although there is really a different refractive index for each different kind of light. It was accordingly the angular position of the plane which was the best representative of a plane of polarization that I sought to determine in this and the subsequent experiments.

In the first experiment the plane of polarization of the diffracted light was determined by six observations for each angle at which the pointer of the polarizer was set. This took a good deal of time, and increased the errors depending on changes in the direction of the light. Accordingly, in a second experiment, I determined the plane of polarization by single observations only, setting the pointer of the polarizer at smaller intervals than before. Both these experiments gave for result that the planes of polarization of the diffracted light were distributed very nearly uniformly. This result already points very decidedly to one of the two hypotheses respecting the direction of vibration. For according to theory the effect of diffraction alone would be, greatly to crowd the planes either in one direction or in the other. It seems very likely that the effect of oblique emergence alone should be to crowd the planes in the manner of refraction, that is, towards the perpendicular to the plane of diffraction. If then we adopt Fresnel's hypothesis, the two effects will be opposed, and may very well be supposed wholly or nearly to neutralize each other. But if we adopt the other hypothesis we shall be obliged to suppose that in the oblique emergence from the glass, or in something else, there exists a powerful cause of crowding towards the plane of diffraction, that is, *in the manner of reflection,* sufficient to neutralize the great crowding in the contrary direction produced by diffraction, which certainly seems almost incredible.

The nearly uniform distribution of the planes of polarization of the diffracted light shews that the two streams of light, polarized in and perpendicular to the plane of diffraction respectively, into which the incident light may be conceived to be decomposed, were diffracted at emergence from the glass in very nearly the same proportion. This result appeared to offer some degree of vague analogy with the depolarization of light produced by such substances as white paper. This analogy, if borne out in other cases, might seem to throw some doubt on the conclusiveness of the experiments with reference to the decision of the question as to the direction of the vibrations of plane-polarized light. For the deviation of the light from its regular course might seem due rather to a sort of scattering than to regular diffraction, though certainly the fact that the observed light was very nearly plane-polarized does not at all harmonize with such a view. Accordingly, I was anxious to obtain a case of diffraction in which the planes of polarization of the diffracted light should be decidedly crowded one way or other. Now, according to the explanation above given, the approximate uniformity of distribution of the planes of polarization in the first two experiments was due to the antagonistic effects of diffraction, (according to Fresnel's hypothesis respecting the direction of vibration), and of oblique emergence from the glass, or irregular refraction, that is, refraction produced wholly by diffraction. If this explanation be correct, a very marked crowding towards the plane of diffraction ought to be produced by diffraction at reflection, since diffraction alone and reflection alone would crowd the planes in the same manner.

To put this anticipation to the test of experiment, I placed the grating with its plane perpendicular to the incident light, and the grooved face towards the polarizer, and observed the light which was diffracted at reflection. Since in this case there would be no crowding of the planes of polarization in the regularly reflected light, any crowding which might be observed would be due either to diffraction directly, or to the irregular reflection due to diffraction, or, far more probably, to a combination of the two.

The experiments indicated indeed a marked crowding towards the plane of diffraction, but the light was so strong at the minimum, for most positions of the pointer of the polarizer, that the observations were very uncertain, and it was evidently only a rough approximation to regard the diffracted light as plane-polarized. The reason of this was evident on consideration. Of the light incident on the grating, a portion is regularly reflected, forming the central image of the system of spectra produced by diffraction at reflection, a portion is diffracted externally at such an angle as to enter the eye, a small portion is scattered, and the greater part enters the glass. Of the light which enters the glass, a portion is diffracted internally at such an angle that after regular reflection and refraction it enters the eye, a portion diffracted at other angles, but the greater part falls perpendicularly on the second surface. A portion of this is reflected to the first surface, and of the light so reflected a portion is diffracted at emergence at such an angle as to enter the eye. Thus there are three principal images, each formed by the light which has been once diffracted and once reflected, the externally diffracted light being considered as both diffracted and reflected, namely, one which has been diffracted internally, and then regularly reflected and refracted, a second in which the light has been regularly refracted and reflected, and then diffracted at emergence, and a third in which the light has been diffracted externally. Any other light which enters the eye must have been at least twice diffracted, or once diffracted and at least three times reflected, and therefore will be comparatively weak, except perhaps when the angle of incidence, or else the angle of diffraction, is very large. Now when the grating is perpendicular to the incident light the second and third of the principal images are necessarily superposed; and as they might be expected to be very differently polarized, it was likely enough that the light arising from the mixture of the two should prove to be very imperfectly polarized.

To separate these images, I placed a narrow vertical slit in front of the grating, between it and the polarizer, and inclined the grating by turning it round a vertical axis so that the normal fell between the polarizer and the analyzer. As soon as the grating was inclined, the image which had been previously observed separated into two, and at a certain inclination the three principal images were seen equidistant. The middle image, which was the second of those above described, was evidently the brightest of the three. The three images were found to be nearly if not perfectly plane-polarized, but polarized in different planes. The third image, and perhaps also the first, did not wholly vanish at the minimum. This might have been due to some subordinate image which then appeared, but it was more probably due to a real defect of polarization.

The planes of polarization of the side images, especially the first, were greatly crowded towards the plane of diffraction, or, which is the same, the plane of incidence. Those of the middle image were decidedly crowded in the same direction, though much less so than those of the side images. The light of the first and second images underwent one regular refraction and one regular reflection besides the diffraction and the accompanying irregular refraction. The crowding of the planes of polarization in one direction or the other produced by the regular refraction and the regular reflection can readily be calculated from the known formulæ*, and thus the crowding due to diffraction and the accompanying irregular refraction can be deduced from the observed result.

The crowding of the planes of polarization of the third image is due solely to diffraction and the accompanying irregular reflection. The crowding in one direction or the contrary, according as one or other hypothesis respecting the direction of vibrations is adopted, is readily calculated from the dynamical theory, and thus is obtained the crowding which is left to be attributed to the irregular reflection. In the absence of an exact theory little or no use can be made of the result in the way of confirming either hypothesis; but it is sufficient to destroy the vague analogy which might have been formed between the effects of diffraction and of irregular scattering.

The crowding of the planes of polarization of the middle image, after the observations had been reduced in the manner which will be explained in the next section, appeared somewhat

---

* It is here supposed that the regularly reflected or refracted light which forms the central colourless image belonging to a system of spectra is affected as to its polarization in the same way as if the surface were free from grooves.

greater than was to have been expected from the first two experiments. This led me to suspect that the crowding in the manner of reflection produced by diffraction accompanying the passage of light from air, across the grooved surface, into the glass plate, might be greater than the crowding had proved to be which was produced by diffraction accompanying the passage from glass, across the grooved surface, into air. I accordingly placed the grating with its plane perpendicular to the incident light, and the grooved face *towards* the polarizer, and placed the analyzer so as to receive the light which was diffracted in passing across the first surface, and then regularly refracted at the second. I soon found that the planes of polarization were very decidedly crowded towards the plane of diffraction, and that, notwithstanding the crowding in the contrary direction which must have been produced by the regular refraction at the second surface of the plate, and the crowding, likewise in the contrary direction, which might naturally be expected to result from the irregular refraction at the first surface, considered apart from diffraction. This result seemed to remove all doubt respecting the hypothesis as to the direction of vibration to which the experiments pointed as the true one.

On account of the decisive character of the result just mentioned, I took several sets of observations on light diffracted in this manner at different angles. I also made two more careful experiments of the same nature as the first two. The result now obtained was, that there was a very sensible crowding towards the plane of diffraction when the grooved face was turned from the polarizer, although there was evidently a marked difference between the two cases, the crowding being much less than when the grooved face was turned towards the polarizer. Even the first two experiments, now that I was aware of the index error of the polarizer, appeared to indicate a small crowding in the same direction.

Before giving the numerical results of the experiments, it may be as well to mention what was observed respecting the defect of polarization. I would here remark that an investigation of the precise nature of the diffracted light was beside the main object of my experiments, and only a few observations were taken which belong to such an investigation. In what follows, ϖ denotes the inclination of the plane of polarization of the light incident on the grating to a vertical plane passing through the ray, that is, to a plane perpendicular to the plane of diffraction. It is given by the reading of the pointer of the polarizer corrected for the index error 25°, and is measured positive in the direction of revolution of the hands of a watch placed with its back towards the incident light.

Whether the diffraction accompanied reflection or refraction, external or internal, the diffracted light was perfectly plane-polarized when ϖ had any one of the values 0°, 90°, 180°, or 270°. The defect of polarization was greatest about 45° from any of the above positions. When the diffracted light observed was red or reddish, on analyzation a blue light was seen at or near the minimum; when the diffracted light was blue or blueish, a red light was seen at or near the minimum. When the angle of diffraction was moderately small, such as 15° or 20°, the defect of polarization was small or insensible; when the angle of diffraction was large, such as 50° or 60°, the defect of polarization was considerable. For equal angles of diffraction, the defect of polarization was much greater when the grooved face was turned towards the polarizer than when it was turned in the contrary direction. By the term *angle of diffraction*, as applied to the case in which the grooved face was turned towards the polarizer, is to be understood the angle measured in air, from which the angle of diffraction within the glass may be calculated, from a knowledge of the refractive index.

The grating being placed perpendicularly to the incident light, with the grooved face towards the polarizer, the light diffracted at a considerable angle, (59° 52′ by measurement,) to the left of the regularly transmitted light was nearly white. When the pointer of the polarizer stood at 70°, so that $\varpi = +45°$, on turning the Nicol's prism of the analyzer in the positive direction through the position of minimum illumination, the light became in succession greenish yellow, blue, plum colour, nearly red. When ϖ was equal to $-45°$, the same appearance was presented on reversing the direction of rotation. Since the colours appeared in the order blue, red, when $\varpi = +45°$, and in the order red, blue, when $\varpi = -45°$, the analyzer being in both cases supposed to turn in the direction of the hands of a watch, the deficiency of colour took place in the order red, blue, when $\varpi = +45°$, and in the order blue, red, when $\varpi = -45°$. Hence the planes of polarization, or approximate polarization, of the blue were more crowded towards the plane of diffraction than those of the red.

On placing a narrow slit so as to allow a small portion only of the diffracted light to pass, and decomposing the light by a prism, in the manner already described, so as to get a spectrum consisting of bright bands with dark intervals, and then analyzing this spectrum with a Nicol's prism, it was found that at a moderate angle of diffraction all the colours were sensibly plane-polarized, though the planes of polarization did not quite coincide. At a large angle of diffraction the bright part of the spectrum did not quite disappear on turning round the Nicol's prism, while the red and blue ends, probably on account of their less intensity, appeared to be still perfectly plane-polarized, though not quite in the same plane. On treating in the same manner the diffracted light produced when the grooved face of the glass plate was turned from the polarizer, all the colours appeared to be sensibly plane-polarized. In the former case the light of the brightest part of the spectrum was made to disappear, or nearly so, by using a thin plate of mica in combination with the Nicol's prism, which shews that the defect of plane polarization was due to a slight elliptic polarization.

The numerical results of the experiments on the rotation of the plane of polarization are contained in the following table. In this table ϖ is the reading of the polarizer corrected for the index error 25°. A reading such as 340° is entered indifferently in the column headed "ϖ" as $+315^{\circ}$ or -45°, that is, $340^{\circ} - 25^{\circ}$ or $-(360^{\circ} - 340^{\circ}) - 25^{\circ}$. α is the reading of the analyzer, determined by one or more observations. The analyzer was graduated only from -90° to $+90^{\circ}$, and any reading such as -20° is entered indifferently as -20°, $+160^{\circ}$, or $+340^{\circ}$, being entered in such a manner as to avoid breaking the sequence of the numbers. On account of the light left at the minimum, the determination of α was very uncertain when the angle of diffraction was large, except when ϖ had very nearly one of the values 0°, 90°, 180°, or 270°. In the most favourable circumstances the mean error in the determination of α was about a quarter of a degree. In some of the experiments a red glass was used to assist in rendering the observations more definite. This had the advantage of stopping all rays except the red, but the disadvantage of considerably diminishing the intensity of the light. The minutes in the given value of θ, the angle of diffraction, cannot be trusted; in fact, during any experiment θ was liable to changes to at least that extent in consequence of the changes in the direction of the light. The same remark applies to i, the angle of incidence, in experiments 11 and 12. In these experiments the three principal images already described were observed separately. The angle of diffraction is measured from the direction of the regularly reflected ray, so that i is the angle of incidence, and $i + \theta$ the angle of reflection, or, in the case of the images which suffered one internal reflection, the angle of emergence.

The eleven experiments which are not found in the following tables consist of five on diffraction by reflection, which did not appear worth giving on account of the superposition of different images ; one on diffraction by refraction, to which the same remark applies, the grating having been placed at a considerable distance from the polarizer, so that the spot illuminated was too large to allow of the separate observation of different images; one on diffraction by reflection, in which the grating was placed perpendicularly to the incident light, with the grooved face turned from the polarizer, but the errors of observation, though much smaller than the whole quantity to be observed, were so large on account of the large angle of diffraction, (about 75°,) with which the observations were attempted, that the details are not worth giving ; one on diffraction by refraction, in which the different observations were so inconsistent that the experiment seemed not worth reducing ; one which was only just begun; and two qualitative experiments, the results of which have been already given. I mention this that I may not appear to have been biassed by any particular theory in selecting the experiments of which the numerical results are given.

The following remarks relate to the particular experiments :

No. 1. In this experiment each value of α was determined by six observations, of which the mean error* ranged from about 15′

* The difference between each individual observation and the mean of the six is regarded as the error of that observation, and the mean of these differences taken positively is what is here called the *mean error*. When two observations only are taken, the mean error is the same thing as the semi-difference between the observations. Since, for a given position of the pointer of the polarizer, the readings of the analyzer were usually taken one immediately after another, the mean error furnishes no criterion by which to judge of the errors produced by the small changes in the direction of the light incident on the grating, but only of those which arise from the vagueness of the object observed. The reader will be much

to 55'. So far the experiment was very satisfactory, but it was vitiated by changes in the direction of the light, sufficient care not having been taken in the adjustment of the mirror.

No. 2. α determined by single observations.

No. 13. α determined by two observations at least, of which the mean error ranged from about 10' to nearly 1°, but was usually decidedly less than 1°. At and about the octants, that is to say, when ϖ was nearly equal to 45°, or an odd multiple of 45°, the light was but very imperfectly polarized in one plane.

No. 14. α determined by two observations. Marked in note book as "a very satisfactory experiment." The mean of the mean errors was only 11'.

No. 15. α determined by three observations at least. The light was very imperfectly polarized, except near the standard points, that is to say when ϖ was equal to 0° or 90°, or a multiple of 90°. This rendered the observations very uncertain. About the octants the mean error in a set of observations taken one immediately after another amounted to near 2°.

No. 17. α determined by two observations. The light was very imperfectly polarized, except near the standard points. Yet the observations agreed very fairly with one another. The mean of the mean errors was 25', and the greatest of them not quite 1°.

No. 18. α determined by two observations, which, generally speaking, agreed well with one another. For $\varpi = -90°$ and $\varpi = +225°$ the light observed was rather scattered than regularly diffracted, the sheet of the cone of illumination having fallen above or below the hole of the analyzer.

No. 21. α determined by two observations at least. In this experiment the polarizer was covered with red glass.

No. 22. α determined by two observations. Marked in note book as "a very satisfactory experiment, though the light was not perfectly polarized."

No. 23. α determined by two observations at least. The hole in a screen placed between the polarizer and the grating was covered with red glass. This appears to have been a good experiment.

better able to judge of the amount of probable error from all causes after examining the reduction of the experiments given in the next section.

No. 11. α determined by two observations, which agreed well with one another. In the table, α (1), α (2), α (3) refer respectively to the first, second, and third of the three principal images already mentioned. In this experiment the polarizer was reversed, that face being turned towards the mirror which in the other experiments was turned towards the grating, which is the reason why α and ϖ increase together, although the light observed suffered one reflection. The same index error as before, namely 25°, is supposed to belong to the polarizer in its reversed position.

No. 12. α determined by three observations. The largeness of the angle of diffraction rendered the determination of α very uncertain.

TABLE I.

Experiment, No. 1.
Grooved face *from* Polarizer.
$\theta = 57°5'$.

ϖ	α
− 115°	− 76°41'
− 92½°	− 52°56'
− 70°	
− 47½°	− 6°52'
− 25°	+ 14°51'
− 2½°	+ 37°51'
+ 20°	+ 61° 5'
+ 42½°	+ 82°54'
+ 65°	+106°46'

Experiment, No. 2.
Grooved face *from* Polarizer.
$\theta = 50°23'$.

ϖ	α
−105°	− 80°
− 95°	− 70°25'
− 85°	− 61°15'
− 75°	− 51°30'
− 65°	− 41°10'
− 55°	− 29°15'
− 45°	− 20° 5'
− 35°	− 9°55'
− 25°	+ 0°20'
− 15°	+ 10°15'

No. 2, continued.

ϖ	α
− 5°	+ 20°20'
+ 5°	+ 30°55'
+ 15°	+ 40°55'
+ 25°	+ 50°45'
+ 35°	+ 61°45'
+ 45°	+ 70°55'
+ 55°	+ 82°15'

Experiment, No. 13.
Grooved face *towards* Polarizer.
$\theta = 39°50'$.

ϖ	α
− 60°	− 6° 5'
− 50°	+ 4°53'
− 40°	+ 15°52'
− 30°	+ 25°
− 20°	+ 33°25'
− 10°	+ 46° 5'
0°	+ 56°35'
+ 10°	+ 67°50'
+ 20°	+ 76°58'
+ 30°	+ 87°55'
+ 40°	+ 99°27'
+ 50°	+108°30'
+ 60°	+120°35'
+ 70°	+129° 2'
+ 80°	+137°42'
+ 90°	+146°57'

Experiment, No. 14.
Grooved face *from* Polarizer.
$\theta = 29°57'$.

ϖ	α
− 50°	+ 22°25'
− 40°	+ 31°15'
− 30°	+ 41°40'
− 20°	+ 51°55'
− 10°	+ 62°37'
0°	+ 71°10'
+ 10°	+ 81°47'
+ 20°	+ 93°47'
+ 30°	+103°10'
+ 40°	+113°15'
+ 50°	+122°42'
+ 60°	+132°42'
+ 70°	+143°
+ 80°	+152°47'
+ 90°	+161°57'
+100°	+171°52'
+110°	+182°52'
+120°	+191°47'
+130°	+202°12'
+140°	+211°42'

Experiment, No. 15.
Grooved face *towards* Polarizer.
$\theta = 59°52'$.

ϖ	α
0°	− 68°10'
− 10°	− 81°
− 20°	− 92°23'

No. 15, continued.

ϖ	α
− 30°	−115°55'
− 40°	−124°25'
− 50°	−133°41'
− 60°	−140°29'
− 70°	−148°18'
− 80°	−152°50'
− 90°	−158°30'

Experiment, No. 17.
Grooved face *towards* Polarizer.
$\theta = 50°45'$.

ϖ	α
− 90°	+ 77°15'
− 80°	+ 85°30'
− 70°	+ 93°12'
− 60°	+101°15'
− 50°	+109°47'
− 40°	+117°12'
− 30°	+129°57'

Experiment, No. 18.
Grooved face *towards* Polarizer.
$\theta = 21°39'$.

ϖ	α
− 90°	−103°23'
− 45°	− 59°53'
0°	− 12°58'
+ 45°	+ 33°37'
+ 90°	+ 77°27'
+135°	+120° 2'
+180°	+167°57'
+225°	+214°10'

TABLE I. (continued).

ϖ	α	ϖ	α	ϖ	α (1)	α (2)	α (3)
Experiment, No. 21. Grooved face *towards* Polarizer. Red glass used. $\theta=28°26'$.		No. 22, continued.		Experiment, No. 11. $i=14°50'$; $\theta=22°30'$.			
		$-135°$	$-140°25'$				
		$-120°$	$-124°45'$				
		$-105°$	$-110°40'$	$-105°$	$-113°35'$	$-117°50'$	
$-90°$	$-29°$	$-90°$	$-96°55'$	$-85°$	$-103°\ 5'$	$-101°$	$-102°20'$
$-75°$	$-16°\ 2'$	$-75°$	$-83°32'$	$-65°$	$-90°$	$-83°\ 5'$	$-89°$
$-60°$	$-\ 2°12'$	$-60°$	$-69°\ 7'$	$-45°$	$-78°40'$	$-63°55'$	$-74°50'$
$-45°$	$+12°35'$	$-45°$	$-54°50'$	$-25°$	$-58°50'$	$-44°$	$-53°19'$
$-30°$	$+27°52'$	$-30°$	$-38°55'$	$-\ 5°$	$-25°\ 5'$	$-21°10'$	$-23°10'$
$-15°$	$+44°47'$	$-15°$	$-22°50'$	$+15°$	$+13°15'$	$+\ 1°25'$	$+\ 7°55'$
$0°$	$+61°40'$			$+35°$	$+38°35'$	$+24°\ 5'$	$+32°$
$+15°$	$+78°25'$	Experiment, No. 23. Grooved face *towards* Polarizer. Red glass used. $\theta=54°53'$.		$+55°$	$+53°50'$	$+43°10'$	$+51°30'$
$+30°$	$+92°18'$						
$+45°$	$+107°25'$			Experiment, No. 12. $i=9°1'$; $\theta=53°39'$.			
$+60°$	$+122°30'$	$0°$	$-\ 6°30'$				
$+75°$	$+137°$	$+15°$	$+11°\ 5'$	$-25°$	$+\ 5°35'$	$-32°$	$-13°45'$
$+90°$	$+151°32'$	$+30°$	$+27°55'$	$-45°$	$+15°$	$-\ 9°40'$	$+\ 2°$
		$+45°$	$+42°30'$	$-90°$	$+26°15'$	$+26°15'$	$+26°15'$
Experiment, No. 22. Grooved face *from* Polarizer. $\theta=55°38'$.		$+60°$	$+58°22'$	$-135°$	$+34°30'$	$+65°$	$+51°15'$
		$+75°$	$+71°\ 5'$				
		$+90°$	$+83°22'$				
$-180°$	$-187°\ 2'$	$+105°$	$+96°12'$				
$-165°$	$-170°37'$	$+120°$	$+108°30'$				
$-150°$	$-154°30'$	$+135°$	$+122°45'$				

SECTION II.

Discussion of the numerical results of the experiments, with reference to theory.

According to the known formulæ which express the laws of the rotation of the plane of polarization of plane-polarized light which has undergone reflection or refraction at the surface of a transparent uncrystallized medium, if ϖ, α' be the azimuths of the planes of polarization of the incident and reflected or refracted light, both measured from planes perpendicular to the plane of incidence, they are connected by the equation

$$\tan \alpha' = m \tan \varpi \quad\quad\quad\quad\quad\quad\quad(48),$$

where m is constant, if the position of the surface and the directions of the rays be given, but is a different constant in the two cases of reflection and refraction. According to the theory developed in this paper, the same law obtains in the case of diffraction in air, or even within an uncrystallized medium, but m has a value distinct from the two former. It seems then extremely likely that the same law should hold good in the case of that combination of diffraction with reflection or refraction which exists when the diffraction takes place at the common surface of two transparent uncrystallized media, such as air and glass. If this be true, it is evident that by combining all the observations belonging to one experiment in such a manner as to get the value of m which best suits that experiment, we shall obtain the crowding of the planes of polarization better than we could from the direct observations, and we shall moreover be able in this way easily to compare the results of different experiments. It seems reasonable then to try in the first instance whether the formula (48) will represent the observations with sufficient accuracy.

In applying this formula to any experiment, there are two unknown quantities to be determined, namely, m, and the index error of the analyzer. Let ϵ be this index error, so that $\alpha = \alpha' + \epsilon$. The regular way to determine ϵ and m would no doubt be to assume an approximate value ϵ_1 of ϵ, put $\epsilon = \epsilon_1 + \Delta\epsilon_1$, where $\Delta\epsilon_1$ is the small error of ϵ_1, form a series of equations of which the type is

$$\tan(\alpha - \epsilon_1) - \sec^2(\alpha - \epsilon_1)\,\Delta\epsilon_1 = m \tan \varpi,$$

and then combine the equations so as to get the most probable values of $\Delta\epsilon_1$ and m. But such a refinement would be wholly unnecessary in the case of the present experiments, which are confessedly but rough. Moreover ϵ can be determined with accuracy, except so far as relates to errors produced by changes in the direction of the light, by means of the observations taken at the standard points, the light being in such cases perfectly polarized. By accuracy is here meant such accuracy as experiments of this sort admit of, where a set of observations giving a mean error of a quarter of a degree would be considered accurate. Besides, whenever the values of ϖ selected for observation are symmetrically taken with respect to one of the standard points, a small error in ϵ would introduce no sensible error into the value of m which would result from the experiment, although it might make the formula appear in fault when the only fault lay in the index error.

Accordingly I have determined the index error of the analyzer in a way which will be most easily explained by an example.

Suppose the values of α to have been determined by experiment corresponding to the following values of ϖ, $- 15^\circ$, 0°, $+ 15^\circ$,... $+ 75^\circ$, $+ 90^\circ$, $+ 105^\circ$. The value of α for $\varpi = 0^\circ$, and the mean of the values for $- 15^\circ$ and $\varpi = + 15^\circ$, furnish two values of ϵ; and the value of α for $\varpi = + 90^\circ$, and the mean of the values for $\varpi = + 75^\circ$ and $\varpi = + 105^\circ$, furnish two values of $\epsilon + 90^\circ$. The mean of the four values of ϵ thus determined is likely to be more nearly correct than any of them. In some few experiments no two values of ϖ were symmetrically taken with respect to the standard points. In such cases I have considered it sufficient to take proportional parts for a small interval. Thus if α_1, α_2 be the readings of the analyzer for $\varpi = - 10^\circ$, $\varpi = + 5^\circ$, assuming

$$\alpha_1 = \epsilon - 10^\circ - 2x, \quad \alpha_2 = \epsilon + 5^\circ + x, \text{ we get } 3x = \alpha_2 - \alpha_1 - 15^\circ,$$

whence ϵ, which is equal to $\alpha_2 - 5^\circ - x$, is known. The index error of the analyzer having been thus determined, it remains to get the most probable value of m from a series of equations of the form (48). For facility of numerical calculation it is better to put this equation under the form

$$\log m = \log \tan \alpha' - \log \tan \varpi \ldots\ldots\ldots\ldots\ldots(49),$$

where it is to be understood that the signs of α and ϖ are to be changed if these angles should lie between 0 and $- 90^\circ$, or their supplements taken if they should lie between $+ 90^\circ$ and $+ 180^\circ$. Now the mean of the values of $\log m$ determined by the several observations belonging to one experiment is not at all the most probable value. For the error in $\log \tan \alpha'$ produced by a small given error in α' increases indefinitely as α' approaches indefinitely to 0° or 90°, so that in this way of combining the observations an infinite weight would be attributed to those which were taken infinitely close to the standard points, although such observations are of no use for the direct determination of $\log m$, their use being to determine ϵ. Let $\alpha' + \Delta\alpha'$ be the true angle of which α' is the approximate value, α' being deduced from the observed angle α corrected for the assumed index error ϵ. Then, neglecting $(\Delta\alpha')^2$, we get for the true equation which ought to replace (49),

$$\log m = \log \tan \alpha' + \frac{2M\Delta\alpha'}{\sin 2\alpha'} - \log \tan \varpi,$$

M being the modulus of the common system of logarithms. Since the effect of the error $\Delta\alpha'$ is increased by the division by $\sin 2\alpha'$, a quantity which may become very small, in combining the equations such as (49) I have first multiplied the several equations by $\sin 2\alpha'$, or the sine of $2(\alpha - \epsilon)$ taken positively, and then added together the equations so formed, and determined $\log m$ from the resulting equation. Perhaps it would have been better to have used for multiplier $\sin^2 2\alpha'$, which is what would have been given by the rule of least squares, if the several observations be supposed equally liable to error; but on the other hand the use of $\sin 2\alpha'$ for multiplier instead of $\sin^2 2\alpha'$ has the effect of diminishing the comparative weight of the observations taken about the octants, where, in consequence of the defect of polarization, the observations were more uncertain.

The following table contains the result of the reduction of the experiments in the way just explained. The value of ϵ used in the reduction, and the resulting value of $\log m$, are written down in each case. The second column belonging to each experiment gives the value of $\alpha' - \varpi$ calculated from (49) with the assumed value of $\log m$, and is put down for the sake of comparison with the value of $\alpha' - \varpi$ deduced from the difference, $\alpha - \varpi$, of the observed angles α, ϖ, corrected for the assumed index error ϵ. In the table, the experiments are arranged in classes, according to their nature, and those belonging to the same class are arranged according to the values of θ. The first three experiments in the table relate to diffraction at refraction, in which the grooved face of the grating was turned from the polarizer, the next six to diffraction at refraction, in which the grooved face was turned towards the polarizer, and the last two to the experiments in which the grating was a little inclined, and the three principal images were observed. The result of Experiment No. 1, is here given separately, on account of the different values of ϖ there employed.

Experiment No. 1. $\theta = 57°5'$;   assumed index error $\epsilon = 40°5'$.

ϖ	$\alpha' - \varpi$
$- 115°$	$- 1°46'$
$- 92\frac{1}{2}°$	$- 0°31'$
$- 70°$	
$- 47\frac{1}{2}°$	$+ 0°33'$
$- 25°$	$- 0°14'$
$- 2\frac{1}{2}°$	$+ 0°16'$
$+ 20°$	$+ 1°$
$+ 42\frac{1}{2}°$	$+ 0°19'$
$+ 65°$	$+ 1°41'$

The values of α' for $\varpi = -115^0$ and $\varpi = +65^0$ ought to differ by 180^0, whereas they differ by $3^027'$ more. This angle is so large compared with the angles $\alpha' - \varpi$ given just above, that it seems best to reject the experiment. The experiment is sufficient however to shew that the crowding of the planes of polarization, be it in what direction it may, is very small. On combining all the observations belonging to this experiment in the manner already described, a small positive value of $\log m$, namely $+ \cdot002$, appeared to result. This value, if exact, would indicate an extremely small crowding in the manner of reflection.

TABLE II.

Experiment, No. 14. $\theta = 29^057'$ $\epsilon = +72^023'$ $\log m = +\cdot009$				Experiment, No. 2. $\theta = 50^023'$ $\epsilon = +24^012'$ $\log m = +\cdot010$			
		$\alpha' - \varpi$				$\alpha' - \varpi$	
ϖ	calc.	obs.	diff.	ϖ	calc.	obs.	diff.
-50^0	$-0^0.6$	$0^0.0$	$+0^0.6$	-105^0	$+0^0.3$	$-0^0.3$	$-0^0.6$
-40^0	$-0^0.6$	$-1^0.1$	$-0^0.5$	-95^0	$+0^0.1$	$-0^0.7$	$-0^0.8$
-30^0	$-0^0.5$	$-0^0.7$	$-0^0.2$	-85^0	$-0^0.1$	$-1^0.5$	$-1^0.4$
-20^0	$-0^0.4$	$-0^0.5$	$-0^0.1$	-75^0	$-0^0.3$	$-1^0.8$	$-1^0.5$
-10^0	$-0^0.2$	$+0^0.2$	$+0^0.4$	-65^0	$-0^0.5$	$-1^0.5$	$-1^0.0$
0^0	$0^0.0$	$-1^0.2$	$-1^0.2$	-55^0	$-0^0.6$	$+0^0.4$	$+1^0.0$
$+10^0$	$+0^0.2$	$-0^0.6$	$-0^0.8$	-45^0	$-0^0.7$	$-0^0.4$	$+0^0.3$
$+20^0$	$+0^0.4$	$+1^0.4$	$+1^0.0$	-35^0	$-0^0.6$	$-0^0.2$	$+0^0.4$
$+30^0$	$+0^0.5$	$+0^0.8$	$+0^0.3$	-25^0	$-0^0.5$	$0^0.0$	$+0^0.5$
$+40^0$	$+0^0.6$	$+0^0.9$	$+0^0.3$	-15^0	$-0^0.3$	$0^0.0$	$+0^0.3$
$+50^0$	$+0^0.6$	$+0^0.3$	$-0^0.3$	-5^0	$-0^0.1$	$0^0.0$	$+0^0.1$
$+60^0$	$+0^0.5$	$+0^0.3$	$-0^0.2$	$+5^0$	$+0^0.1$	$+0^0.6$	$+0^0.5$
$+70^0$	$+0^0.4$	$+0^0.6$	$+0^0.2$	$+15^0$	$+0^0.3$	$+0^0.6$	$+0^0.3$
$+80^0$	$+0^0.2$	$+0^0.4$	$+0^0.2$	$+25^0$	$+0^0.5$	$+0^0.4$	$-0^0.1$
$+90^0$	$0^0.0$	$-0^0.4$	$-0^0.4$	$+35^0$	$+0^0.6$	?	?
$+100^0$	$-0^0.2$	$-0^0.5$	$-0^0.3$	$+45^0$	$+0^0.7$	$+0^0.4$	$-0^0.3$
$+110^0$	$-0^0.4$	$+0^0.5$	$+0^0.9$	$+55^0$	$+0^0.6$	$+1^0.9$	$+1^0.3$
$+120^0$	$-0^0.5$	$-0^0.6$	$-0^0.1$				
$+130^0$	$-0^0.6$	$-0^0.2$	$+0^0.4$				
$+140^0$	$-0^0.6$	$-0^0.7$	$-0^0.1$				

TABLE II. (*continued*).

Experiment, No. 22.
θ = 55°38'
ε = − 7°27'
log m = + ·035

ϖ	a' − ϖ		
	calc.	obs.	diff.
− 180°	0°·0	0°·0	0°·0
− 165°	+ 1°·2	+ 1°·4	+ 0°·2
− 150°	+ 2°·0	+ 2°·5	+ 0°·5
− 135°	+ 2°·3	+ 1°·6	− 0°·7
− 120°	+ 2°·0	+ 2°·2	+ 0°·2
− 105°	+ 1°·1	+ 1°·3	+ 0°·2
− 90°	0°·0	+ 0°·1	+ 0°·1
− 75°	− 1°·1	− 1°·5	− 0°·4
− 60°	− 2°·0	− 2°·1	− 0°·1
− 45°	− 2°·3	− 2°·8	~ 0°·5
− 30°	− 2°·0	− 1°·9	+ 0°·1
− 15°	− 1°·2	− 0°·8	+ 0°·4

Experiment, No. 18.
θ = 21°39'
ε = − 12°44'
log m = + ·029

ϖ	a' − ϖ		
	calc.	obs.	diff.
− 90°	0°·0	− 0°·6	− 0°·6
− 45°	− 1°·9	− 2°·1	− 0°·2
0°	0°·0	− 0°·2	− 0°·2
+ 45°	+ 1°·9	+ 1°·3	− 0°·6
+ 90°	0°·0	+ 0°·2	+ 0°·2
+ 135°	− 1°·9	− 2°·2	− 0°·3
+ 180°	0°·0	+ 0°·7	+ 0°·7
+ 225°	+ 1°·9	+ 1°·9	0°·0

Experiment, No. 21.
θ = 28°26'
ε = 60°49'
log m = + ·039

ϖ	a' − ϖ		
	calc.	obs.	diff.
− 90°	0°·0	+ 0°·2	+ 0°·2
− 75°	− 1°·2	− 1°·6	− 0°·4
− 60°	− 2°·2	− 3°·0	− 0°·8
− 45°	− 2°·6	− 3°·2	− 0°·6
− 30°	− 2°·3	− 2°·9	− 0°·6
− 15°	− 1°·3	− 1°·0	+ 0°·3
0°	0°·0	+ 0°·8	+ 0°·8
+ 15°	+ 1°·3	+ 2°·6	+ 1°·3
+ 30°	+ 2°·3	+ 1°·5	− 0°·8
+ 45°	+ 2°·6	+ 1°·6	− 1°·0
+ 60°	+ 2°·2	+ 1°·7	− 0°·5
+ 75°	+ 1°·2	+ 1°·2	0°·0
+ 90°	0°·0	+ 0°·7	+ 0°·7

Experiment, No. 13.
θ = 39°50'
ε = 56°50'
log m = + ·034

ϖ	a' − ϖ		
	calc.	obs.	diff.
− 60°	− 1°·9	− 0°·9	+ 1°·0
− 50°	− 2°·2	− 1°·9	+ 0°·3
− 40°	− 2°·2	− 1°·0	+ 1°·2
− 30°	− 2°·0	− 1°·8	+ 0°·2
− 20°	− 1°·5	− 3°·4	− 1°·9
− 10°	− 0°·8	− 0°·7	+ 0°·1
0°	0°·0	− 0°·2	− 0°·2
+ 10°	+ 0°·8	+ 1°·0	+ 0°·2
+ 20°	+ 1°·5	+ 0°·1	− 1°·4
+ 30°	+ 2°·0	+ 1°·1	− 0°·9
+ 40°	+ 2°·2	+ 2°·6	+ 0°·4
+ 50°	+ 2°·2	+ 1°·7	− 0°·5
+ 60°	+ 1°·9	+ 3°·7	+ 1°·8
+ 70°	+ 1°·4	+ 2°·2	+ 0°·8
+ 80°	+ 0°·7	+ 0°·9	+ 0°·2
+ 90°	0°·0	+ 0°·1	+ 0°·1

Experiment, No. 17.
θ = 50°45'
ε = + 167°15'
log m = + ·122

ϖ	a' − ϖ		
	calc.	obs.	diff.
− 90°	0°·0	0°·0	0°·0
− 80°	− 2°·4	− 1°·7	+ 0°·7
− 70°	− 4°·6	− 4°·0	+ 0°·6
− 60°	− 6°·4	− 6°·0	+ 0°·4
− 50°	− 7°·6	− 7°·5	+ 0°·1
− 40°	− 8°·0	− 10°·0	− 2°·0
− 30°	− 7°·4	− 7°·3	+ 0°·1

Experiment, No. 23.
θ = 54°53'
ε = − 7°27'
log m = + ·082

ϖ	a' − ϖ		
	calc.	obs.	diff.
0°	0°·0	+ 0°·2	+ 0°·2
+ 15°	+ 2°·9	+ 2°·7	− 0°·2
+ 30°	+ 4°·3	+ 4°·6	+ 0°·3
+ 45°	+ 5°·4	+ 4°·2	− 1°·2
+ 60°	+ 4°·4	+ 5°·0	+ 0°·6
+ 75°	+ 2°·5	+ 3°·5	+ 1°·0
+ 90°	0°·0	0°·0	0°·0
+ 105°	− 2°·5	− 3°·1	− 0°·6
+ 120°	− 4°·4	− 4°·8	− 0°·4
+ 135°	− 5°·4	− 5°·6	− 0°·2

TABLE II. (continued).

Experiment, No. 15.

$\theta = 59^0 52'$

$\epsilon = -68^0 15'$

$\log m = + \cdot 225$

ϖ	$a' - \varpi$ calc.	$a' - \varpi$ obs.	diff.
0^0	$0^0 \cdot 0$	$+ 0^0 \cdot 1$	$+ 0^0 \cdot 1$
$- 10^0$	$- 6^0 \cdot 5$	$- 2^0 \cdot 7$	$+ 3^0 \cdot 8$
$- 20^0$	$- 11^0 \cdot 4$	$- 4^0 \cdot 1$	$+ 7^0 \cdot 3$
$- 30^0$	$- 14^0 \cdot 1$	$- 17^0 \cdot 7$	$- 3^0 \cdot 6$
$- 40^0$	$- 14^0 \cdot 6$	$- 16^0 \cdot 2$	$- 1^0 \cdot 6$
$- 50^0$	$- 13^0 \cdot 4$	$- 15^0 \cdot 4$	$- 2^0 \cdot 0$
$- 60^0$	$- 11^0 \cdot 0$	$- 12^0 \cdot 2$	$- 1^0 \cdot 2$
$- 70^0$	$- 7^0 \cdot 8$	$- 10^0 \cdot 0$	$- 2^0 \cdot 2$
$- 80^0$	$- 4^0 \cdot 0$	$- 4^0 \cdot 6$	$- 0^0 \cdot 6$
$- 90^0$	$0^0 \cdot 0$	$- 0^0 \cdot 2$	$- 0^0 \cdot 2$

Experiment, No. 11.

$i = 14^0 50'$; $\theta = 22^0 30'$; $\epsilon = - 15^0 30'$.

ϖ	First Image. $\log m = + \cdot 289$. $a'-\varpi$ calc.	obs.	diff.	Second Image. $\log m = + \cdot 061$. $a'-\varpi$ calc.	obs.	diff.	Third Image. $\log m = + \cdot 209$. $a'-\varpi$ calc.	obs.	diff.
$- 105^0$	$+ 7^0 \cdot 1$	$+ 6^0 \cdot 9$	$- 0^0 \cdot 2$	$+ 1^0 \cdot 9$	$+ 2^0 \cdot 7$	$+ 0^0 \cdot 8$			
$- 85^0$	$- 2^0 \cdot 4$	$- 2^0 \cdot 6$	$- 0^0 \cdot 2$	$- 0^0 \cdot 7$	$- 0^0 \cdot 5$	$+ 0^0 \cdot 2$	$- 1^0 \cdot 9$	$- 1^0 \cdot 8$	$+ 0^0 \cdot 1$
$- 65^0$	$- 11^0 \cdot 5$	$- 9^0 \cdot 5$	$+ 2^0 \cdot 0$	$- 2^0 \cdot 9$	$- 2^0 \cdot 6$	$+ 0^0 \cdot 3$	$- 8^0 \cdot 9$	$- 8^0 \cdot 5$	$+ 0^0 \cdot 4$
$- 45^0$	$- 17^0 \cdot 8$	$- 18^0 \cdot 2$	$- 0^0 \cdot 4$	$- 4^0 \cdot 0$	$- 3^0 \cdot 4$	$+ 0^0 \cdot 6$	$- 13^0 \cdot 3$	$- 14^0 \cdot 3$	$- 1^0 \cdot 0$
$- 25^0$	$- 17^0 \cdot 2$	$- 18^0 \cdot 3$	$- 1^0 \cdot 1$	$- 3^0 \cdot 2$	$- 3^0 \cdot 5$	$- 0^0 \cdot 3$	$- 12^0 \cdot 0$	$- 12^0 \cdot 8$	$- 0^0 \cdot 8$
$- 5^0$	$- 4^0 \cdot 6$	$- 4^0 \cdot 6$	$0^0 \cdot 0$	$- 0^0 \cdot 7$	$- 0^0 \cdot 7$	$0^0 \cdot 0$	$- 3^0 \cdot 0$	$- 2^0 \cdot 7$	$+ 0^0 \cdot 3$
$+ 15^0$	$+ 12^0 \cdot 5$	$+ 13^0 \cdot 7$	$+ 1^0 \cdot 2$	$+ 2^0 \cdot 1$	$+ 1^0 \cdot 9$	$- 0^0 \cdot 2$	$+ 8^0 \cdot 4$	$+ 8^0 \cdot 4$	$0^0 \cdot 0$
$+ 35^0$	$+ 18^0 \cdot 7$	$+ 19^0 \cdot 1$	$+ 0^0 \cdot 4$	$+ 3^0 \cdot 9$	$+ 4^0 \cdot 6$	$+ 0^0 \cdot 7$	$+ 13^0 \cdot 6$	$+ 12^0 \cdot 5$	$- 1^0 \cdot 1$
$+ 55^0$	$+ 15^0 \cdot 2$	$+ 14^0 \cdot 3$	$- 0^0 \cdot 9$	$+ 3^0 \cdot 7$	$+ 3^0 \cdot 7$	$0^0 \cdot 0$	$+ 11^0 \cdot 6$	$+ 12^0 \cdot 0$	$+ 0^0 \cdot 4$

Experiment, No. 12.

$i = 9^0 1'$; $\theta = 53^0 39'$; $\epsilon = - 63^0 45'$.

ϖ	First Image. $\log m = + \cdot 756$. $a'-\varpi$ calc.	obs.	diff.	Second Image. $\log m = + \cdot 122$. $a'-\varpi$ calc.	obs.	diff.	Third Image. $\log m = + \cdot 366$. $a'-\varpi$ calc.	obs.	diff.
$+ 25^0$	$+ 44^0 \cdot 4$	$+ 44^0 \cdot 3$	$- 0^0 \cdot 1$	$+ 6^0 \cdot 7$	$+ 6^0 \cdot 7$	$0^0 \cdot 0$	$+ 22^0 \cdot 3$	$+ 25^0 \cdot 0$	$+ 2^0 \cdot 7$
$+ 45^0$	$+ 35^0 \cdot 1$	$+ 33^0 \cdot 7$	$- 1^0 \cdot 4$	$+ 7^0 \cdot 9$	$+ 9^0 \cdot 1$	$+ 1^0 \cdot 2$	$+ 21^0 \cdot 7$	$+ 21^0 \cdot 7$	$0^0 \cdot 0$
$+ 90^0$	$0^0 \cdot 0$	$0^0 \cdot 0$	$0^0 \cdot 0$	$0^0 \cdot 0$	$0^0 \cdot 0$	$0^0 \cdot 0$	$0^0 \cdot 0$	$0^0 \cdot 0$	$0^0 \cdot 0$
$+ 135^0$	$- 35^0 \cdot 1$	$- 36^0 \cdot 7$	$- 1^0 \cdot 6$	$- 7^0 \cdot 9$	$- 6^0 \cdot 2$	$+ 1^0 \cdot 7$	$- 21^0 \cdot 7$	$- 20^0 \cdot 0$	$+ 1^0 \cdot 7$

A nearly constant error appearing in the table of differences would indicate merely that the value of ϵ used in the reduction was slightly erroneous. A slight error in ϵ, it is to be remembered, produces no sensible error in $\log m$, whenever the observations are balanced with respect to one of the standard points.

In the first two experiments entered in the table, the crowding of the planes of polarization is so small that it is masked by errors of observation, and it is only by combining all the observations that a slight crowding towards the plane of diffraction can be made out. In all the other experiments, however, a glance at the numbers in the third column is sufficient to shew in what direction the crowding takes place. From an inspection of the numbers found in the columns headed "diff." it seems pretty evident that if the formula (49) be not exact the error cannot be made out without more accurate observations. In the case of experiment No. 15, the errors are unusually large, and moreover appear to follow something of a regular law. In this experiment the observations were extremely uncertain on account of the large angle of diffraction and the great defect of polarization of the light observed, but besides this there appears to have been some confusion in the entry of the values of ϖ. This confusion affecting one or two angles, or else some unrecorded change of adjustment, was probably the cause of the apparent break in the second column between the third and fourth numbers. Since the value of $\log m$ is deduced from all the observations combined, there seems no occasion to reject the experiment, since even a large error affecting one angle would not produce a large error in the value of $\log m$ resulting from the whole series. In the entry of experiment No. 12 the signs of ϖ have been changed, to allow for the reversion produced by reflection. This change of sign was unnecessary in No. 11, because in that experiment the polarizer was actually reversed. The results of experiment No. 12 would be best satisfied by using slightly different values of the index error of the analyzer for the three images, adding to the assumed index error about $-1\frac{1}{2}°$, $+1\frac{1}{2}°$, $+2°$, for the first, second, and third images respectively. The largest error in the third columns, $2·7°$, is for $\varpi = +25°$, third image. The three readings by which α was determined in this case were $-15°$, $-13°30'$, $-12°$? Hence the error $+2·7°$, even if no part of it were due to an index error, would hardly be too large to be attributed to errors of observation.

Since the formula (49), even if it be not strictly true, represents the experiments with sufficient accuracy, we may consider the value of $\log m$ which results from the combination of all the observations belonging to one experiment as itself the result of direct observation, and proceed to discuss its magnitude. Let us consider first the experiments on diffraction at refraction, in which the light was incident perpendicularly on the grating.

Although the theory of this paper does not meet the case in which diffraction takes place at the confines of air and glass, it leads to a definite result on each of the three following suppositions:

First, that the diffraction takes place in air, before the light reaches the glass:

Second, that the diffraction takes place in glass, after the light has entered the first surface perpendicularly:

Third, that the diffraction takes place in air, after the light has passed perpendicularly through the plate.

On the first supposition let α_1, α_2, α be the azimuths of the plane of polarization of the light after diffraction, after the first refraction, and after the second refraction respectively, and θ' the angle of refraction corresponding to the angle of incidence θ, so that $\sin \theta = \mu \sin \theta'$, μ being the refractive index of the plate: and first, let us suppose the vibrations of plane-polarized light to be perpendicular to the plane of polarization. Then by the theory of this paper we have $\tan \alpha_1 = \sec \theta \tan \varpi$, and by the known formula applying to refraction we have $\tan \alpha_2 = \cos (\theta - \theta') \tan \alpha_1$, $\tan \alpha = \cos (\theta - \theta') \tan \alpha_2$, whence $\tan \alpha = m \tan \varpi$, where

$$m = \sec \theta \cos^2 (\theta - \theta').$$

On the second supposition, if α_1 be the azimuth after diffraction at an angle θ' within the glass, we have $\tan \alpha_1 = \sec \theta' \tan \varpi$, $\tan \alpha = \cos (\theta - \theta') \tan \alpha_1$, whence $\tan \alpha = m \tan \varpi$, where

$$m = \sec \theta' \cos (\theta - \theta').$$

On the third supposition we have $\tan \alpha = m \tan \varpi$, where

$$m = \sec \theta.$$

If we suppose the vibrations parallel to the plane of polarization, we shall obtain the same formulæ except that $\cos \theta$, $\cos \theta'$ will come in place of $\sec \theta$, $\sec \theta'$, the factor $\cos (\theta - \theta')$ being unaltered.

Theory would lead us to expect to find the value of $\log m$ deduced from observations in which the grooved face was turned from the polarizer lying between the values obtained on the second and third of the suppositions respecting the place of diffraction, or at most not much differing from one of these limits. Similarly, we should expect from theory to find the value of $\log m$ deduced from observations in which the grooved face was turned towards the polarizer lying between the values obtained on the first and second suppositions, or at most not lying far beyond one of these values.

The following table contains the values of $\log m$ calculated from theory on each of the hypotheses respecting the direction of vibration, and on each of the three suppositions respecting the place of diffraction. The numerals refer to these suppositions. The table extends from $\theta = 0$ to $\theta = 90°$, at intervals of 5°. When $\theta = 0$, $m = 1$, and $\log m = 0$, in all cases. In calculating the table, I have supposed $\mu = 1\cdot52$, or rather equal to the number, $(1\cdot5206,)$ whose common logarithm is $\cdot182$. This table is followed by another containing the values of $\log m$ deduced from experiment.

TABLE III.   Values of $\log m$ from theory, μ being supposed equal to $1\cdot5206$.

θ	Vibrations supposed perpendicular to the plane of polarization.			Vibrations supposed parallel to the plane of polarization.		
	I	II	III	I	II	III
5°	+ ·001	+ ·001	+ ·002	− ·002	− ·001	− ·002
10°	+ ·005	+ ·002	+ ·007	− ·008	− ·004	− ·007
15°	+ ·011	+ ·004	+ ·015	− ·019	− ·008	− ·015
20°	+ ·020	+ ·008	+ ·027	− ·034	− ·015	− ·027
25°	+ ·032	+ ·012	+ ·043	− ·053	− ·023	− ·043
30°	+ ·047	+ ·017	+ ·062	− ·078	− ·033	− ·062
35°	+ ·065	+ ·022	+ ·087	− ·109	− ·044	− ·087
40°	+ ·086	+ ·028	+ ·116	− ·146	− ·058	− ·116
45°	+ ·111	+ ·033	+ ·150	− ·190	− ·073	− ·150
50°	+ ·139	+ ·037	+ ·192	− ·244	− ·090	− ·192
55°	+ ·173	+ ·040	+ ·241	− ·310	− ·109	− ·241
60°	+ ·214	+ ·040	+ ·301	− ·388	− ·129	− ·301
65°	+ ·262	+ ·039	+ ·374	− ·486	− ·151	− ·374
70°	+ ·324	+ ·034	+ ·466	− ·608	− ·175	− ·466
75°	+ ·408	+ ·022	+ ·587	− ·766	− ·202	− ·587
80°	+ ·533	+ ·005	+ ·760	− ·987	− ·231	− ·760
85°	+ ·773	− ·022	+1·060	−1·347	− ·265	−1·060
90°	+ ∞	− ·059	+ ∞	− ∞	− ·305	− ∞

TABLE IV.   Values of $\log m$ from observation.

Nature of Experiment.	No.	θ	$\log m$
Diffraction at refraction. Incidence perpendicular. Grooved face of glass plate turned *from* the incident light.	14	$29^0 57'$	$+ \cdot 009$
	2	$50^0 23'$	$+ \cdot 010$
	22	$54^0 38'$	$+ \cdot 035$
Diffraction at refraction. Incidence perpendicular. Grooved face of glass plate turned *towards* the incident light.	18	$21^0 39'$	$+ \cdot 029$
	21	$28^0 26'$	$+ \cdot 039$
	13	$39^0 50'$	$+ \cdot 034$
	17	$50^0 45'$	$+ \cdot 122$
	23	$54^0 53'$	$+ \cdot 082$
	15	$59^0 52'$	$+ \cdot 225$

A comparison of the two tables will leave no reasonable doubt
that the experiments are decisive in favour of Fresnel's hypothesis, if the theory be considered well founded. In considering
the conclusiveness of the experiments, it is to be remembered
that on either the first or second supposition respecting the place
of diffraction, (and the third certainly cannot apply to the case
in which the grooved face is turned towards the incident light,)
the planes of polarization of the diffracted light are crowded by
refraction towards the perpendicular to the plane of diffraction,
and therefore the observed crowding towards the plane of diffraction does not represent the whole effect of the cause, be it what
it may, of crowding in that direction.

If β be the value of $\alpha' - \varpi$ for $\varpi = 45^0$, $\beta = 1^0$ when $\log m =$
$\cdot 015$, nearly; and when $\log m$ is not large, β is nearly proportional to $\log m$. In this case β is nearly the maximum value
of $\alpha' - \varpi$. Hence the greatest value of $\alpha' - \varpi$, expressed in degrees,
may be obtained approximately from Table IV, and, within the
range of observation, from Table III, by regarding the decimals
as integers and dividing by 15. Thus, for $\log m = - \cdot 388$ the
real maximum is $24^0 \cdot 8$, and the approximate rule gives $25^0 \cdot 9$, so
that this rule is abundantly sufficient to allow us to judge of the
magnitude of the quantity by which the two theories differ. For
$\theta = 60^0$, the two columns in Table III headed "I", as well as
those headed "III", differ by $\cdot 602$, and those headed "II", differ
by $\cdot 169$, so that the values assigned to β by the two theories differ
by about 40^0 or 11^0, according as we suppose the diffraction to
take place in air or in glass. For $\theta = 40^0$, the corresponding
differences are 15^0 and 6^0, nearly. These differences, even those

which belong to diffraction within the glass plate, are large compared with the errors of observation; for the probable cause of the large errors in experiment No. 15, has been already mentioned.

In the following figure the abscissæ of the curves represent the angle of diffraction, and the ordinates the values of log m calculated from theory. The numerals refer to the three suppositions respecting the place of diffraction, and the letters E, A, (the first vowels in the words *perpendicular* and *parallel*,) to the two hypotheses respecting the direction of vibration. The dots represent the results of the experiments in which the grooved face of the glass plate was turned towards the polarizer, and the crosses those of the experiments in which it was turned in the contrary direction.

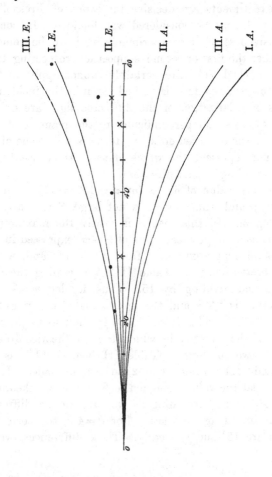

The smallness of log m in experiment No. 23, to which the 5th dot belongs, is probably due in part to the use of the red glass, since, as has been already remarked, the planes of polarization of the blue were more crowded towards the plane of diffraction than those of the red. On this account the dot ought to be slightly raised to make this experiment comparable with its neighbours. On the other hand it will be seen by referring to Table II, that No. 23 was a much better experiment than No. 15, which is represented by the 6th dot, and apparently also better than No. 17, which is represented by the 4th dot. No. 21, represented by the 2nd dot, seems to have been decidedly better than No. 13, which is represented by the 3rd. Nos. 14 and 22, represented by the 1st and 3rd crosses respectively, were probably much better, especially the latter of them, than No. 2, which is represented by the 2nd cross. Now, bearing in mind the character of the experiments, conceive two curves drawn with a free hand, both starting from the origin, where they touch the axis, and passing, the one among the dots, and the other among the crosses. The former of these would apparently lie a little below the curve marked I. E, and the latter a very little below the curve II. E.

Hence the observations are very nearly represented by adopting Fresnel's hypothesis respecting the direction of vibration, and, whether the grooved face be turned towards or from the incident light, supposing the wave broken up *before* it reaches the grooves.

I think a physical reason may be assigned why the supposition of the wave's being broken up before it reaches the grooves should be a better representation of the actual state of things than the supposition of its being broken up after it has passed between them. Till it reaches the grooves, the wave is regularly propagated, and, according to what has been already remarked in the introduction, we have a perfect right to conceive it broken up at any distance we please in front of the grooves. Let the figure represent a section of the grooves, &c., by the plane of diffraction. Let aA, bB be sections of two consecutive grooves, AB being the polished interval. Let eh be the plane at which a wave incident in the direction represented by the arrow is conceived to be broken up. Let O be any point in eh, and from O draw ORS in the direction of a ray proceeding regularly from O and entering the eye; so that OR, RS are inclined to the normal at angles θ, θ', or θ', θ, according as the light is passing from air into glass or from glass into air. The latter case is represented in the figure. Of a secondary wave diverging spherically from O, which is only partly represented in the figure, those rays which are situated between the limits OA, OB, and are not inclined at a small angle to either of these limiting directions, may be regarded as regularly refracted across AB. In a direction inclined at a small angle only to OA or OB, it would be necessary to take account of the diffraction at the edge A or B. Let γ be a small angle such that if OR be inclined to OA and OB at angles greater than γ the ray OR may be regarded as regularly refracted, and draw Ae, Bg inclined at angles γ to OR, and Af, Bh inclined at angles $-\gamma$. Then, in finding the illumination in the direction RS, all the secondary waves except those which come from points situated in portions such as ef, gh of the plane eh may be regarded as regularly refracted, or else completely stopped, those which come from points in fg and similar portions being regularly refracted, and those which come from points to the left of e, between e and the point which bears to a the same relation that h bears to b, as well as those which come from similar portions of the plane eh, being completely stopped. Now the whole of the aperture AB is not effective in producing illumination in the direction RS. For let C be the centre of AB, and through C draw a plane perpendicular to RS, and then draw a pair of parallel planes each at a distance $\frac{1}{2}\lambda$ from the former plane, cutting AB in M_1, N_1, another pair at a distance λ, and cutting AB in M_2, N_2, and so on as long as the points of section fall between A and B. Let M, N be the last points of section. Then the vibrations proceeding from MN in the direction RS neutralize each other by interference, so that the effective portions of the aperture are reduced to AM, NB. Now the distance between the feet of the perpendiculars let fall from A, M on RS may have any value from 0 to $\frac{1}{2}\lambda$, and for the angle of diffraction actually employed AM was equal to about twice that distance on the average, or rather less. Hence AM may be regarded as ranging from 0 to λ; and since for the brightest part of a band forming that portion of a spectrum of the first class which belongs to light of given refrangibility AM has just half its greatest value, we may suppose $AM = \frac{1}{2}\lambda$. But if the distance between the planes eh, ab be a small multiple of λ, and γ be small, ef will be small compared with λ, and therefore compared with AM. Hence the breadth of the portions of the plane eh, such as ef, for which we are not at liberty to regard the light as first diffracted and then regularly refracted, is small compared with the breadth of the portions of the aperture, such as AM, which are really effective; and therefore, so far as regards the main part of the illumination, we are at liberty to make the supposition just mentioned. But we must not suppose the wave to be first regularly refracted and then diffracted, because the regular refraction presupposes the continuity of the wave.

The above reasoning is not given as perfectly satisfactory, nor could we on the strength of it venture to predict with confidence the result; but the result having been obtained experimentally, the explanation which has just been given seems a plausible way of accounting for it. According to this view of the subject, the result is probably not strictly exact, but only a very near approximation to the fact. For, if we suppose the distance between the planes eh, ab to be only a small multiple of λ, we cannot apply the regular law of refraction, except as a near approximation. Moreover, the dynamical theory of diffraction points to the existence of terms which, though small, would not be wholly insensible at the distance of the plane ab. Lastly, when the radius of a secondary wave which passes the edge A or B is only a small multiple of λ, we cannot regard γ as exceedingly small.

Let us consider now the results of experiments Nos. 11 and 12. In diffraction at refraction, the amount of crowding with respect to which the theory leaves us in doubt vanishes along with $\mu - 1$; and although this amount is far from insensible in the actual experiments, it is still not sufficiently large to prevent the results from being decisive in favour of one of the two hypotheses respecting the direction of vibration. Thus the curves marked "A" in the first figure are well separated from those marked "E", and if μ were to approach indefinitely to 1, the curves I. A and II. A would approach indefinitely to III. A, and I. E, and II. E to III. E. In diffraction at reflection, however, the case is quite different, and in the absence of a precise theory little can be made of the experiments, except that they tend to confirm the law expressed by the equation (49). In the case of the first and second images the diffraction accompanied refraction, and so far the experiments were of the same nature as those which have been just discussed, but the angle of incidence was not equal to zero, and in that respect they differ.

Let i', ρ be the angles of refraction corresponding to the angles of incidence, $i, i + \theta$. Then in the case of the first image the tangent of the azimuth of the plane of polarization is multiplied by $\cos(i + \theta - \rho) \sec(i + \theta + \rho)$ in consequence of reflection, and by $\cos(i + \theta - \rho)$ in consequence of refraction; and in the case of the second image by $\cos(i - i')$ in consequence of refraction, and by $\cos(i - i') \sec(i + i')$ in consequence of reflection. Hence if m' be the factor corresponding to diffraction and the accompanying refraction, m the factor got from observation, and regarded as correct, we have

for 1st image, $\log m' = \log m + \log \cos(i + \theta + \rho) - 2\log \cos(i + \theta - \rho)$,
for 2nd image, $\log m' = \log m + \log \cos(i + i') - 2\log \cos(i - i')$.

In the case of the first image, m' relates to diffraction at refraction from air into glass, where i is the angle of incidence in air, and $\rho - i'$ the angle of diffraction in glass. In the case of the second image, m' relates to diffraction from glass into air, where i' is the angle of incidence in glass, and θ the angle of diffraction in air.

In experiment No. 11, 1st image, we have from Table II, $\log m = + \cdot289$; for the 2nd image $\log m = + \cdot061$. In this experiment $i = 14^\circ 50'$, $\theta = 22^\circ 30'$, whence $i' = 9^\circ 41'$, $\rho = 23^\circ 30'$. We thus get

for 1st image, $\log m' = + \cdot289 - \cdot286 = + \cdot003$,
for 2nd image, $\log m' = + \cdot061 - \cdot037 = + \cdot024$.

The positive values of $\log m'$ which result from these experiments, notwithstanding the refraction which accompanied the diffraction, bear out the results of the experiments already discussed, and confirm the hypothesis of Fresnel. It may be remarked that $\log m'$ comes out larger for the second image, in which diffraction accompanied refraction from air into glass, than for the first image, in which diffraction accompanied refraction from glass into air. This also agrees with the experiments just referred to.

In experiment No. 12, the light which entered the eye came in a direction not much different from that in which light regularly reflected would have been perfectly polarized. Since in regularly reflected light the amount of crowding of the planes of polarization changes rapidly about the polarizing angle, it is probable that small errors in μ, i, and θ would produce large errors in m. Hence little can be made of this experiment beyond confirming the formula (49).

I will here mention an experiment of Fraunhofer's, which, when the whole theory is made out, will doubtless be found to have a most intimate connexion with those here described. In this experiment the light observed was reflected from the grooved face of a glass-grating; the reflection from the second surface was stopped by black varnish. In Fraunhofer's notation ϵ is the interval from one groove to the corresponding point of its consecutive, and is measured in parts of a French inch, σ is the angle of incidence, τ the inclination of the light observed to the plane of the grating, $(E\tau)$ the value of τ for the fixed line E, and the numerals mark the order of the spectrum, reckoned from the axis, or central colourless image, the order being reckoned positive on the side of the acute angle made by the regularly reflected light with the plane of the grating. The following is a translation of Fraunhofer's description of the experiment.

" It is very remarkable that, under a certain angle of incidence, a part of a spectrum arising from reflection consists of *perfectly polarized light*. This angle of incidence is very different for the different spectra, and even very sensibly different for the different colours of one and the same spectrum. With the glass-grating $\epsilon = 0\cdot0001223$ there is polarized: $(E\tau)^{(+\mathrm{I})}$, that is, the *green* part of this first spectrum, when $\sigma = 49°$; $(E\tau)^{(+\mathrm{II})}$, or the green part in the second spectrum lying on the same side of the axis, when $\sigma = 40°$; lastly, $(E\tau)^{(-\mathrm{I})}$, or the green part of the first spectrum lying on the opposite side of the axis, when $\sigma = 69°$. When $(E\tau)^{(+\mathrm{I})}$ is polarized perfectly, the remaining colours of this spectrum are still but imperfectly polarized. This is less the case with $(E\tau)^{(+\mathrm{II})}$, and σ can be sensibly changed while this colour still remains polarized. $(E\tau)^{(-\mathrm{I})}$ is under no angle of incidence so completely polarized (so ganz vollständig polarisirt) as $(E\tau)^{(+\mathrm{I})}$. With a grating in which ϵ is greater than in that here spoken of, the angle of incidence would have to be quite different in order that the above-mentioned spectra should be polarized*."

* Gilbert's *Annalen der Physik*, B. xiv. (1823) S. 364.

If we suppose σ_ν a function of ν such that $\sigma_{-1} = 69$, $\sigma_{+1} = 49$, $\sigma_{+2} = 40$, we get by interpolation $\sigma_0 = 58\cdot33$; so that if we suppose the central colourless image, which arises from light reflected according to the regular law, to have been polarized at the polarizing angle for light reflected at a surface free from grooves, we get $\mu = \tan 58^\circ 40' = 1\cdot64$, from which it would result that the grating was made of flint glass.   The inclination of E in the spectrum of the order ν to the plane of the grating may be calculated from the formula $\cos \tau = \sin \sigma + \nu\lambda/\epsilon^*$, given by Fraunhofer, and obtained from the theory of interference; and $\theta = 90^\circ - \tau - \sigma$, where θ is the angle of diffraction.   We thus get for green light polarized by reflection and the accompanying diffraction,

order of spectrum	σ	θ	$\sigma + \theta$
-1	69°	$-18^\circ 13'$	$50^\circ 47'$
0	$58^\circ 40'$	0	$58^\circ 40'$
$+1$	49°	$+17^\circ 1'$	$66^\circ 1'$
$+2$	40°	$+33^\circ 52'$	$73^\circ 52'$.

If we suppose the formula (49) to hold good in this case, m becomes infinite for the angles of incidence σ and the corresponding angles of reflection $\sigma + \theta$ contained in the preceding table.

Another observation of Fraunhofer's described in the same paper deserves to be mentioned in connexion with the present investigation, because at first sight it might seem to invalidate the conclusions which have been built on the results of the experiments.   On examining the spectra produced by refraction in another glass-grating on which the light was incident perpendicularly, Fraunhofer found that the spectra on one side of the axis were more than twice as bright as those on the other †.   To account for this phenomenon, he supposed that in ruling the grating the diamond had had such a position with respect to the plate that one side of each groove was sharp, the other less defined. This view was confirmed by finding that a glass plate covered with a thin coat of grease, and purposely ruled in such a manner, gave similar results.   Now with reference to the present investigation the question might naturally be asked, If such material changes in intensity are capable of being produced by such slight modifications in the diffracting edge, how is it possible to build any certain con-

---

* In Fraunhofer's notation the wave length is denoted by w.
† Gilbert's *Annalen der Physik*, B. xiv. p. 353.

clusions on an investigation in which the nature of the diffracting edge is not taken into account ?

To facilitate the explanation of the apparent cause of the above-mentioned want of symmetry, suppose the diffraction produced by a wire grating in which the section of each wire is a right-angled triangle, with one side of the right angle parallel to the plane of the grating, and perpendicular to the incident light, and the equal acute angles all turned the same way. The triangles ABC, DEF in the figure represent sections of two consecutive wires, and GB, HD, IE represent incident rays, or normals to the incident waves, which are supposed plane. Let $BE = \epsilon$, and $BD : DE :: n : 1 - n$. Draw BK, DL, EM parallel to one another in the direction of the spectrum of the order ν

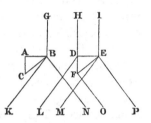

on the one side of the axis, so that $\nu\lambda$ is the retardation of the ray EM relatively to BK, and therefore $\sin\theta = \nu\lambda/\epsilon$, θ being the angle of diffraction, or the inclination of BK to GB produced. Draw BN, FO, EP at an inclination θ on the other side of the axis, and let $\angle DBF = \alpha$. Then the retardation of DL relatively to BK is equal to $n\nu\lambda$, or $n\epsilon\sin\theta$, and that of BN relatively to FO is equal to $n\epsilon\sin\theta + n\epsilon\tan\alpha\cos\theta - n\epsilon\tan\alpha$, so that if we denote these retardations by

$$R_1, R_2, \quad R_1 = n\epsilon\sin\theta, \quad R_2 = n\epsilon\sin\theta - n\epsilon\tan\alpha \text{ versin } \theta.$$

Let p_1, p_2 be the greatest integers contained in the quotients of R_1, R_2 divided by λ, and let $R_1 = p_1\lambda + r_1$, $R_2 = p_2\lambda + r_2$. Then the relative intensities of the two spectra of the order $+\nu$ and $-\nu$ depend on r_1, r_2: in fact, we find for the ratio of intensities, on the theory of interference, $\sin^2 \pi r_1/\lambda : \sin^2 \pi r_2/\lambda$. Now this ratio may have any value, and we may even have a bright spectrum on one side of the axis answering to an evanescent spectrum on the other side. It appears then in the highest degree probable that the want of symmetry of illumination in Fraunhofer's experiment was due to *a different mode of interference* on opposite sides of the axis. But this has nothing whatsoever to do with the nature of the polarization of the incident light, and consequently does not in the slightest degree affect the ratio of the intensities, or rather the ratio of the coefficients of vibration, of the two streams of light *belonging to the same spectrum* corresponding to the two streams of oppositely polarized light into which we may conceive the incident light decomposed, and consequently does not affect the law of the rotation of the plane of polarization of the diffracted light.

P. S.   Since the above was written, Professor Miller has determined for me the refractive index of the glass plate by means of the polarizing angle.   Four observations, made by candle-light, of which the mean error was only $1'\frac{1}{2}$, gave for the double angle $113^{\circ} 20'$, whence $\mu = \tan 56^{\circ} 40' = 1\cdot52043$, which agrees almost exactly with the value I had assumed.   In two of these observations the light was reflected at the ruled, and in two at the plane surface.   The accordance of the results bears out the supposition made in Part II, that the light belonging to the central colourless image, which is reflected or refracted according to the regular laws, is also affected as to its polarization in the same manner as if the surface were free from grooves.   The refractive index of the plate being now known for certain, the experiments described in this paper render it probable that the crowding of the planes of polarization which actually takes place is rather less than that which results from theory on the supposition (which is in a great measure empirical), that the diffraction takes place before the light reaches the grooves.   The difference is however so small that more numerous and more accurate experiments would be required before we could affirm with confidence that such is actually the case.

When a stream of light is incident obliquely on an aperture, it is sometimes necessary to conceive each wave broken up as its elements arrive in succession at the plane of the aperture.   In applying the formula (46) to such a case, it will be sufficient to substitute for dS the projection of an element of the aperture on the wave's front, θ being measured as before from the normal to the wave, which no longer coincides with the normal to the plane of the aperture.

Before concluding, it will be right to say a few words respecting M. Cauchy's dynamical investigation of the problem of diffraction, if it be only to shew that I have not been anticipated in the results which I here lay before the Society. This investigation is referred to in Moigno's *Repertoire d'Optique moderne*, p. 190, and will be found in the fifteenth Volume of the *Comptes Rendus*, where two short memoirs of M. Cauchy's on the subject are printed, the first of which begins at p. 605, and the second at p. 670. The first contains the analysis which M. Cauchy had some years before applied to the problem. This solution he afterwards, as it appears, saw reason to abandon, or at least greatly to restrict; and he has himself stated (p. 675), that it is only applicable when certain conditions are fulfilled, and when moreover the nature of the medium is such that normal and transversal vibrations are propagated with equal velocity. This latter condition, as Green has shewn, is incompatible with the stability of the medium. In the second memoir M. Cauchy has explained the principles of a new solution of the problem which he had obtained, without giving any of the analysis. The principal result, it would appear, at which he has arrived is, that light incident on an aperture in a screen is capable of being reflected, so to speak, *by the aperture itself* (p. 675); and he proposes seeking, by the use of very black screens, for these new rays which are reflected and diffracted. But it follows from reasoning similar to that of Art. 34, or even from the general formula (45) or (46), that such rays would be wholly insensible in all ordinary cases of diffraction, even were the screen to reflect absolutely no light. The only way apparently of rendering them sensible would be, to construct a grating of actual threads, so fine as to allow of observations at a large angle of diffraction. Such a grating I believe has never been made; and even if it could be made it would apparently be very difficult, if not impossible, to separate the effect to be investigated from the effect of reflection at the threads of the grating.

[A few years after the appearance of the above Paper, the question was re-examined experimentally by M. Holtzmann*, who at first employed glass gratings, but without getting consistent results (though there seemed some indication of a conclusion the same as that which I had obtained), and afterwards had recourse

---

* Poggendorff's *Annalen*, Vol. 99 (1856) p. 446, or *Philosophical Magazine*, Vol. 13, p. 135.

to a Schwerd's lampblack grating. With the latter consistent results were obtained. But the crowding of the planes of polarization was *towards* the plane of diffraction; and when instead of measuring the azimuths of the planes of polarization of the incident and diffracted light, the incident light was polarized at an azimuth of 45° to the lines of the grating, and the diffracted light was divided by a double-image prism into two beams polarized in and perpendicularly to the plane of diffraction, it was the *latter* that was the brighter. From these experiments the conclusion seemed to follow that in polarized light the vibrations are *in* the plane of polarization. The amount of rotation did not very well agree with theory. The subject was afterwards more elaborately investigated by M. Lorenz*. He commences by an analytical investigation which he substitutes for that which I had given, which latter he regards as incomplete, apparently from not having seized the spirit of my method. He then gives the results of his experiments, which were made with gratings of various kinds, especially smoke gratings. His results with these do not confirm those of Holtzmann, and he points out an easily over-looked source of error, which he himself had not for some time perceived, which he thinks may probably have affected Holtzmann's observations. Lorenz's results like mine were decisively in favour of the supposition that in polarized light the vibrations are *perpendicular* to the plane of polarization. He found as I had done that the results of observation as to the azimuth of the plane of polarization of the diffracted light agreed very approximately with the theoretical result, provided we imagine the diffraction to take place *before* the light reaches the ruled lines.]

* Poggendorff's *Annalen*, Vol. 111 (1860) p. 315, or *Philosophical Magazine*, Vol. 21, p. 321.

[From the *Transactions of the Cambridge Philosophical Society*. Vol. IX. Part I.]

ON THE NUMERICAL CALCULATION OF A CLASS OF DEFINITE INTEGRALS AND INFINITE SERIES.

[Read *March* 11, 1850.]

IN a paper " On the Intensity of Light in the neighbourhood of a Caustic*," Mr Airy the Astronomer Royal has shewn that the undulatory theory leads to an expression for the illumination involving the square of the definite integral $\int_0^\infty \cos \frac{\pi}{2} (w^3 - mw) \, dw$, where m is proportional to the perpendicular distance of the point considered from the caustic, and is reckoned positive towards the illuminated side. Mr Airy has also given a table of the numerical values of the above integral extending from $m = -4$ to $m = +4$, at intervals of 0·2, which was calculated by the method of quadratures. In a Supplement to the same paper† the table has been re-calculated by means of a series according to ascending powers of m, and extended to $m = \pm 5·6$. The series is convergent for all values of m, however great, but when m is at all large the calculation becomes exceedingly laborious. Thus, for the latter part of the table Mr Airy was obliged to employ 10-figure logarithms, and even these were not sufficient for carrying the table further. Yet this table gives only the first two roots of the equation $W = 0$, W denoting the definite integral, which answer to the theoretical places of the first two dark bands in a system of spurious rainbows, whereas Professor Miller was able to observe 30 of these bands. To attempt the computation of 30 roots of the equation $W = 0$ by

---

* *Camb. Phil. Trans.* Vol. VI. p. 379.   † Vol. VIII. p. 595.

means of the ascending series would be quite out of the question, on account of the enormous length to which the numerical calculation would run.

After many trials I at last succeeded in putting Mr Airy's integral under a form from which its numerical value can be calculated with extreme facility when m is large, whether positive or negative, or even moderately large. Moreover the form of the expression points out, without any numerical calculation, the law of the progress of the function when m is large. It is very easy to deduce from this expression a formula which gives the ith root of the equation $W = 0$ with hardly any numerical calculation, except what arises from merely passing from $(m/3)^{\frac{3}{2}}$, the quantity given immediately, to m itself.

The ascending series in which W may be developed belongs to a class of series which are of constant occurrence in physical questions. These series, like the expansions of ϵ^{-x}, $\sin x$, $\cos x$, are convergent for all values of the variable x, however great, and are easily calculated numerically when x is small, but are extremely inconvenient for calculation when x is large, give no indication of the law of progress of the function, and do not even make known what the function becomes when $x = \infty$. These series present themselves, sometimes as developments of definite integrals to which we are led in the first instance in the solution of physical problems, sometimes as the integrals of linear differential equations which do not admit of integration in finite terms. Now the method which I have employed in the case of the integral W appears to be of very general application to series of this class. I shall attempt here to give some sort of idea of it, but it does not well admit of being described in general terms, and it will be best understood from examples.

Suppose then that we have got a series of this class, and let the series be denoted by y or $f(x)$, the variable according to ascending powers of which it proceeds being denoted by x. It will generally be easy to eliminate the transcendental function $f(x)$ between the equation $y = f(x)$ and its derivatives, and so form a linear differential equation in y, the coefficients in which involve powers of x. This step is of course unnecessary if the differential equation is what presented itself in the first instance, the series being only an integral of it. Now by taking the terms of this differential equation in pairs, much as in Lagrange's method of expanding implicit functions which is given by Lacroix[*], we shall easily find what terms are of most importance when x is large: but this step will be best understood from examples. In this way we shall be led to assume for the integral a circular or exponential function multiplied by a series according to descending powers of x, in which the coefficients and indices are both arbitrary. The differential equation will determine the indices, and likewise the coefficients in terms of the first, which remains arbitrary. We shall thus have the complete integral of the differential equation, expressed in a form which admits of ready computation when x is large, but containing a certain number of arbitrary constants, according to the order of the equation, which have yet to be determined.

For this purpose it appears to be generally requisite to put the infinite series under the form of a definite integral, if the series be not itself the developement of such an integral which presented itself in the first instance. We must now endeavour to determine by means of this integral the leading term in $f(x)$ for indefinitely large values of x, a process which will be rendered more easy by our previous knowledge of the *form* of the term in question, which is given by the integral of the differential equation. The arbitrary constants will then be determined by comparing the integral just mentioned with the leading term in $f(x)$.

There are two steps of the process in which the mode of proceeding must depend on the particular example to which the method is applied. These are, first, the expression of the ascending series by means of a definite integral, and secondly, the determination thereby of the leading term in $f(x)$ for indefinitely large values of x. Should either of these steps be found impracticable, the method does not on that account fall to the ground. The arbitrary constants may still be determined, though with more trouble and far less elegance, by calculating the numerical value of $f(x)$ for one or more values of x, according to the number of arbitrary constants to be determined, from the ascending and descending series separately, and equating the results.

---

[*] *Traité du Calcul*, &c. Tom. I. p. 104.

In this paper I have given three examples of the method just described. The first relates to the integral W, the second to an infinite series which occurs in a great many physical investigations, the third to the integral which occurs in the case of diffraction with a circular aperture in front of a lens. The first example is a good deal the most difficult. Should the reader wish to see an application of the method without involving himself in the difficulties of the first example, he is requested to turn to the second and third examples.

FIRST EXAMPLE.

1.   Let it be required to calculate the integral

$$W = \int_0^\infty \cos \frac{\pi}{2} \left(w^3 - mw\right) dw \dots\dots\dots\dots (1),$$

for different values of m, especially for large values, whether positive or negative, and in particular to calculate the roots of the equation $W = 0$.

2.   Consider the integral

$$u = \int_0^\infty \epsilon^{-(\cos 3\theta + \sqrt{-1} \sin 3\theta)(x^3 - nx)} \, dx \dots\dots\dots\dots (2),$$

where θ is supposed to lie between $-\pi/6$ and $+\pi/6$, in order that the integral may be convergent.

Putting $x = (\cos \theta - \sqrt{-1} \sin \theta) z,$

we get $dx = (\cos \theta - \sqrt{-1} \sin \theta) \, dz$, and the limits of z are 0 and ∞; whence, writing for shortness

$$p = (\cos 2\theta + \sqrt{-1} \sin 2\theta) \, n \dots\dots\dots\dots (3),$$

we get $u = (\cos \theta - \sqrt{-1} \sin \theta) \int_0^\infty \epsilon^{-(z^3 - pz)} \, dz* \dots\dots\dots\dots (4).$

---

* The legitimacy of this transformation rests on the theorem that if $f(x)$ be a continuous function of x, which does not become infinite for any real or imaginary, but finite, value of x, we shall obtain the same result for the integral of $f(x) \, dx$ between two given real or imaginary limits through whatever series of real or imaginary values we make x pass from the inferior to the superior limit. It is unnecessary here to enunciate the theorem which applies to the case in which $f(x)$ becomes infinite for one or more real or imaginary values of x. In the present case

3. Let now θ, which hitherto has been supposed less than $\pi/6$, become equal to $\pi/6$. The integral obtained from (2) by putting $\theta = \pi/6$ under the integral sign may readily be proved to be convergent. But this is not sufficient in order that we may be at liberty to assert the equality of the results obtained from (2), (4) by putting $\theta = \pi/6$ before integration. It is moreover necessary that the convergency of the integral (2) should not become infinitely slow when θ approaches indefinitely to $\pi/6$, in other words, that if X be the superior limit to which we must integrate in order to render the remainder, or rather its modulus, less than a given quantity which may be as small as we please, X should not become infinite when θ becomes equal to $\pi/6$*. This may be readily proved in the present case, since the integral (2) is even more convergent than the integral

$$\int_0^\infty \epsilon^{-\sqrt{-1}\sin 3\theta\,(x^3 - nx)}\, dx,$$

which may be readily proved to be convergent.

Putting then $\theta = \pi/6$ in (2) and (4), we get

$$u = \int_0^\infty \cos(x^3 - nx)\, dx - \sqrt{-1} \int_0^\infty \sin(x^3 - nx)\, dx \ldots\ldots (5),$$

$$u = \left(\cos\frac{\pi}{6} - \sqrt{-1}\sin\frac{\pi}{6}\right)\int_0^\infty \epsilon^{-(z^3 - pz)}\, dz \ldots\ldots (6),$$

where $$p = \left(\cos\frac{\pi}{3} + \sqrt{-1}\sin\frac{\pi}{3}\right)n \ldots\ldots\ldots (7).$$

Let $$u = U - \sqrt{-1}\, U',$$

and in the expression for U got from (5) put

$$x = \left(\frac{\pi}{2}\right)^{\frac{1}{3}} w, \quad n = \left(\frac{\pi}{2}\right)^{\frac{2}{3}} m \ldots\ldots\ldots (8);$$

then we get $$W = \left(\frac{\pi}{2}\right)^{-\frac{1}{3}} U \ldots\ldots\ldots\ldots (9).$$

the limits of x are 0 and real infinity, and accordingly we may first integrate with respect to z from 0 to a large real quantity z_1, θ' (which is supposed to be written for θ in the expression for x) being constant, then leave z equal to z_1, make θ' vary, and integrate from θ to 0, and lastly make z_1 infinite. But it may be proved without difficulty (and the proof may be put in a formal shape as in Art. 8), that the second integral vanishes when z_1 becomes infinite, and consequently we have only to integrate with respect to z from 0 to real infinity.

* See Section III. of a paper "On the Critical Values of the sums of Periodic Series." *Camb. Phil. Trans.* Vol. VIII. p. 561. [*Ante*, Vol. I. p. 279.]

4. By the transformation of u from the form (5) to the form (6), we are enabled to differentiate it as often as we please with respect to n by merely differentiating under the integral sign. By expanding the exponential ϵ^{pz} in (6) we should obtain u, and therefore U, in a series according to ascending powers of n. This series is already given in Mr Airy's Supplement. It is always convergent, but is not convenient for numerical calculation when n is large.

We get from (6)

$$\frac{d^2u}{dp^2} - \frac{pu}{3} = \left(\cos\frac{\pi}{6} - \sqrt{-1}\sin\frac{\pi}{6}\right)\int_0^\infty \epsilon^{-(z^3-pz)}\left(z^2 - \frac{p}{3}\right)dz$$

$$= \frac{1}{3}\left(\cos\frac{\pi}{6} - \sqrt{-1}\sin\frac{\pi}{6}\right),$$

which becomes by (7)

$$\frac{d^2u}{dn^2} + \frac{n}{3}u \tfrac{1}{3}\sqrt{-1} \quad\dots\dots\dots\dots\dots (10).$$

Equating to zero the real part of the first member of this equation, we get

$$\frac{d^2U}{dn^2} + \frac{n}{3}U = 0 \quad\dots\dots\dots\dots\dots\dots(11).$$

5. We might integrate this equation by series according to ascending powers of n, and we should thus get, after determining the arbitrary constants, the series which have been already mentioned. What is required at present is, to obtain for U an expression which shall be convenient when n is large.

The form of the differential equation (11) already indicates the general form of U for large values of n. For, suppose n large and positive, and let it receive a small increment δn. Then the proportionate increment of the coefficient $n/3$ will be very small; and if we regard this coefficient as constant, and δn as variable, we shall get for the integral of (11)

$$U = N\cos\left\{\sqrt{\left(\frac{n}{3}\right)}\cdot\delta n\right\} + N'\sin\left\{\sqrt{\left(\frac{n}{3}\right)}\cdot\delta n\right\} \dots (12),$$

where N, N' are regarded as constants, δn being small, which does not prevent them from being in the true integral of (11) slowly varying functions of n. The approximate integral (12) points out the existence of circular functions such as $\cos f(n)$, $\sin f(n)$ in the true integral; and since $\sqrt{(n/3)} \cdot \delta n$ must be the small increment of $f(n)$, we get $f(n) = \frac{2}{3}\sqrt{(n^3/3)}$, omitting the constant, which it is unnecessary to add. When n is negative, and equal to $-n'$, the same reasoning would point to the existence of exponentials with $\pm \frac{2}{3}\sqrt{(n'^3/3)}$ in the index. Of course the exponential with a positive index will not appear in the particular integral of (11) with which we are concerned, but both exponentials would occur in the complete integral. Whether n be positive or negative, we may, if we please, employ exponentials, which will be real or imaginary as the case may be.

6. Assume then to satisfy (11)

$$U = \epsilon^{\frac{2}{3}\sqrt{-\frac{n^3}{3}}} \{An^\alpha + Bn^\beta + Cn^\gamma + \dots\} * \dots \dots (13),$$

where $A, B, C \dots \alpha, \beta, \gamma \dots$ are constants which have to be determined. Differentiating, and substituting in (11), we get

$$\alpha(\alpha - 1)An^{\alpha-2} + \beta(\beta -)1\, Bn^{\beta-2} + \dots$$

$$+ \frac{\sqrt{-1}}{2\sqrt{3}}\{(4\alpha + 1)An^{\alpha-\frac{1}{2}} + (4\beta + 1)Bn^{\beta-\frac{1}{2}} + \dots\} = 0.$$

As we want a series according to descending powers of n, we must put

$$4\alpha + 1 = 0, \quad \beta = \alpha = \tfrac{3}{2}, \quad \gamma = \beta - \tfrac{3}{2} \dots$$

$$B = 2\sqrt{-3}\,\frac{\alpha(\alpha-1)}{4\beta + 1}A, \quad C = 2\sqrt{-3}\,\frac{\beta(\beta-1)}{4\gamma + 1}B \dots$$

---

* The idea of multiplying the circular functions by a series according to descending powers of n was suggested to me by seeing in Moigno's *Repertoire d'optique moderne*, p. 189, the following formulæ which M. Cauchy has given for the calculation of Fresnel's integrals for large, or moderately large, values of the superior limit:

$$\int_0^m \cos\frac{\pi}{2}z^2 dz = \tfrac{1}{2} - N\cos\frac{\pi}{2}m^2 + M\sin\frac{\pi}{2}m^2;$$

$$\int_0^m \sin\frac{\pi}{2}z^2 dz = \tfrac{1}{2} - M\cos\frac{\pi}{2}m^2 - N\sin\frac{\pi}{2}m^2;$$

where $\quad M = \dfrac{1}{m\pi} - \dfrac{1.3}{m^5\pi^3} + \dfrac{1.3.5.7}{m^9\pi^5} - \dots; \quad N = \dfrac{1}{m^3\pi^2} - \dfrac{1.3.5}{m^7\pi^4} + \dots$

The demonstration of these formulæ will be found in the 15th Volume of the *Comptes Rendus*, pp. 554 and 573. They may be readily obtained by putting $\pi z^2 = 2x$, and integrating by parts between the limits $\frac{1}{2}\pi m^2$ and ∞ of x.

whence

$$U = An^{-\frac{1}{4}} \epsilon^{\frac{2}{3}\sqrt{-\frac{n^3}{3}}} \left\{ 1 - \frac{1.5}{1} \frac{\sqrt{-1}}{16\sqrt{(3n^3)}} + \frac{1.5.7.11}{1.2} \left(\frac{\sqrt{-1}}{16\sqrt{(3n^3)}} \right)^2 \right.$$

$$\left. - \frac{1.5.7.11.13.17}{1.2.3} \left(\frac{\sqrt{-1}}{16\sqrt{(3n^3)}} \right)^3 + \ldots \right\} \ldots \ldots (14).$$

By changing the sign of $\sqrt{(-1)}$ both in the index of ϵ and in the series, writing B for A, and adding together the results, we shall obtain the complete integral of (11) with its two arbitrary constants. The integral will have different forms according as n is positive or negative.

First, suppose n positive. Putting the function of n of which A is the coefficient at the second side of (14) under the form $P + \sqrt{(-1)}\, Q$, and observing that an expression of the form

$$A\left(P + \sqrt{-1}\, Q\right) + B\left(P - \sqrt{-1}\, Q\right),$$

where A and B are imaginary arbitrary constants, and which is supposed to be real, is equivalent to $AB + BQ$, where A and B are real arbitrary constants, we get

$$U = An^{-\frac{1}{4}} \left(R \cos \tfrac{2}{3} \sqrt{\frac{n^3}{3}} + S \sin \tfrac{2}{3} \sqrt{\frac{n^3}{3}} \right)$$

$$+ Bn^{-\frac{1}{4}} \left(R \sin \tfrac{2}{3} \sqrt{\frac{n^3}{3}} - S \cos \tfrac{2}{3} \sqrt{\frac{n^3}{3}} \right) \ldots \ldots (15),$$

where

$$\left. \begin{aligned} R &= 1 - \frac{1.5.7.11}{1.2.16^2.3n^3} + \frac{1.5.7.11.13.17.19.23}{1.2.3.4.16^4.3^2 n^6} \ldots \ldots \\[2mm] S &= \frac{1.5}{1.16\,(3n^3)^{\frac{1}{2}}} - \frac{1.5.7.11.13.17}{1.2.3.16^3\,(3n^3)^{\frac{3}{2}}} + \ldots \ldots \end{aligned} \right\} \ldots (16).$$

Secondly, suppose n negative, and equal to $-n'$. Then, writing $-n'$ for n in (14), and changing the arbitrary constant, and the sign of the radical, we get

$$U = Cn'^{-\frac{1}{4}} \epsilon^{-\frac{2}{3}\sqrt{\frac{n'^3}{3}}} \left\{ 1 - \frac{1.5}{1.16\,(3n^3)^{\frac{1}{2}}} + \frac{1.5.7.11}{1.2.16.3n^3} - \ldots \right\} \ldots (17).$$

It is needless to write down the part of the complete integral of (11) which involves an exponential with a positive index, because, as has been already remarked, it does not appear in the particular integral with which we are concerned.

7. When n or n' is at all large, the series (16) or (17) are at first rapidly convergent, but they are ultimately in all cases hypergeometrically divergent. Notwithstanding this divergence, we may employ the series in numerical calculation, provided we do not take in the divergent terms. The employment of the series may be justified by the following considerations.

Suppose that we stop after taking a finite number of terms of the series (16) or (17), the terms about where we stop being so small that we may regard them as insensible; and let U_1 be the result so obtained. From the mode in which the constants A, B, C,... α, β, γ... in (13) were determined, it is evident that if we form the expression

$$\frac{d^2 U_1}{dn^2} + \frac{n}{3} U_1, \quad \text{or} \quad \frac{d^2 U_1}{dn'^2} - \frac{n'}{3} U_1,$$

according as n is positive or negative, the terms will destroy each other, except one or two at the end, which remain undestroyed. These terms will be of the same order of magnitude as the terms at the part of the series (16) or (17) where we stopped, and therefore will be insensible for the value of n or n' for which we are calculating the series numerically, and, much more, for all superior values. Suppose the arbitrary constants A, B in (16) determined by means of the ultimate form of U for $n = \infty$, and C in (17) by means of the ultimate form of U for $n' = \infty$. Then U_1 satisfies exactly a differential equation which differs from (11) by having the zero at the second side replaced by a quantity which is insensible for the value of n or n' with which we are at work, and which is still smaller for values comprised between that and the particular value, (namely ∞), by means of which the arbitrary constants were determined so as to make U_1 and U agree. Hence U_1 will be a near approximation to U. But if we went too far in the series (16) or (17), so as, after having gone through the insensible terms, to take in some terms which were not insensible, the differential equation which U_1 would satisfy exactly would differ sensibly from (11), and the value of U_1 obtained would be faulty.

8.   It remains to determine the arbitrary constants A, B, C. For this purpose consider the integral

$$Q = \int_0^\infty \epsilon^{-x^3 + 3q^2 x}\, dx \dots \dots \dots \dots \dots (18),$$

where q is any imaginary quantity whose amplitude does not lie beyond the limits $-\pi/6$ and $+\pi/6$. Since the quantity under the integral sign is finite and continuous for all finite values of x, we may, without affecting the result, make x pass from its initial value 0 to its final value ∞ through a series of imaginary values. Let then $x = q + y$, and we get

$$Q = \epsilon^{2q^3} \int_{-q}^\infty \epsilon^{-y^3 - 3qy^2}\, dy,$$

where the values through which y passes in the integration are not restricted to be such as to render x real. Putting $y = (3q)^{-\frac{1}{2}} t$, where that value of the radical is supposed to be taken which has the smallest amplitude, we get

$$Q = (3q)^{-\frac{1}{2}} \epsilon^{2q^3} \int \epsilon^{-(3q)^{-\frac{3}{2}} t^3 - t^2}\, dt \dots \dots \dots \dots (19).$$

The limits of t are $-3^{\frac{1}{2}} q^{\frac{3}{2}}$ and an imaginary quantity with an infinite modulus and an amplitude equal to $\frac{1}{2}\alpha$, where α denotes the amplitude of q. But we may if we please integrate up to a real quantity ρ, and then, putting $t = \rho \epsilon^{\theta \sqrt{(-1)}}$, and leaving ρ constant, integrate with respect to θ from 0 to $\frac{1}{2}\alpha$, and lastly put $\rho = \infty$. The first part of the integral will be evidently convergent at the limit ∞, since amplitude of the coefficient of t^3 in the index does not lie beyond the limits $-\frac{1}{2}\pi$ and $+\frac{1}{2}\pi$; and calling the two parts of the integral with respect to t in (19) T, T_4, we get

$$T = \int_{-3^{\frac{1}{2}} q^{\frac{3}{2}}}^\infty \epsilon^{-(3q)^{-\frac{3}{2}} t^3 - t^2}\, dt \dots \dots \dots \dots \dots \dots (20),$$

$$T_4 = \text{limit}_{(\rho = \infty)} \rho \sqrt{-1} \int_0^{\frac{1}{2}\alpha} \epsilon^{-(3q)^{-\frac{3}{2}} \rho^3 \epsilon^{3\theta \sqrt{-1}} - \rho^2 \epsilon^{2\theta \sqrt{-1}} + \theta \sqrt{-1}}\, d\theta \dots (21).$$

We shall evidently obtain a superior limit to either the real or the imaginary part of T_4 by reducing the expression under the integral sign to its modulus. The modulus is $\epsilon^{-\Theta}$ where

$$\Theta = (3c)^{-\frac{3}{2}} \rho^3 \cos (3\theta - \tfrac{3}{2}\alpha) + \rho^2 \cos 2\theta,$$

c being the modulus of q. The first term in this expression is never negative, being only reduced to zero in the particular case in which $\theta = 0$ and $\alpha = \pm \pi/6$. The second term is never less than $\rho^2 \cos \frac{1}{3}\pi$ or $\frac{1}{2}\rho^2$, and is in general greater. Hence both the real and the imaginary parts of the expression of which T_4 is the limit are numerically less than $\frac{1}{2}\alpha\rho e^{-\frac{1}{2}\rho^2}$, which vanishes when $\rho = \infty$, and therefore $T_4 = 0$. Hence we have rigorously

$$Q = (3q)^{-\frac{1}{4}} \epsilon^{2q^{\frac{3}{2}}} T \dots\dots\dots\dots\dots(22).$$

Let us now seek the limit to which T tends when c becomes infinite. For this purpose divide the integral T into three parts T_1, T_2, T_3, where T_1 is the integral taken from $-3^{\frac{1}{2}}q^{\frac{1}{2}}$ to a real negative quantity $-a$, T_2 from $-a$ to a real positive quantity $+b$, and T_3 from b to ∞; and suppose c first to become infinite, a and b remaining constant, and lastly make a and b infinite.

Changing the sign of t in T_1, and the order of the limits, we get

$$T_1 = \int_a^{3^{\frac{1}{3}}c^{\frac{2}{3}}} \epsilon^{(3q)^{-\frac{1}{2}}t^3 - t^2} dt.$$

Put $t = \rho e^{\theta \sqrt{(-1)}}$. Then we may integrate first from $\rho = a$ to $\rho = 3^{\frac{1}{3}}c^{\frac{2}{3}}$ while θ remains equal to 0, and afterwards from $\theta = 0$ to $\theta = \alpha$ while $\frac{3}{2}\rho$ remains equal to $3^{\frac{1}{3}}c^{\frac{2}{3}}$. Let the two parts of the integral be denoted by T', T''. We shall evidently obtain a superior limit to T' by making the following changes in the integral: first, replacing the quantity under the integral sign by its modulus; secondly, replacing t^3 in the index by the product of t^2 and the greatest value (namely $3^{\frac{1}{3}}c^{\frac{2}{3}}$) which t receives in the integration; thirdly, replacing α by the smallest quantity (namely 0) to which it can be equal, and, fourthly, extending the superior limit to ∞. Hence the real and imaginary parts of T' are both numerically less than $\int_a^\infty \epsilon^{-\frac{1}{3}t^2} dt$, a quantity which vanishes in the limit, when a becomes infinite.

We shall obtain a superior limit to the real or imaginary part of T'' by reducing the quantity under the integral sign to its modulus, and omitting $\sqrt{(-1)}$ in the coefficient. Hence L will be such a limit if

$$L = 3^{\frac{1}{3}}c^{\frac{2}{3}} \int_0^{\frac{2}{3}\alpha} \epsilon^{-c^2 f(\theta)} d\theta, \text{ where } f(\theta) = 3\cos 2\theta - \cos (3\theta - \tfrac{3}{2}\alpha).$$

We may evidently suppose α to be positive, if not equal to zero, since the case to which it is negative may be reduced to the case in which it is positive by changing the signs of α and θ. When $\theta = \pi/6$, the first term in $f(\theta)$ is equal to $\frac{3}{2}$, which, being greater than 1, determines the sign of the whole, and therefore $f(\theta)$ is positive; and $f(\theta)$ is evidently positive from $\theta = 0$ to $\theta = \pi/6$, since for such values $\cos 2\theta > \frac{1}{2}$. Also in general $f'(\theta) = -6 \sin 2\theta + 3 \sin(3\theta - \frac{3}{2}\alpha)$, which is evidently positive from $\theta = \pi/6$ to $\theta = \pi/4$, and the latter is the largest value we need consider, being the extreme value of θ when α has its extreme value $\pi/6$. When θ has its extreme value $\frac{3}{2}\alpha$, $f(\theta) = 2 \cos 3\alpha$, which is positive when $\alpha < \pi/6$, and vanishes when $\alpha = \pi/6$. Hence $f(\theta)$ is positive when $\theta < \frac{3}{2}\alpha$; for it has been shewn to be positive when $\theta < \pi/6$, which meets the case in which $\alpha < \pi/9$ or $= \pi/9$, and to be constantly decreasing from $\theta = \pi/6$ to $\theta = \frac{3}{2}\alpha$, which meets the case in which $\theta > \pi/9$. Hence when $\alpha < \pi/6$ the limit of L for $c = \infty$ is zero, inasmuch as the coefficient of c^3 in the index of ϵ is negative and finite; and when $\alpha = \pi/6$ the same is true, for the same reason, if it be not for a range of integration lying as near as we please to the superior limit. In this case put for shortness $f(\theta) = \delta$, regard $\frac{3}{2}\alpha - \theta$ as a function of δ, $F(\delta)$, and integrate from $\delta = 0$ to $\delta = \beta$, where β is a constant which may be as small as we please. By what precedes, $F'(\delta)$ will be finite in the integration, and may be made as nearly as we please equal to the constant $F'(0)$ by diminishing β. Hence the integral ultimately becomes

$$3^{\frac{1}{2}} F'(0) \, c^{\frac{3}{2}} \int_0^\beta \epsilon^{-c^3\delta} \, d\delta,$$

which vanishes when c becomes infinite. Hence the limit of T_1 is zero.

We have evidently $T_3 < \int_b^\infty \epsilon^{-t^2} \, dt$,

which vanishes when b becomes infinite. Hence the limit of T is equal to that of T_2. Now making c first infinite and afterwards a and b, we get

$$\text{limit of } T_2 = \text{limit of } \int_{-a}^b \epsilon^{-t^2} \, dt = \int_{-\infty}^{+\infty} \epsilon^{-t^2} \, dt = \sqrt{\pi},$$

and therefore we have ultimately, for very large values of c,

$$Q = \left(\frac{\pi}{3q}\right)^{\frac{1}{2}} \epsilon^{2q^3} \quad \dots\dots\dots\dots\dots\dots\dots (22).$$

In order to apply this expression to the integral u given by (6), we must put

$$3q^2 = n\epsilon^{\frac{\pi}{3}\sqrt{-1}}, \quad \text{whence} \quad q = \left(\frac{n}{3}\right)^{\frac{1}{2}} \epsilon^{\frac{\pi}{6}\sqrt{-1}},$$

$$\epsilon^{-\frac{\pi}{6}\sqrt{-1}}\left(\frac{\pi}{3q}\right)^{\frac{1}{2}} = \frac{\pi^{\frac{1}{2}}}{(3n)^{\frac{1}{4}}} \epsilon^{-\frac{\pi}{4}\sqrt{-1}}, \quad 2q^3 = 2\left(\frac{n}{3}\right)^{\frac{3}{2}}\sqrt{-1},$$

whence we get ultimately

$$u = \frac{\pi^{\frac{1}{2}}}{(3n)^{\frac{1}{4}}} \epsilon^{\left\{2\left(\frac{n}{3}\right)^{\frac{3}{2}} - \frac{\pi}{4}\right\}\sqrt{-1}}, \quad U = \frac{\pi^{\frac{1}{2}}}{(3n)^{\frac{1}{4}}} \cos\left\{2\left(\frac{n}{3}\right)^{\frac{3}{2}} - \frac{\pi}{4}\right\}^{*} \dots\dots (23).$$

Comparing with (15) we get

$$A = B = \frac{\pi^{\frac{1}{2}}}{2^{\frac{1}{2}}3^{\frac{1}{4}}} \dots\dots\dots\dots\dots\dots (24).$$

9. We cannot make n pass from positive to negative through a series of real values, so long as we employ the series according to descending powers, because these series become illusory when n is small. When n is imaginary we cannot speak of the integrals which appear at the right-hand side of (5), because the exponential with a positive index which would appear under the integral signs would render each of these integrals divergent. If however we take equation (6) as the definition of u, and suppose U always derived from u by changing the sign of $\sqrt{(-1)}$ in the coefficient of the integral and in the value of p, but not in the expression for n, and taking half the sum of the results, we may regard u and U as certain functions of n whether n be real or imaginary. According to this definition, the series involving ascending integral powers of n, which is convergent for all values of n, real or imaginary, however great be the modulus, will continue to represent u

---

* This result might also have been obtained from the integral U in its original shape, namely, $\int_0^\infty \cos(x^3 - nx)\,dx$, by a method similar to that employed in Art. 21. If x_1 be the positive value of x which renders $x^3 - nx$ a minimum, we have $x_1 = 3^{-\frac{1}{2}}n^{\frac{1}{2}}$. Let the integral U be divided into three parts, by integrating separately from $x=0$ to $x=x_1-a$, from $x=x_1-a$ to $x=x_1+b$, and from $x=x_1+b$ to $x=\infty$; then make n infinite while a and b remain finite, and lastly, let a and b vanish. In this manner the second of equations (23) will be obtained, by the assistance of the known formulæ

$$\int_{-\infty}^\infty \cos x^2 dx = \int_{-\infty}^\infty \sin x^2 dx = 2^{-\frac{1}{2}}\pi^{\frac{1}{2}}.$$

when n is imaginary. The differential equation (11), and consequently the descending series derived from it, will also hold good when n is imaginary; but since this series contains radicals, while U is itself a rational function of n, we might expect beforehand that in passing from one imaginary value of n to another it should sometimes be necessary to change the sign of a radical, or make some equivalent change in the coefficients A, B. Let $n = n_1 e^{\nu \sqrt{(-1)}}$ where n_1 is positive. Since both values of $2 (n/3)^{\frac{3}{2}}$ are employed in the series, with different arbitrary constants, we may without loss of generality suppose that value of $n^{\frac{3}{2}}$ which has $\frac{3}{2}\nu$ for its amplitude to be employed in the circular functions or exponentials, as well as in the expression for S. In the multiplier we may always take $-\nu/4$ for the amplitude of $n^{-\frac{1}{4}}$ by including in the constant coefficients the factor by which one fourth root of n differs from another; but then we must expect to find the arbitrary constants discontinuous. In fact, if we observe the forms of R and S, and suppose the circular functions in (15) expanded in ascending series, it is evident that the expression for U will be of the form

$$An^{-\frac{1}{4}} N + Bn^{\frac{1}{4}} N' \quad\dots\dots\dots\dots\dots\dots(25),$$

where N and N' are rational functions of n. At least, this will be the case if we regard as a rational function a series involving descending integral powers of n, and which is at first rapidly convergent, though ultimately divergent, or rather, if we regard as such the function to which the convergent part of the series is a very close approximation when the modulus of n is at all large. Now, if A and B retained the same values throughout, the above expression would not recur till ν was increased by 8π, whereas U recurs when ν is increased by 2π. If we write $\nu + 2\pi$ for ν, and observe that N and N' recur, the expression (25) will become

$$- \sqrt{-1}An^{-\frac{1}{4}} N + \sqrt{-1}Bn^{\frac{1}{4}} N';$$

and since U recurs it appears that A, B become $\sqrt{(-1)} A, -\sqrt{(-1)}B$, respectively, when ν is increased by 2π. Also the imaginary part of the expression (25) changes sign with ν, as it ought; so that, in order to know what A and B are generally, it would be sufficient to know what they are from $\nu = 0$ to $\nu = \pi$.

If we put $n_1 e^{\pi \sqrt{(-1)}}$ for n in the second member of equation (15), and write β for $2 . 3^{-\frac{3}{2}} n_1^{\frac{3}{2}}$, and R_1, S_1 for what R, S become when

n_1 is put for n in the second members of equations (16) and ·all the terms are taken positively, we shall get as our result

$$\tfrac{1}{2}\epsilon^{-\frac{\pi}{4}\sqrt{-1}} n_1^{-\frac{1}{4}} \{(A-\sqrt{-1}B)(R_1+S_1)\epsilon^\beta + (A+\sqrt{-1}B)(R_1-S_1)\epsilon^{-\beta}\}.$$

Now the part of this expression which contains $(R_1+S_1)\epsilon^\beta$ ought to disappear, as appears from (17). If we omit the first part of the expression, and in the second part put for A and B their values given by (24), we shall obtain an expression which will be identical with the second member of (17) provided

$$C = \frac{\pi^{\frac{1}{2}}}{2 \cdot 3^{\frac{1}{4}}} \quad \dots\dots\dots\dots\dots\dots\dots\dots \text{(26)}.$$

This mode of determining the constant C is anything but satisfactory. I have endeavoured in vain to deduce the leading term in U for n negative from the integral itself, whether in the original form in which it appears in (5), or in the altered form in which it is obtained from (6) *. The correctness of the above value of C will however be verified further on.

10. Expressing n, U in terms of m, W by means of (8) and (9), putting for shortness

$$\phi = 2\left(\frac{n}{3}\right)^{\frac{3}{2}} = \pi\left(\frac{m}{3}\right)^{\frac{3}{2}} \quad \dots\dots\dots\dots\dots \text{(27)},$$

where the numerical values of m and n are supposed to be taken when these quantities are negative, observing that $16\sqrt{(3n^3)} = 72\phi$, and reducing, we get when m is positive

$$W = 2^{\frac{1}{2}}(3m)^{-\frac{1}{4}}\left\{R\cos\left(\phi-\frac{\pi}{4}\right) + S\sin\left(\phi-\frac{\pi}{4}\right)\right\} \dots\text{(28)},$$

where

$$\left.\begin{array}{l} R = 1 - \dfrac{1 \cdot 5 \cdot 7 \cdot 11}{1 \cdot 2\,(72\phi)^2} + \dfrac{1 \cdot 5 \cdot 7 \cdot 11 \cdot 13 \cdot 17 \cdot 19 \cdot 23}{1 \cdot 2 \cdot 3 \cdot 4\,(72\,\phi)^4} - \dots \\[2ex] S = \dfrac{1 \cdot 5}{1 \cdot 72\phi} - \dfrac{1 \cdot 5 \cdot 7 \cdot 11 \cdot 13 \cdot 17}{1 \cdot 2 \cdot 3\,(72\phi)^3} + \dots \end{array}\right\} \dots\text{(29)}.$$

When m is negative, so that W is the integral expressed by writing $-m$ for m in (1), we get

$$W = 2^{-\frac{1}{2}}(3m)^{-\frac{1}{4}}\epsilon^{-\phi}\left\{1 - \frac{1 \cdot 5}{1 \cdot 72\phi} + \frac{1 \cdot 5 \cdot 7 \cdot 11}{1 \cdot 2\,(72\phi)^2} - \dots\right\}\dots\text{(30)}.$$

[* The difficulty was overcome in a later paper entitled "On the discontinuity of arbitrary constants which appear in divergent developments." (*Transactions of the Cambridge Philosophical Society*, Vol. x. p. 105.)]

11.   Reducing the coefficients of ϕ^{-1}, ϕ^{-2}... in the series (29) for numerical calculation, we have, not regarding the signs,

order   (i)   (ii)   (iii)
logarithm $\bar{2}$·841638; $\bar{2}$·569766; $\bar{2}$·579704;
coefficient   ·0694444;   ·0371335;   ·0379930;

(iv)   (v)   (vi)
$\bar{2}$·760793; $\bar{1}$·064829; $\bar{1}$·464775;
·0576490;   ·116099;   ·291592.

Thus, for $m = 3$, in which case $\phi = \pi$, we get for the successive terms after the first, which is 1,

·022105, ·003762, ·001225, ·000592, ·000379, ·000303.

We thus get for the value of the series in (30), by taking half the last term but one and a quarter of its first difference, ·980816; whence for $m = 3$, $W = 6^{-\frac{1}{2}} \times \cdot 980816 \epsilon^{-\pi} = \cdot 0173038$, of which the last figure cannot be trusted.   Now the number given by Mr Airy to 5 decimal places, and calculated from the ascending series and by quadratures separately, is ·01730, so that the correctness of the value of C given by (26) is verified.

For $m = +3$ we have from (28)

$$W = - 3^{-\frac{1}{2}} (R - S) = - 3^{-\frac{1}{2}} (\cdot 9965 - \cdot 0213) = - \cdot 5632,$$

which agrees with Mr Airy's result $- \cdot 56322$ or $- \cdot 56323$.   As m increases, the convergency of the series (29) or (30) increases rapidly.

12.   The expression (28) will be rendered more easy of numerical calculation by assuming $R = M \cos \psi$, $S = M \sin \psi$, and expanding M and $\tan \psi$ in series to a few terms.   These series will evidently proceed, the first according to even, and the second according to odd inverse powers of ϕ.   Putting the several terms, taken positively, under the form 1, $a\phi^{-1}$, $ab\phi^{-2}$, $abc\phi^{-3}$, $abcd\phi^{-4}$, &c., and proceeding to three terms in each series, we get

$$M = 1 - a\left(b - \frac{a}{2}\right) \phi^{-2} + a\left\{ bc(d - a) + \frac{a^2}{2}\left(b - \frac{a}{4}\right)\right\} \phi^{-4}...(31),$$

$$\tan \psi = a\phi^{-1} - ab(c - a) \phi^{-3} + ab\{cd(e - a) - ab(c - a)\} \phi^{-5}...(32).$$

The roots of the equation $W = 0$ are required for the physical problem to which the integral W relates.   Now equations (28),

(29) shew that when m is at all large the roots of this equation are given very nearly by the formula $\phi = (i - \frac{1}{4})\,\pi$, where i is an integer. From the definition of ψ it follows that the root satisfies exactly the equation

$$\phi = (i - \tfrac{1}{4})\,\pi + \psi \dots\dots\dots\dots\dots (33).$$

By means of this equation we may expand ϕ in a series according to descending powers of Φ, where $\Phi = (i - \frac{1}{4})\,\pi$. For this purpose it will be convenient first to expand ψ in a series according to descending powers of ϕ, by means of the expansion of $\tan^{-1} x$ and the equation (32), and having substituted the result in (33) to expand by Lagrange's theorem. The result of the expansion carried as far as to Φ^{-5} is

$$\phi = \Phi + a\Phi^{-1} - \{ab\,(c - a) + \tfrac{1}{3}a^3 + a^2\}\,\Phi^{-3}$$
$$+ \{ab\,[cd\,(e - a) - ab\,(c - a)] + a^3 b\,(c - a) + \tfrac{1}{5}a^5$$
$$+ 4a\,[ab\,(c - a) + \tfrac{1}{3}a^3] + 2a^3\}\,\Phi^{-5}\dots\dots\dots(34).$$

13.   To facilitate the numerical calculation of the coefficients let

$$a = \frac{a'}{1 \cdot D};\ \ b = \frac{b'}{2 \cdot D};\ \ c = \frac{c'}{3 \cdot D};\ \ \&c.,$$

and let the coefficients of ϕ^{-2}, ϕ^{-4} in (31) be put under the forms $-\dfrac{A_2}{1 \cdot 2 D^2}$, $\dfrac{A_4}{1 \cdot 2 \cdot 3 \cdot 4 D^4}$, and similarly with respect to (32), (34). Then to calculate W for a given value of m, we have

$$W = 2^{\frac{1}{2}}(3m)^{-\frac{1}{4}} M \cos\left(\phi - \frac{\pi}{4} - \psi\right)\dots\dots\dots(35),$$

where $$M = 1 - \frac{A_2}{1 \cdot 2 D^2}\,\phi^{-2} + \frac{A_4}{1 \cdot 2 \cdot 3 \cdot 4 D^4}\,\phi^{-4}\dots\dots\dots(36),$$

$$\tan\psi = \frac{C_1}{1 \cdot D}\,\phi^{-1} - \frac{C_3}{1 \cdot 2 \cdot 3 D^3}\,\phi^{-3} + \frac{C_5}{1 \cdot 2 \cdot 3 \cdot 4 \cdot 5 D^5}\,\phi^{-5}\dots(37),$$

and for calculating the roots of the equation $W = 0$, we have

$$\phi = \Phi + \frac{E_1}{1 \cdot D}\,\Phi^{-1} - \frac{E_3}{1 \cdot 2 \cdot 3 D^3}\,\Phi^{-3} + \frac{E_5}{1 \cdot 2 \cdot 3 \cdot 4 \cdot 5 D^5}\,\Phi^{-5}\dots(38).$$

The coefficients in these formulæ are given by the equations

$$\left.\begin{aligned} A_2 &= a'\,(b' - a');\quad A_4 = a'\,\{b'c'\,(d' - 4a') + 3a'^2\,(2b' - a')\} \\ C_1 &= a';\quad C_3 = a'b'\,(c' - 3a');\quad C_5 = a'b'\,\{c'd'\,(e' - 5a') - 10\,C_3\} \\ E_1 &= a';\quad E_3 = C_3 + 2a'^2\,(3D + a') \\ E_5 &= C_5 + 20a'\,(4D + a')\,C_3 + 24a'^5 + 80a'^3\,D\,(3D + 2a') \end{aligned}\right\} \dots(39).$$

14. Putting in these formulæ

$$a' = 1\,.\,5;\; b' = 7\,.\,11;\; c' = 13\,.\,17;\; d' = 19\,.\,23;\; e' = 25\,.\,29;\; D = 72;$$

we get

$$A_2 = 5.72;\; A_4 = 3.5.72^2.457;\; C_1 = 5;\; C_3 = 2.5.7.11.103;$$
$$C_5 = 4^2.5^3.7^2.11.23861;\; E_1 = 5;\; E_3 = 72.1255;\; E_5 = 4.5^3.72^2.10883;$$

whence we obtain, on substituting in (36), (37), (38),

$$M = 1 - \frac{5}{144}\,\phi^{-2} + \frac{2285}{41472}\,\phi^{-4},$$

$$\tan\psi = \frac{5}{72}\,\phi^{-1} - \frac{39655}{1119744}\,\phi^{-3} + \frac{321526975}{2902376448}\,\phi^{-5},$$

$$\phi = \Phi + \frac{5}{72}\,\Phi^{-1} - \frac{1255}{31104}\,\Phi^{-3} + \frac{272075}{2239488}\,\Phi^{-5}.$$

Reducing to decimals, having previously divided the last equation by π, and put for Φ its value $(i - \frac{1}{4})\,\pi$, we get

$$M = 1 - \cdot034722\,\phi^{-2} + \cdot055097\,\phi^{-4}\dots\dots\dots\dots(40),$$

$$\tan\psi = \cdot069444\,\phi^{-1} - \cdot035414\,\phi^{-3} + \cdot110781\,\phi^{-5}\dots\dots(41),$$

$$\frac{\phi}{\pi} = i - \cdot25 + \frac{\cdot028145}{4i - 1} - \frac{\cdot026510}{(4i - 1)^3} + \frac{\cdot129402}{(4i - 1)^5}\dots\dots\dots(42).$$

15. Supposing $i = 1$ in (42), we get

$$\frac{\phi}{\pi} = \cdot75 + \cdot0094 - \cdot0010 + 0005 = \cdot7589;$$

whence $m = 3\,(\phi/\pi)^{\frac{2}{3}} = 2\cdot496$.   The descending series obtained in this paper fail for small values of m; but it appears from Mr Airy's table that for such values the function W is positive, the first change of sign occurring between $m = 2\cdot4$ and $m = 2\cdot6$. Hence the integer i in (42) is that which marks the order of the root. A more exact value of the first root, obtained by interpolation from Mr Airy's table, is $2\cdot4955$. For $i = 1$ the series (42) is not convergent enough to give the root to more than three places of decimals, but the succeeding roots are given by this series with great accuracy. Thus, even in the case of the second root the value of the last term in (42) is only ·000007698. It appears then that this term might have been left out altogether.

16. To determine when W is a maximum or minimum we must put $dW/dm = 0$. We might get dW/dm by direct differentiation, but the law of the series will be more easily obtained from the differential equation. Resuming equation (11), and putting V for dU/dn, we get by dividing by n and then differentiating

$$\frac{d^2V}{dn^2} - \frac{1}{n}\frac{dV}{dn} + \frac{n}{3}V = 0.$$

This equation may be integrated by descending series just as before, and the arbitrary constants will be determined at once by comparing the result with the derivative of the second member of (15), in which A, B are given by (24). As the process cannot fail to be understood from what precedes, it will be sufficient to give the result, which is

$$V = 3^{-\frac{3}{4}}\pi^{\frac{1}{2}}n^{\frac{1}{4}}\left\{R'\cos\left(\phi + \frac{\pi}{4}\right) + S'\sin\left(\phi + \frac{\pi}{4}\right)\right\}\ \ldots\ldots (43),$$

where

$$\left.\begin{array}{l}R' = 1 - \dfrac{-1.7.5.13}{1.2\,(72\phi)^2} + \dfrac{-1.7.5.13.11.19.17.25}{1.2.3.4\,(72\phi)^4} - \cdots\\[2mm] S' = \dfrac{-1.7}{1.72\phi} - \dfrac{-1.7.5.13.11.19}{1.2.3\,(72\phi)^3} + \cdots.\end{array}\right\}\ \ldots (44).$$

17. The expression within brackets in (43) may be reduced to the form $M\cos(\phi + \frac{1}{4}\pi - \psi)$ just as before, and the formulæ of Art. 13 will apply to this case if we put

$$a' = -1.7;\ b' = 5.13;\ c' = 11.19;\ \&c.,\ D = 72.$$

The roots of the equation $dW/dm = 0$ are evidently the same as those of $V = 0$. They are given approximately by the formula $\phi = (i - \frac{3}{4})\pi$, and satisfy exactly the equation $\phi = (i - \frac{3}{4})\pi + \psi$. The root corresponding to any integer i may be expanded in a series according to the inverse odd powers of $4i - 3$ by the formulæ of Art. 13. Putting $(i - \frac{3}{4})\pi$ for Φ, and taking the series to three terms only, we get

$$E_1 = -7; \quad E_3 = -84168;$$

whence
$$\phi = \Phi - \tfrac{7}{72}\Phi^{-1} + \tfrac{1169}{31104}\Phi^{-3};$$

or, reducing as before,

$$\frac{\phi}{\pi} = i - \cdot75 - \frac{\cdot039403}{4i - 3} + \frac{\cdot024693}{(4i - 3)^3} \dots\dots\dots\dots(45).$$

This series will give only a rough approximation to the first root, but will answer very well for the others.

For $i = 1$ the series gives $\pi^{-1}\phi = \cdot25 - \cdot039 + \cdot025$, which becomes on taking half the second term and a quarter of its first difference $\cdot25 - \cdot019 - \cdot004 = \cdot227$, whence $m = 1\cdot12$. The value of the first root got by interpolation from Mr Airy's table is $1\cdot0845$. For the second and third roots we get from (45)

for $i = 2$, $\pi^{-1}\phi = 1\cdot25 - \cdot00788 + \cdot00020 = 1\cdot24232$;

for $i = 3$, $\pi^{-1}\phi = 2\cdot25 - \cdot00438 + \cdot00003 = 2\cdot24565$.

For higher values of i the last term in (45) may be left out altogether.

18.   The following table contains the first fifty roots of the equation $W = 0$, and the first ten roots of the derived equation. The first root in each case was obtained by interpolation from Mr Airy's table; the series (42) and (45) were sufficiently convergent for the other roots. In calculating the second root of the derived equation, a rough value of the first term left out in (45) was calculated, and its half taken since the next term would be of opposite sign. The result was only $-\cdot000025$, so that the series (45) may be used even when i is as small as 2. By far the greater part of the calculation consisted in passing from the values of $\pi^{-1}\phi$ to the corresponding values of m. In this part of the calculation 7-figure logarithms were used in obtaining the value of $\frac{1}{3}m$, and the result was then multiplied by 3.

A table of differences is added, for the sake of exhibiting the decrease indicated by theory in the interval between the consecutive dark bands seen in artificial rainbows. This decrease will be readily perceived in the tables which contain the results

of Professor Miller's observations*. The table of the roots of the
derived equation, which gives the maxima of W^2, is calculated for
the sake of meeting any observations which may be made on the
supernumerary bows accompanying a natural rainbow, since in
that case the maximum of the red appears to be what best admits
of observation.

i	m	diff.	i	m	diff.
1	2·4955	1·8676	26	26·1602	·6730
2	4·3631	1·5291	27	26·8332	·6647
3	5·8922	1·3514	28	27·4979	·6567
4	7·2436	1·2352	29	28·1546	·6491
5	8·4788	1·1512	30	28·8037	·6419
6	9·6300	1·0861	31	29·4456	·6349
7	10·7161	1·0335	32	30·0805	·6284
8	11·7496	·9899	33	30·7089	·6219
9	12·7395	·9529	34	31·3308	·6159
10	13·6924	·9208	35	31·9467	·6100
11	14·6132	·8927	36	32·5567	·6043
12	15·5059	·8676	37	33·1610	·5989
13	16·3735	·8452	38	33·7599	·5936
14	17·2187	·8250	39	34·3535	·5885
15	18·0437	·8065	40	34·9420	·5836
16	18·8502	·7897	41	35·5256	·5788
17	19·6399	·7740	42	36·1044	·5742
18	20·4139	·7597	43	36·6786	·5698
19	21·1736	·7463	44	37·2484	·5655
20	21·9199	·7337	45	37·8139	·5612
21	22·6536	·7221	46	38·3751	·5572
22	23·3757	·7111	47	38·9323	·5532
23	24·0868	·7008	48	39·4855	·5494
24	24·7876	·6909	49	40·0349	·5456
25	25·4785	·6817	50	40·5805	
1	1·0845	2·3824	6	9·0599	1·1175
2	3·4669	1·6777	7	10·1774	1·0590
3	5·1446	1·4336	8	11·2364	1·0111
4	6·5782	1·2903	9	12·2475	·9710
5	7·8685	1·1914	10	13·2185	

* *Cambridge Philosophical Transactions*, Vol. VII. p. 277.

SECOND EXAMPLE.

19. Let us take the integral

$$u = \frac{2}{\pi} \int_0^{\frac{\pi}{2}} \cos\left(x \cos \theta\right) d\theta = 1 - \frac{x^2}{2^2} + \frac{x^4}{2^2 4^2} - \frac{x^6}{2^2 4^2 6^2} + \dots * \quad \dots \dots (46)$$

which occurs in a great many physical investigations. If we perform the operation $x \cdot d/dx$ twice in succession on the series we get the original series multiplied by $-x^2$, whence

$$\frac{d^2u}{dx^2} + \frac{1}{x}\frac{du}{dx} + u = 0 \quad \dots \dots \dots \dots (47).$$

20. The form of this equation shews that when x is very large, and receives an increment δx, which, though not necessarily a very small fraction itself, is very small compared with x, u is expressed by $A \cos \delta x + B \sin \delta x$, where under the restrictions specified A and B are sensibly constant†. Assume then, according to the plan of Art. 5,

$$u = \epsilon^{x\sqrt{-1}} \{A x^\alpha + B x^\beta + C x^\gamma + \dots\} \quad \dots \dots \dots (48).$$

On substituting in (47) we get

$$\sqrt{-1} \left\{ (2\alpha + 1) A x^{\alpha-1} + (2\beta + 1) B x^{\beta-1} + \dots \right\}$$
$$+ \alpha^2 A x^{\alpha-2} + \beta^2 B^{\beta-2} + \dots = 0.$$

Since we want a descending series, we must put

$$2\alpha + 1 = 0; \; \beta = \alpha - 1; \; \gamma = \beta - 1 \dots;$$

$$(2\beta + 1) B = \sqrt{-1}\, \alpha^2 A \;;\; (2\gamma + 1) C = \sqrt{-1}\beta^2 B \dots;$$

---

* This integral has been tabulated by Mr Airy from $x=0$ to $x=10$, at intervals of 0·2. The table will be found in the 18th Volume of the *Philosophical Magazine*, page 1.

† That the 1st and 3rd terms in (47) are ultimately the important terms, may readily be seen by trying the terms two and two in the way mentioned in the introduction. Thus, if we suppose the first two to be the important terms, we get ultimately $U = A$ or $U = B \log x$, either of which would render the last term more important than the 1st or 2nd, and if we suppose the 2nd and 3rd to be the important terms, we get ultimately $u = A\epsilon^{-x^2/2}$, which would render the first term more important than either of the others.

whence $\qquad \alpha = -\tfrac{1}{2}; \; \beta = -\tfrac{3}{2}; \; \gamma = -\tfrac{5}{2} \ldots;$

$$B = -\frac{1^2}{1.8}\sqrt{-1}A; \quad C = +\frac{1^2.3^2}{1.2.8^2}(\sqrt{-1})^2A;$$

$$D = -\frac{1^2.3^2.5^2}{1.2.3.8^3}(\sqrt{-1})^3A\ldots.$$

Substituting in (48), reducing the result to the form

$$A(P + \sqrt{-1}Q),$$

adding another solution of the form $B(P - \sqrt{-1}Q)$, and changing the arbitrary constants, we get

$$u = Ax^{-\frac{1}{2}}(R\cos x + S\sin x) + Bx^{-\frac{1}{2}}(R\sin x - S\cos x)\ldots.(49),$$

where
$$\left. \begin{aligned} R &= 1 - \frac{1^2.3^2}{1.2\,(8x)^2} + \frac{1^2.3^2.5^2.7^2}{1.2.3.4\,(8x)^4} \cdots \\ S &= \frac{1^2}{1.8x} - \frac{1^2.3^2.5^2}{1.2.3\,(8x)^3} + \cdots \end{aligned} \right\} \ldots\ldots\ldots\ldots(50).$$

21. It remains to determine the arbitrary constants A, B. In equation (46) let $\cos\theta = 1 - \mu$, whence

$$d\theta = \frac{d\mu}{\sin\theta} = \frac{d\mu}{(2\mu - \mu^2)^{\frac{1}{2}}} = \frac{d\mu}{(2\mu)^{\frac{1}{2}}} + Md\mu,$$

where $\qquad M = (2\mu - \mu^2)^{-\frac{1}{2}} - (2\mu)^{-\frac{1}{2}},$

a quantity which does not become infinite between the limits of μ. Substituting in (46) we get

$$u = \frac{\sqrt{2}}{\pi}\int_0^1 \cos\{(1 - \mu)\,x\}\,\mu^{-\frac{1}{2}}\,d\mu + \frac{2}{\pi}\int_0^1 \cos\{(1 - \mu)\,x\}\,Md\mu\ldots(51).$$

By considering the series whose n^{th} term is the part of the latter integral for which the limits of μ are $n\pi x^{-1}$ and $(n + 1)\,\pi x^{-1}$ respectively, it would be very easy to prove that the integral has a superior limit of the form Hx^{-1}, where H is a finite constant, and therefore this integral does not furnish any part of the leading terms in u. Putting $\mu x = \nu$ in the first integral in (51), so that

$$\mu^{-\frac{1}{2}}\,d\mu = x^{-\frac{1}{2}}\,\nu^{-\frac{1}{2}}\,d\nu,$$

observing that the limits of ν are 0 and x, of which the latter ultimately becomes ∞, and that

$$\int_0^\infty \cos \nu \,.\, \nu^{-\frac{1}{2}}\, d\mu = 2\int_0^\infty \cos \lambda^2 d\lambda = \sqrt{\frac{\pi}{2}}$$

$$= 2\int_0^\infty \sin \lambda^2 d\lambda = \int_0^\infty \sin \nu \,.\, \nu^{-\frac{1}{2}}\, d\nu,$$

we get ultimately for very large values of x

$$u = (\pi x)^{-\frac{1}{2}} (\cos x + \sin x).$$

Comparing with (49) we get

$$A = B = \pi^{-\frac{1}{2}},$$

whence $u = \left(\dfrac{2}{\pi x}\right)^{\frac{1}{2}} R \cos \left(x - \dfrac{\pi}{4}\right) + \left(\dfrac{2}{\pi x}\right)^{\frac{1}{2}} S \sin \left(x - \dfrac{\pi}{4}\right)^{*} \dots (52).$

For example, when $x = 10$ we have, retaining 5 decimal places in the series,

$$R = 1 - {\cdot}00070 + {\cdot}00001 = {\cdot}99931 ; \quad S = {\cdot}01250 - {\cdot}00010 = {\cdot}01240$$

$$\text{Angle } x - \frac{\pi}{4} = 527^0 \,{\cdot}95780 = 3 \times 180^0 - 12^0 \, 2' \, 32'' ;$$

whence $u = -\,{\cdot}24594$, which agrees with the number $(-\,{\cdot}2460)$ obtained by Mr Airy by a far more laborious process, namely, by calculating from the original series.

22.   The second member of equation (52) may be reduced to the same form as that of (28), and a series obtained for calculating the roots of the equation $u = 0$ just as before. The formulæ of Art. 13 may be used for this purpose on putting

$$a' = 1^2; \quad b' = 3^2; \quad c' = 5^2; \quad \&c.; \quad D = 8,$$

and writing x, X for ϕ, Φ, where $X = (i - \frac{1}{4})\,\pi$. We obtain

$$A_2 = 8; \quad A_4 = 3\,.\,8^2\,.\,53; \quad C_1 = 1; \quad C_3 = 2\,.\,3^2\,.\,11;$$

$$C_5 = 3^2\,.\,4^2\,.\,5\,.\,1139; \quad E_1 = 1; \quad E_3 = 8\,.\,31; \quad E_5 = 4^4\,.\,3779;$$

---

* This expression for u, or rather an expression differing from it in nothing but notation and arrangement, has been already obtained in a different manner by Sir William R. Hamilton, in a memoir " On Fluctuating Functions." See *Transactions of the Royal Irish Academy*, Vol. xix. p. 313.

whence we get for calculating u for a given value of x

$$M = 1 - \tfrac{1}{16} x^{-2} + \tfrac{53}{512} x^{-4},$$

$$\tan \psi = \tfrac{1}{8} x^{-1} - \tfrac{33}{512} x^{-3} + \tfrac{3417}{16384} x^{-5},$$

$$u = \left(\frac{2}{\pi x} \right)^{\tfrac{1}{2}} M \cos \left(x - \frac{\pi}{4} - \psi \right) \dots \dots (53).$$

For calculating the roots of the equation $u = 0$ we have

$$x = X + \tfrac{1}{8} X^{-1} - \tfrac{31}{384} X^{-3} + \tfrac{3779}{15360} X^{-5}.$$

Reducing to decimals as before, we get

$$M = 1 - \cdot 0625 \, x^{-2} + \cdot 103516 \, x^{-4} \dots\dots\dots\dots\dots(54),$$

$$\tan \psi = \cdot 125 \, x^{-1} - \cdot 064453 \, x^{-3} + \cdot 208557 \, x^{-5} \dots\dots\dots(55),$$

$$\frac{x}{\pi} = i - \cdot 25 + \frac{\cdot 050661}{4i - 1} - \frac{\cdot 053041}{(4i-1)^3} + \frac{\cdot 262051}{(4i-1)^5} \dots\dots(56).$$

As before, the series (56) is not sufficiently convergent when $i = 1$ to give a very accurate result. In this case we get

$$\pi^{-1} x = \cdot 75 + \cdot 017 - \cdot 002 + \cdot 001 = \cdot 766,$$

whence $x = 2 \cdot 41$. Mr Airy's table gives $u = + \cdot 0025$ for $x = 2 \cdot 4$, and $u = - \cdot 0968$ for $x = 2 \cdot 6$, whence the value of the root is $2 \cdot 4050$ nearly.

The value of the last term in (56) is $\cdot 0000156$ for $i = 2$, and $\cdot 00000163$ for $i = 3$, so that all the roots after the first may be calculated very accurately from this series.

THIRD EXAMPLE.

23. Consider the integral

$$v = \frac{2}{\pi} \int_0^x \int_0^{\frac{\pi}{2}} \cos \left(x \cos \theta \right) x \, dx \, d\theta$$

$$= \int_0^x u x \, dx = \frac{x^2}{2} - \frac{x^4}{2^2 \cdot 4} + \frac{x^6}{2^2 \cdot 4^2 \cdot 6} - \dots * \dots\dots\dots(57),$$

---

* The series $1 - \dfrac{x^2}{2 \cdot 4} + \dfrac{x^4}{2 \cdot 4^2 \cdot 6} \dots$ or $\dfrac{2v}{x^2}$ has been tabulated by Mr Airy from $x = 0$ to $x = 12$ at intervals of $0 \cdot 2$. See *Camb. Phil. Trans.* Vol. v. p. 291. The same function has also been calculated in a different manner and tabulated by M. Schwerd which occurs in investigating the diffraction of an object-glass with a circular aperture.

By performing on the series the operation denoted by $x \cdot d/dx \cdot x^{-1} \cdot d/dx$, we get the original series with the sign changed, whence

$$\frac{d^2v}{dx^2} - \frac{1}{x}\frac{dv}{dx} + v = 0 \quad\ldots\ldots\ldots\ldots\ldots\ldots\ldots(58).$$

We may obtain the integral of this equation in a form similar to (49). As the process is exactly the same as before, it will be sufficient to write down the result, which is

$$v = A' x^{\frac{1}{2}} (R \cos x + S \sin x) + B' x^{\frac{1}{2}} (R \sin x - S \cos x) \ldots\ldots (59),$$

where

$$\left. \begin{aligned} R &= 1 - \frac{-1 \cdot 3 \cdot 1 \cdot 5}{1 \cdot 2 \; (8x)^2} + \frac{-1 \cdot 3 \cdot 1 \cdot 5 \cdot 3 \cdot 7 \cdot 5 \cdot 9}{1 \cdot 2 \cdot 3 \cdot 4 \; (8x)^4} - \ldots \\ S &= \frac{-1 \cdot 3}{1 \cdot 8x} - \frac{-1 \cdot 3 \cdot 1 \cdot 5 \cdot 3 \cdot 7}{1 \cdot 2 \cdot 3 \; (8x)^3} + \ldots \end{aligned} \right\} \ldots(60),$$

the last two factors in the numerator of any term being formed by adding 2 to the last two factors respectively in the numerator of the term of the preceding order.

The arbitrary constants may be easily determined by means of the equation

$$\frac{dv}{dx} = ux \ldots\ldots\ldots\ldots\ldots\ldots\ldots\ldots\ldots(61).$$

Writing down the leading terms only in this equation, we have

$$x^{\frac{1}{2}} (-A' \sin x + B' \cos x) = \pi^{-\frac{1}{2}} x^{\frac{1}{2}} (\cos x + \sin x),$$

whence

$$-A' = B' = \pi^{-\frac{1}{2}},$$

$$v = \left(\frac{2x}{\pi}\right)^{\frac{1}{2}} \left\{ R \cos\left(x - \frac{3\pi}{4}\right) + S \sin\left(x - \frac{3\pi}{4}\right) \right\} \ldots\ldots\ldots(62).$$

24. Putting in the formulæ of Art. 13,

$$a' = -1 \cdot 3; \; b' = 1 \cdot 5; \; c' = 3 \cdot 7; \; d' = 5 \cdot 9; \; e' = 7 \cdot 11; \; D = 8;$$

in his work on diffraction. The argument in the latter table is the angle $180^0/\pi \cdot x$, and the table extends from 0^0 to 1125^0 at intervals of 15^0, that is, from $x=0$ to $x=19 \cdot 63$ at intervals of $0 \cdot 262$ nearly.

we get

$$A_2 = -3 \cdot 8; \quad A_4 = -3^3 \cdot 8^2 \cdot 11; \quad C_1 = -3; \quad C_3 = -2 \cdot 3^2 \cdot 5^2;$$

$$C_5 = -3^3 \cdot 4^2 \cdot 5^2 \cdot 127; \quad E_1 = -3; \quad E_3 = -3^2 \cdot 8; \quad E_5 = -3^3 \cdot 4 \cdot 8^2 \cdot 131;$$

whence we get for the formulæ answering to those of Art. 22,

$$M = 1 + \tfrac{3}{16} x^{-2} - \tfrac{99}{512} x^{-4},$$

$$\tan \psi = -\tfrac{3}{8} x^{-1} + \tfrac{75}{512} x^{-3} - \tfrac{5715}{16384} x^{-5},$$

$$x = X - \tfrac{3}{8} X^{-1} + \tfrac{3}{128} X^{-3} + \tfrac{1179}{5120} X^{-5},$$

X being in this case equal to $(i + \tfrac{1}{4}) \pi$.

Reducing to decimals as before, we get for the calculation of v for a given value of x,

$$M = 1 + \cdot 1875 x^{-2} + \cdot 193359 x^{-4} \dots\dots\dots\dots(63),$$

$$\tan \psi = -\cdot 375 x^{-1} + \cdot 146484 x^{-3} - \cdot 348817 x^{-5} \dots\dots(64),$$

$$v = \left(\frac{2x}{\pi}\right)^{\frac{1}{2}} M \cos\left(x - \frac{3\pi}{4} - \psi\right) \dots\dots\dots\dots (65);$$

and for calculating the roots of the equation $v = 0$,

$$\frac{x}{\pi} = i + \cdot 25 - \frac{\cdot 151982}{4i + 1} + \frac{\cdot 015399}{(4i + 1)^3} - \frac{\cdot 245835}{(4i + 1)^5} \dots\dots(66).$$

25. The following table contains the first 12 roots of each of the equations $u = 0$, and $x^{-2} v = 0$. The first root of the former

i	$\dfrac{x}{\pi}$ for $u = 0$	diff.	$\dfrac{x}{\pi}$ for $v = 0$	diff.
1	·7655		1·2197	
2	1·7571	·9916	2·2330	1·0133
3	2·7546	·9975	3·2383	1·0053
4	3·7534	·9988	4·2411	1·0028
5	4·7527	·9993	5·2428	1·0017
6	5·7522	·9995	6·2439	1·0011
7	6·7519	·9997	7·2448	1·0009
8	7·7516	·9997	8·2454	1·0006
9	8·7514	·9998	9·2459	1·0005
10	9·7513	·9999	10·2463	1·0004
11	10·7512	·9999	11·2466	1·0003
12	11·7511	·9999	12·2469	1·0003

was got by interpolation from Mr Airy's table, the others were calculated from the series (56). The roots of the latter equation were all calculated from the series (66), which is convergent enough even in the case of the first root. The columns which contain the roots are followed by columns which contain the differences between consecutive roots, which are added for the purpose of shewing how nearly equal these differences are to 1, which is what they ultimately become when the order of the root is indefinitely increased.

26.   The preceding examples will be sufficient to illustrate the general method. I will remark in conclusion that the process of integration applied to the equations (11), (47), and (58) leads very readily to the complete integral in finite terms of the equation

$$\frac{d^2y}{dx^2} - \left\{ q^2 + \frac{i(i+1)}{x^2} \right\} y = 0 \quad ....(67),$$

where i is an integer, which without loss of generality may be supposed positive. The form under which the integral immediately comes out is

$$y = A\epsilon^{qx} \left\{ 1 - \frac{i(i+1)}{1.2qx} + \frac{(i-1)\,i\,(i+1)\,(i+2)}{1.2\,(2qx)^2} - \cdots \right\},$$

$$+ B\epsilon^{-qx} \left\{ 1 + \frac{i(i+1)}{1.2qx} + \frac{(i-1)\,i\,(i+1)\,(i+2)}{1.2\,(2qx)^2} + \cdots \right\},$$

where each series will evidently contain $i+1$ terms. It is well known that (67) is a general integrable form which includes as a particular case the equation which occurs in the theory of the figure of the earth, for q in (67) is any quantity real or imaginary, and therefore the equation formed from (67) by writing $+ q^2 y$ for $- q^2 y$ may be supposed included in the form (67).

It may be remarked that the differential equations discussed in this paper can all be reduced to particular cases of the equation obtained by replacing $i(i+1)$ in (67) by a general constant. By taking $gn^{\frac{2}{3}}$, where g is any constant, for the independent variable in place of n in the differential equations which U, V in the first example satisfy, these equations are reduced to the form

$$\frac{d^2y}{dx^2} + \frac{2a}{x}\frac{dy}{dx} + \left(\frac{b}{x^2} + c\right)y = 0,$$

and (47), (58) are in this form already. Putting now $y = x^{-a}z$, we shall reduce the last equation to the form required.

[The four following are from the *Report of the British Association* for 1850, Part II. p. 19.]

ON THE MODE OF DISAPPEARANCE OF NEWTON'S RINGS IN PASSING THE ANGLE OF TOTAL INTERNAL REFLEXION.

WHEN Newton's rings are formed between the under surface of a prism and the upper surface of a lens, there is no difficulty in increasing the angle of incidence so as to pass through the angle of total internal reflexion. When the rings are observed with the naked eye in the ordinary way, they appear to break in the upper part on approaching the angle of total internal reflexion, and pass nearly into semicircles when that angle is reached, the upper edges of the semicircles, which are in all cases indistinct, being slightly turned outwards when the curvature of the lens is small.

The cause of the indistinctness will be evident from the following considerations. The *order* of the ring (a term here used to denote a number not necessarily integral) to which a ray reflected at a given obliquity from a given point of the thin plate of air belongs, depends partly on the obliquity and partly on the thickness of the plate at that point. When the angle of incidence is small, or even moderately large, the rings would not be seen, or at most would be seen very indistinctly, if the glasses were held near the eye, and the eye were adapted to distinct vision of distant objects, because in that case the rays brought to a focus at a given point of the retina would correspond to a pencil reflected at a given obliquity from an area of the plate of air, the size of which would correspond to the pupil of the eye; and the order of the rays reflected from this area would vary so much in passing from the point of contact outwards that the rings would be altogether confused. When, however, as in the usual mode of observation, the eye is adapted to distinct vision of an object at the distance of the plate of air, the rings are seen distinctly, because in this case the rays proceeding from a given point of the plate of air, and entering the pupil of the eye, are brought to a focus on the retina, and the variation in the obliquity of the rays forming this pencil is so small that it may be neglected.

When, however, the angle of incidence becomes nearly equal to that of total internal reflexion, a small change of obliquity produces a great change in the order of the ring to which the reflected ray belongs, and therefore the rings are indistinct to an eye adapted to distinct vision of the surfaces of the glass. They are also indistinct, for the same reason as before, if the eye be adapted to distinct vision of distant objects.

To see distinctly the rings in the neighbourhood of the angle of total internal reflexion, the author used a piece of blackened paper in which a small hole was pierced with the point of a needle. When the rings were viewed through the needle-hole, in the light of a spirit-lamp, the appearance was very remarkable. The first dark band seen within the bright portion of the field of view where the light suffered total internal reflexion was somewhat bow-shaped towards the point of contact, the next still more so, and so on, until at last one of the bands made a great bend and passed under the point of contact and the rings which surrounded it, the next band passing under it, and so on. As the incidence was gradually increased, the outermost ring united with the bow-shaped band next above it, forming for an instant a curve with a loop and two infinite branches, or at least branches which ran out of the field of view: then the loop broke, and the curve passed into a bulging band similar to that which had previously surrounded the rings. In this manner the rings, one after another, joined the corresponding bands till all had disappeared, and nothing was left but a system of bands which had passed completely below the point of contact, and the central black spot which remained isolated in the bright field where the light suffered total internal reflexion. Corresponding appearances were seen with daylight or candlelight, but in these cases the bands were of course coloured, and not near so many could be seen at a time.

On Metallic Reflexion.

THE effect which is produced on plane-polarized light by reflexion at the surface of a metal, shews that if the incident light be supposed to be decomposed into two streams, polarized in and perpendicularly to the plane of reflexion respectively, the *phases* as well as the intensities of the two streams are differently affected by the reflexion. It remains a question whether the phase of vibration of the stream polarized in the plane of reflexion is accelerated or retarded relatively to that of the stream polarized perpendicularly to the plane of reflexion. This question was first decided by the Astronomer Royal, by means of a phænomenon relating to Newton's rings when formed between a speculum and a glass plate. Mr Airy's paper is published in the *Cambridge Philosophical Transactions*. M. Jamin has since been led to the same result, apparently by a method similar in principle to that of Mr Airy. In repeating Mr Airy's experiment, the author experienced considerable difficulty in observing the phænomenon. The object of the present communication was to point out an extremely easy mode of deciding the question experimentally. Light polarized at an azimuth of about 45° to the plane of reflexion at the surface of the metal was transmitted, after reflexion, through a plate of Iceland spar, cut perpendicular to the axis, and analysed by a Nicol's prism. When the angle of incidence was the smallest with which the observation was practicable, on turning the Nicol's prism properly the dark cross was formed almost perfectly; but on increasing the angle of incidence it passed into a pair of hyperbolic brushes. This modification of the ring is very well known, having been described and figured by Sir D. Brewster in the *Philosophical Transactions* for 1830. Now the question at issue may be imme-. diately decided by observing in which pair of opposite quadrants it is that the brushes are formed, an observation which does not present the slightest difficulty. In this way the author was led to Mr Airy's result, namely, that as the angle of incidence increases from zero, the phase of vibration of light polarized in the plane of incidence is *accelerated* relatively to that of light polarized in a plane perpendicular to the plane of incidence.

ON A FICTITIOUS DISPLACEMENT OF FRINGES OF INTERFERENCE.

THE author remarked that the mode of determining the refractive index of a plate by means of the displacement of a system of interference fringes, is subject to a theoretical error depending upon the dispersive power of the plate. It is an extremely simple consequence (as the author shewed) of the circumstance that the bands are broader for the less refrangible colours, that the point of symmetry, or nearest approach to symmetry, in the system of displaced fringes, is situated *in advance* of the position calculated in the ordinary way for rays of mean refrangibility. Since an observer has no other guide than the symmetry of the bands in fixing on the centre of the system, he would thus be led to attribute to the plate a refractive index which is slightly too great.

The author has illustrated this subject by the following experiment. A set of fringes, produced in the ordinary way by a flat prism, were viewed through an eye-piece, and bisected by its cross wires. On viewing the whole through a prism of moderate angle, held in front of the eye-piece with its edge parallel to the fringes, an indistinct prismatic image of the wires was seen, together with a *distinct* set of fringes which lay *quite at one side* of the cross wires, the dispersion produced by the prism having thus occasioned an *apparent* displacement of the fringes in the direction of the general deviation.

In conclusion, the author suggested that it might have been the fictitious displacement due to the dispersion accompanying eccentrical refraction, which caused some philosophers to assert that the central band was black, whereas, according to theory, it ought to be white. A fictitious displacement of half an order, which might readily be produced by eccentrical refraction through the lens or eye-piece with which the fringes were viewed, would suffice to cause one of the two black bands of the first order to be the band with respect to which the system was symmetrical.

ON HAIDINGER'S BRUSHES.

It is now several years since these brushes were discovered, and they have since been observed by various philosophers, but the author has not met with any observations made with a view of investigating the action of different colours in producing them. The author's attention was first called to the subject, by observing that a green tourmaline, which polarized light very imperfectly, enabled him to see the brushes very distinctly, while he was unable to make them out with a brown tourmaline which transmitted a much smaller quantity of unpolarized light. He then tried the effect of combining various coloured glasses with a Nicol's prism. A red glass gave no trace of brushes. A brownish yellow glass, which absorbed only a small quantity of light, rendered the brushes very indistinct. A green glass enabled the author to see the brushes rather more distinctly than they were seen in the light of the clouds viewed without a coloured glass. A deep blue glass gave brushes of remarkable intensity, notwithstanding the large quantity of light absorbed. With the green and blue glasses, the brushes were not coloured, but simply darker than the rest of the field.

To examine still further the office of the different colours in producing the brushes seen with ordinary daylight, the author used a telescope and prism mounted for shewing the fixed lines of the spectrum. The sun's light having been introduced into a darkened room through a narrow slit, it was easy, by throwing the eye-piece a little out of focus, to form a pure spectrum on a screen of white paper, placed a foot or two in front of the eye-piece. On examining this spectrum with a Nicol's prism, which was suddenly turned round from time to time through about a right angle, the author found that the red and yellow did not present the least trace of brushes. The brushes began to be visible in the green, about the fixed line E of Fraunhofer. They became more distinct on passing into the blue, and were particularly strong about the line F. The author was able to trace them about as far as the line G; and when they were no longer visible, the cause appeared to be merely the feebleness of the light, not the incapacity of the greater part of the violet to produce them. With homogeneous light, the brushes, when they were formed at all, were simply darker than the rest of the field, and, as might have been expected, did not appear of a different tint. In the blue, where the brushes were most distinct, it appeared to the author that they were somewhat shorter than usual. The contrast between the more and less refrangible portions of the spectrum, in regard to their capability of producing brushes, was most striking. The most brilliant part of the spectrum gave no brushes; and the intensity of the orange and more refrangible portion of the red, where not the slightest trace of brushes was discoverable, was much greater than that of the more refrangible portion of the blue, where the brushes were formed with great distinctness, although *cæteris paribus* a considerable degree of intensity is favourable to the exhibition of the brushes.

These observations account at once for the colour of the brushes seen with ordinary daylight. Inasmuch as no brushes are seen with the less refrangible colours, and the brushes seen with the more refrangible colours consist in the removal of a certain quantity of light, the tint of the brushes ought to be made up of red, yellow, and perhaps a little green, the yellow predominating, on account of its greater brightness in the solar spectrum. The mixture would give an impure yellow, which is the colour observed. The blueness of the side patches may be merely the effect of contrast, or the cause may be more deeply seated. If the total illumination perceived be independent of the brushes, the light withdrawn from the brushes must be found at their sides, which would account, independently of contrast, both for the comparative brightness and for the blue tint of the side patches.

The observations with homogeneous light account likewise for a circumstance with which the author had been struck, namely, that the brushes were not visible by candle-light, which is explained by the comparative poverty of candle-light in the more refrangible rays. The brushes ought to be rendered visible by absorbing a certain quantity of the less refrangible rays, and accordingly the author found that a blue glass, combined with a Nicol's prism, enabled him to see the brushes very distinctly when looking at the flame of a candle. The specimen of blue glass which shewed them best, which was of a tolerably deep colour, gave brushes which were decidedly red, and were only comparatively dark, so that the difference of tint between the brushes and side patches was far more conspicuous than the difference of intensity. This is accounted for by the large quantity of extreme red rays which such a glass transmits. That the same glass gave red brushes with candle-light, and dark brushes with daylight, is accounted for by the circumstance, that the ratio which the intensity of the transmitted red rays bears to the intensity of the transmitted blue rays is far larger with candle-light than with daylight.

INDEX TO VOL. II.

CAMBRIDGE : PRINTED BY C. J. CLAY AND SON, AT THE UNIVERSITY PRESS.

Printed in the United States
By Bookmasters